Progress in Systems and Control Theory
Volume 15

Series Editor
Christopher I. Byrnes, Washington University

Associate Editors

Computation and Control III

Proceedings of the Third Bozeman Conference, Bozeman, Montana, August 5-11, 1992

K. L. Bowers

J. Lund

Editors

Springer Science+Business Media, LLC

Kenneth Bowers
Department of Mathematics
Montana State University
Bozeman, Montana 59717

John Lund
Department of Mathematics
Montana State University
Bozeman, Montana 59717

Library of Congress Cataloging In-Publication Data

Computation and control III : proceedings of the Third Bozeman
    Conference, Bozeman, Montana, August 5-11, 1992 / K.L. Bowers, J.
    Lund.
        p.   cm. -- (Porgress in systems and control theory ; v. 15)
    Includes bibliographical references.
    ISBN 978-1-4612-6706-5   ISBN 978-1-4612-0321-6 (eBook)
    DOI 10.1007/978-1-4612-0321-6
    1. Engineering mathematics--Congresses. 2. System analysis-
-Congresses.   3. Control theory--Congresses.   4. Differential
equations, Partial--Congresses.   I. Bowers, K. (Kenneth)   II. Lund,
J. (John)   III. Series.
TA329.C6442   1993   93-14021
629.8'312--dc20   CIP

ISBN 978-1-4612-6706-5
Typeset by the Authors in LaTEX.

9 8 7 6 5 4 3 2 1

CONTENTS

Preface

Theoretical and Computational Aspects of Feedback in
Structural Systems with Piezoceramic Controllers
  H. T. Banks, K. Ito and B. B. King   1

Modeling and Approximation of a Coupled 3-D Structural
Acoustics Problem
  H. T. Banks and R. C. Smith   29

Parameter Identification in the Frequency Domain
  H. T. Banks and Y. Wang   49

On Model Identification of Gaussian Reciprocal Processes
from the Eigenstructure of Their Covariances
  C. F. Borges and R. Frezza   63

An Inverse Problem in Thermal Imaging
  K. Bryan   73

Optimal Fixed-Finite-Dimensional Compensator for Burgers'
Equation with Unbounded Input/Output Operators
  J. A. Burns and H. Marrekchi   83

Boundary Control and Stabilization for a Viscous
Burgers' Equation
  C. I. Byrnes and D. S. Gilliam   105

A Sinc-Galerkin Method for Convection Dominated Transport
  T. S. Carlson, J. Lund and K. L. Bowers   121

Discrete Observability of the Wave Equation on Bounded
Domains in Euclidean Space
  A. DeStefano   141

A New Algorithm for Nonlinear Filtering
  G. B. DiMasi, D. B. Hernández and T. J. Taylor   153

CONTENTS

Continuation Methods for Nonlinear Eigenvalue Problems via a
Sinc-Galerkin Scheme
    J. D. Dockery and N. J. Lybeck   165

On the Kalman-Yacubovich-Popov Lemma for Nonlinear Systems
    K. A. Doll and C. I. Byrnes   181

Robust Control of Distributed Parameter Systems with
Structured Uncertainty
    R. H. Fabiano, A. J. Kurdila and T. Strganac   193

On the Phase Portrait of the Karmarkar's Flow
    L. E. Faybusovich   203

The Reduced Basis Method in Control Problems
    M. D. Gunzburger and J. S. Peterson   211

Numerical Treatment of Oscillating Integrals Appearing in
Heat Conduction Problems
    S.-Å. Gustafson   219

Root Locus for Control Systems with Completely Separated
Boundary Conditions
    J. He   229

On the Problem of Parameter Identification in Perspective
Systems and its Application to Motion Estimation
Problems in Computer Vision
    E. P. Loucks and B. K. Ghosh   241

Over-Regularization of Ill-Posed Problems
    B. A. Mair   255

A Model for the Optimal Control of a Measles Epidemic
    C. Martin, L. Allen, M. Stamp, M. Jones and R. Carpio   265

Condition Numbers for the Sinc Matrices Associated with
Discretizing the Second-Order Differential Operator
    K. M. McArthur   285

Computational Models for Lattice Structures
    R. E. Miller   301

The Partial Differential Equations of Controlled Invariance
    V. Ramakrishna   313

CONTENTS

What is the Distance Between Two Autoregressive Systems?
J. Rosenthal and X. Wang — 333

Sinc Convolution Approximate Solution of Burgers' Equation
F. Stenger, B. Barkey and R. Vakili — 341

Sinc-Galerkin Collocation Method for Parabolic Equations
in Finite Space-Time Regions
M. Stromberg and X. L. Gilliam — 355

A Modified Levenberg-Marquardt Algorithm for Large-Scale
Inverse Problems
C. R. Vogel and J. G. Wade — 367

A Local Sampling Scheme for Invariant Evolution Equations
on a Compact Symmetric Space, Especially the Sphere
D. I. Wallace — 379

Hasse Diagram and Dynamic Feedback of Linear Systems
X. Wang and J. Rosenthal — 391

Point Placement for Observation of the Heat Equation
on the Sphere
J. A. Wolf — 399

CONTRIBUTORS

Linda Allen, Department of Mathematics, Texas Tech University, Lubbock,
Texas 79409

Tom Banks, Center for Research in Scientific Computation, North Carolina
State University, Raleigh, North Carolina 27695-8205

Brian Barkey, Department of Computer Science, University of Utah,
Salt Lake City, Utah 84112

Carlos Borges, Department of Mathematics, Naval Postgraduate School,
Monterey, California 93943

Ken Bowers, Department of Mathematical Sciences, Montana State
University, Bozeman, Montana 59717-0240

Kurt Bryan, Institute for Computer Applications in Science and
Engineering, NASA Langley Research Center, Hampton,
Virginia 23681

John Burns, Interdisciplinary Center for Applied Mathematics,
Department of Mathematics, Virginia Polytechnic Institute and
State University, Blacksburg, Virginia 24061

Chris Byrnes, Department of Systems Science and Mathematics,
Washington University, St. Louis, Missouri, 63130

Tim Carlson, Department of Mathematical Sciences, Montana State
University, Bozeman, Montana 59717-0240

R. Carpio, Department of Mathematics, Texas Tech University, Lubbock,
Texas 79409

Alisa DeStefano, Department of Mathematics, College of the Holy Cross,
Worcester, Massachusetts 01610

Giovanni DiMasi, Dipartimento di Matematica Pura ed Applicata,
Universitá di Padova and LADSEB-CNR, 35100 Padova, Italy

Jack Dockery, Department of Mathematical Sciences, Montana State
University, Bozeman, Montana 59717-0240

CONTRIBUTORS

Kenneth Doll, Department of Systems Science and Mathematics, Washington University, St. Louis, Missouri, 63130

Rich Fabiano, Department of Mathematics, Texas A&M University, College Station, Texas 77843

Leonid Faybusovich, Department of Mathematics, University of Notre Dame, Notre Dame, Indiana 46556

Ruggero Frezza, Universitá di Padova, 35131 Padova, Italy

Bijoy Ghosh, Department of Systems Science and Mathematics, Washington University, St. Louis, Missouri 63130

Dave Gilliam, Department of Mathematics, Texas Tech University, Lubbock, Texas 79409

Xiaoning Gilliam, Department of Mathematics, Texas Tech University, Lubbock, Texas 79409

Max Gunzburger, Interdisciplinary Center for Applied Mathematics, Department of Mathematics, Virginia Polytechnic Institute and State University, Blacksburg, Virginia 24061

Sven-Åke Gustafson, Høgskolesenteret i Rogaland, N-4004 Stavanger, Norway

Jianqiu He, Department of Mathematics, Texas Tech University, Lubbock, Texas 79409

Diego Hernández, Centro de Investigación en Matemáticas, A. C., 36000 Guanajuato, Gto., Apartado Postal 402, Mexico

Kazufumi Ito, Center for Research in Scientific Computation, North Carolina State University, Raleigh, North Carolina 27695-8205

M. Jones, Department of Mathematics, Texas Tech University, Lubbock, Texas 79409

Belinda King, Center for Research in Scientific Computation, North Carolina State University, Raleigh, North Carolina 27695-8205

Andrew Kurdila, Department of Aerospace Engineering, Texas A&M University, College Station, Texas 77843

CONTRIBUTORS

Edward Loucks, Department of Systems Science and Mathematics, Washington University, St. Louis, Missouri 63130

John Lund, Department of Mathematical Sciences, Montana State University, Bozeman, Montana 59717-0240

Nancy Lybeck, Department of Mathematical Sciences, Montana State University, Bozeman, Montana 59717-0240

Bernard Mair, Department of Mathematics, University of Florida, Gainesville, Florida 32611

Hamadi Marrekchi, Interdisciplinary Center for Applied Mathematics, Department of Mathematics, Virginia Polytechnic Institute and State University, Blacksburg, Virginia 24061

Clyde Martin, Department of Mathematics, Texas Tech University, Lubbock, Texas 79409

Kelly McArthur, Department of Mathematics, Colorado State University, Fort Collins, Colorado 80523

Robert Miller, Department of Mathematical Sciences, University of Arkansas, Fayetteville, Arkansas 72701

Janet Peterson, Interdisciplinary Center for Applied Mathematics, Department of Mathematics, Virginia Polytechnic Institute and State University, Blacksburg, Virginia 24061

Viswanath Ramakrishna, Frick Laboratories, Princeton University, Princeton, New Jersey 08544

Joachim Rosenthal, Department of Mathematics, University of Notre Dame, Notre Dame, Indiana 46556

Ralph Smith, Institute for Computer Applications in Science and Engineering, NASA Langley Research Center, Hampton, Virginia 23681

Mark Stamp, Department of Mathematics, Worcester Polytechnic Institute, Worcester, Massachusetts 01609

CONTRIBUTORS

Frank Stenger, Department of Computer Science, University of Utah, Salt Lake City, Utah 84112

Thomas Strganac, Department of Aerospace Engineering, Texas A&M University, College Station, Texas 77843

Marc Stromberg, Department of Mathematics, Texas Tech University, Lubbock, Texas 79409

Tom Taylor, Department of Mathematics, Arizona State University, Tempe, Arizona 85287-1804

Reza Vakili, Department of Computer Science, University of Utah, Salt Lake City, Utah 84112

Curt Vogel, Department of Mathematical Sciences, Montana State University, Bozeman, Montana 59717-0240

Gordon Wade, Institute for Scientific Computation, Texas A&M University, College Station, Texas 77843

Dorothy Wallace, Department of Mathematics and Computer Science, Dartmouth College, Hanover, New Hampshire 03755

Xiaochang Wang, Department of Mathematics, Texas Tech University, Lubbock, Texas 79409

Yun Wang, Center for Research in Scientific Computation, North Carolina State University, Raleigh, North Carolina 27695-8205

Joe Wolf, Department of Mathematics, University of California at Berkeley, Berkeley, California 94720

PREFACE

The third Conference on Computation and Control was held at Montana State University in Bozeman, Montana from August 5-11, 1992 and this proceedings represents the evolution that the conference has taken since its 1988 and 1990 predecessors. The first conference and proceedings (Volume 1 in PSCT) nurtured a dialogue between researchers in control theory and the area of numerical computation. This cross-fertilization was continued with the 1990 conference and proceedings (Volume 11 in PSCT) while forecasting the theme for this conference.

The present volume contains a collection of papers addressing issues ranging from noise abatement via smart material technology, robotic vision, and parameter identification to feedback design challenges in fluid control and other areas of topical interest. The area of feedback design in fluid control spawns computational challenges in the form of Burgers' equation which is addressed both with standard numerical methods as well as new computational procedures. Applications which involve inverse problems include material parameter estimation and sampling in observability. Whether motivated by the plant or arising as the distributed system in the design of a feedback compensator for problems in nonlinear control, the theme of this conference placed an emphasis on the use of partial differential equations in control theory. Through challenges initiated via the control problem or the subsequent computational problem, the joint efforts of experts from the respective disciplines enhance the development of both.

There are many parties responsible for the success of the conference and these proceedings. Outside of the university community, the Bozeman Chamber of Commerce and the Sweet Pea Organizers provided a warm welcome for all participants. The Cannery hospitality orchestrated by Bob Fletcher and Candy Bartholomew provides a respite from the rigor in true western ambiance. Within the university community the list is lengthy. The continued support of the Montana State University Foundation and the Office of the Vice-President for Research provided the monetary seed which afforded the Department of Mathematical Sciences the opportunity to, once again, sponsor this biennial conference. The in-house efforts of the staff, faculty and graduate students provided the infrastructure so important to successfully bring the conference from inception to fruition. To all of these people and to the conference participants we extend with a debt of gratitude our warmest thank you. Finally, the editors wish, in particular, to thank Edwin Beschler of Birkhäuser for his presence at the conference and his assistance and encouragement in the preparation of this volume.

Kenneth L. Bowers and John Lund
Bozeman, April 5, 1993

THEORETICAL AND COMPUTATIONAL ASPECTS OF FEEDBACK IN STRUCTURAL SYSTEMS WITH PIEZOCERAMIC CONTROLLERS

H.T. Banks *, Kazufumi Ito * and Belinda B. King *

Center for Research in Scientific Computation
North Carolina State University
Raleigh, North Carolina 27695-8205

1 Introduction

An emerging technology for the 90s that offers new challenges for the distributed parameter control community involves the design, development and use of smart material structures. These structures may involve one or more of several materials such as shape memory alloys, piezoceramics, magnetostrictives, or electrorheological fluids. In this paper, we present a brief summary of recent results of some importance in mathematical and computational issues that are fundamental to the development of smart material technology. To illustrate certain ideas and findings, we apply these results to a specific example of a clamped beam with piezoceramic actuators. We emphasize that such a structure is not a "smart material" (usually defined as one with combined sensing and actuation units embedded along with control elements) but that with proper circuitry (to render the piezoceramic element a "self-sensing actuator" (see [16])) such a structure could be an example of a smart material beam.

In discussing various aspects of our example, we first summarize modeling concepts for piezoceramic actuators. The resulting models involve unbounded control or input operators that require careful mathematical formulation and treatment. We discuss a weak formulation that permits rigorous mathematical discussions of well-posedness, approximation and feedback control issues. After a brief presentation of results on these issues, computational ideas are outlined and applied to our example to demonstrate their usefulness in developing control strategies for piezoceramic actuation of structures.

2 Modeling Piezoceramic Patches as Actuators

We discuss approximation schemes and convergence results for unbounded input operators in the context of a linear beam model with piezoceramic patch actuators. Specifically, as a motivating example we consider

*Research supported in part by the Air Force Office of Scientific Research under grant AFOSR-90-0091.

the Euler-Bernoulli beam model with Kelvin-Voigt damping which can be written mathematically as

$$\rho(x)w_{tt}(t,x) + D^2\left[c_D I(x)D^2 w_t(t,x) + EI(x)D^2 w(t,x)\right] = g(t,x), \quad (2.1)$$

where $w(t,x)$ is the transverse displacement of the beam, $\rho(x)$ is the linear mass density of the beam material, $c_D I(x)$ is the damping coefficient, $EI(x)$ is the flexural rigidity, $g(t,x)$ is the external forcing function or control, the subscripts denote partial differentiation, and the notation $D = \frac{\partial}{\partial x}$ is adopted. Note that the usual density, stiffness and damping parameters are a function of the spatial variable. It was seen in [11, 12, 13], that when using piezoceramic patches as actuators, the structural parameters should be discontinuous over the region where the patch is bonded. Thus, the standard beam parameters ρ, $c_D I$, and EI are "step functions", each having one value along the beam alone and another value along the region of the beam to which the patch is bonded. These will be designated as ρ_b, $c_D I_b$, and EI_b for the parameters along the beam, and ρ_p, $c_D I_p$, and EI_p for the parameters along the beam/patch region.

The form of g is determined by the use of piezoceramic patches as actuators. We summarize salient points for mathematical models for this type of controller as developed in [8, 11, 12]. Consider a beam of length ℓ, thickness h and width 1 as depicted in Figure 2.1.

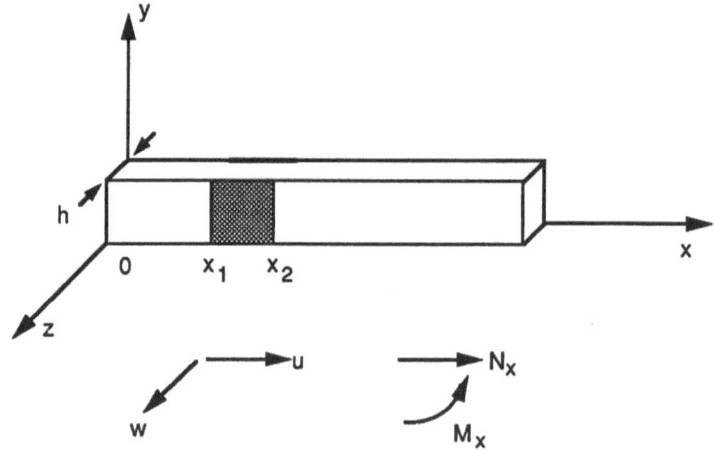

Figure 2.1: Clamped-clamped beam with piezoceramic patch actuators.

Assume that a pair of patches is bonded to the beam at $x_1 \leq x \leq x_2$ along the beam which is clamped at both ends, i.e., at $x = 0$ and at $x = \ell$. Each piezoceramic patch is inherently an electro-mechanical transducer which, when excited by an electric field, induces a strain in the material in the axial direction of the beam if the patch is poled across the thickness as depicted in Figure 2.2. Further assume that the patches have identical polarization, each of which can be excited independently with an applied voltage to produce elongation or contraction (see, e.g. [14, 16, 17] along with the extensive reference list of [12] for more specific details). If the patches are excited "out-of phase" (with equal magnitude voltages which are in opposite sense across the polarizations as depicted Figure 2.3a), one produces contraction in one patch and elongation in the other. This induces surface strains in the beam as shown in Figure 2.3b. The net result on the beam is a pure bending moment about the neutral axis as illustrated in Figure 2.3c.

Figure 2.2: Induced piezoceramic patch strain.

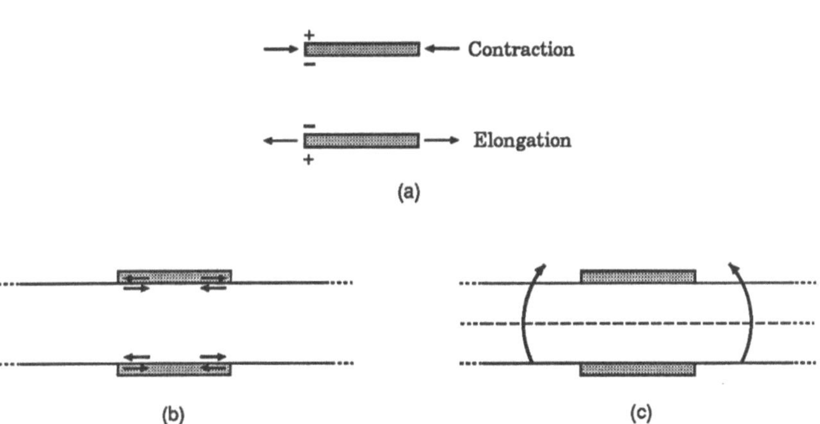

Figure 2.3: (a) Out-of-phase patch excitation; (b) Surface strains and stresses; (c) Pure bending moment about neutral axis of beam.

The patches may also be excited "in-phase", resulting in in-plane compressional or extensional forces along the neutral axis of the beam, although that aspect will not be the focus in this paper. For details on in-plane excitations, see [11, 12].

Quantitative expressions for excitation of the patches can be derived as special cases of the general patch/thin shell interaction models developed in [11, 12]. Consider transverse and axial displacements of the beam, denoted by $w(t, x)$ and $u(t, x)$, respectively. It will be seen that for the case of "out-of-phase" excitation, one need not model the axial displacements of the beam, and the model in (2.1) is sufficient. However, to summarize the patch model in the general setting, we will temporarily consider the axial displacements. Referring to [11, 12], one finds that for an undamped beam, the dynamic equations of motion are given by

$$
\begin{aligned}
\rho(x)\frac{\partial^2 u}{\partial t^2} - \frac{\partial}{\partial x}\left(Eh(x)\frac{\partial u}{\partial x}\right) &= -S_{1,2}(x)\frac{\partial}{\partial x}\,[N_x]_p \\
\rho(x)\frac{\partial^2 w}{\partial t^2} + \frac{\partial^2}{\partial x^2}\left(EI(x)\frac{\partial^2 w}{\partial x^2}\right) &= \hat{q}_n + \frac{\partial}{\partial x}\left(-\frac{\partial}{\partial x}\,[M_x]_p\right)
\end{aligned}
\tag{2.2}
$$

for $0 \leq x \leq \ell$, $t > 0$. We remark that damping can be added to the model in the standard fashion so long as one takes into account appropriate discontinuities of the damping coefficients due to the presence of the bonded patches. The discontinuities for mass density and stiffness can be expressed in terms of material properties of the patches and the beam. For patch pairs with edges at x_1 and x_2 (see Figure 2.1), the linear mass density is given by

$$
\rho(x) = \rho_b h + 2\rho_p T \chi_p(x), \quad \chi_p(x) = \begin{cases} 1, & x_1 \leq x \leq x_2 \\ 0, & \text{otherwise} \end{cases}
$$

where T is the thickness of the patch (throughout this section, we assume that the beam and patches have width 1). Similarly the terms Eh and EI are given by

$$
\begin{aligned}
Eh(x) &= E_b h + 2E_p T \chi_p(x), \\
EI(x) &= E_b \frac{h^3}{12} + E_p\left[hT^2 + \frac{1}{2}h^2 T + \frac{2}{3}T^3\right]\chi_p(x),
\end{aligned}
$$

where E_b and E_p are the Young's moduli for the beam and patch, respectively.

The force components are given in terms of $\hat{q}_n$, the total external normal (transverse) surface load, and $[N_x]_p$, the piezoceramic induced line forces in the x-direction (see [12] for a discussion concerning the relationship between the induced line forces $[N_x]_p$, having units of force, and the longitudinal force $S_{1,2}\frac{\partial}{\partial x}\,[N_x]_p$ having units of force per unit length along the neutral axis). The indicator function, $S_{1,2}(x)$, appearing in the longitudinal force has the values 1 for $x < (x_1 + x_2)/2$, 0 for $x = (x_1 + x_2)/2$, and -1 for $x > (x_1 + x_2)/2$, and determines the direction of the applied force. The quantity $[M_x]_p$ represents the piezoceramic induced line moments (about the neutral axis) about the x-axis and has units of moment.

As shown in [12], the external moments and forces resulting from the excitation of the patches can be written as

$$[N_x]_p = -E_p d_{31} \left[H(x - x_1) - H(x - x_2) \right] S_{1,2}(x) \left(V_1 + V_2 \right)$$

$$[M_x]_p = -E_p \frac{d_{31}}{2} \left[h + T \right] \left[H(x - x_1) - H(x - x_2) \right] \left(V_1 - V_2 \right) \tag{2.3}$$

where H is the Heaviside function, and V_1 and V_2 are the voltages into the patches (note that this allows for different voltages into the individual patches in the pair). The piezoelectric charge coefficient d_{31} relates the applied electric field with the resulting mechanic strain. For notational brevity, we write $H_i = H(x - x_i), i = 1, 2$.

By choosing $V_1 = -V_2 = V$, one excites the patches *out-of-phase* and obtains

$$[N_x]_p = 0$$

$$[M_x]_p = -E_p d_{31} \left[h + T \right] \left[H_1 - H_2 \right] V \tag{2.4}$$

which results in pure bending of the beam.

Therefore, when the patches are used as bending moment actuators and there is no external surface load on the beam, g has the form

$$g(t, x) = D^2 E_p d_{31} \left[h + T \right] \left[H_1 - H_2 \right] u(t) \tag{2.5}$$

where $u(t) = V(t)$ is the voltage applied to the patch. This situation will be the focus of our considerations in this paper.

Since the beam was assumed to be clamped at both ends and of length ℓ, the following boundary conditions hold for $t \geq 0$:

$$w(t, 0) = 0, \quad Dw(t, 0) = 0,$$

$$w(t, \ell) = 0, \quad Dw(t, \ell) = 0. \tag{2.6}$$

In addition, we assume initial conditions of the form

$$w(0, x) = w_0 \quad w_t(0, x) = w_1 \tag{2.7}$$

for $0 \leq x \leq \ell$.

The system (2.1), (2.5), (2.6), (2.7) represents our piezoceramically excited clamped beam model in strong form. We observe that this involves second derivatives of Heaviside functions or derivatives of Dirac delta functions. Consequently, the variational form of this model is most conveniently used as a basis for approximation and computation. To reformulate, we take V and H to be Hilbert spaces with (see [21])

$$V \hookrightarrow H \simeq H^* \hookrightarrow V^*.$$

Then by choosing the state $z(t) = w(t, \cdot) \in H = L_2[0, \ell]$, one can write the system (2.1), (2.5), (2.6), (2.7) as

$$\ddot{z}(t) + A_2 \dot{z}(t) + A_1 z(t) = Bu(t) \quad in \ V^*$$
$$z(0) = z_0, \ \dot{z}(0) = z_1.$$
(2.8)

The variational form of (2.8) can be derived by choosing the Hilbert space $V = H_0^2[0, \ell] = \{z \in H^2[0, \ell] : z(0) = 0, Dz(0) = 0, z(\ell) = 0, Dz(\ell) = 0\}$. Then we seek $z(t) \in V$ satisfying

$$\langle \rho z_{tt}, \phi \rangle_{V^*, V} + \sigma_2(z_t, \phi) + \sigma_1(z, \phi) = \langle g, \phi \rangle_{V^*, V},$$
$$z(0) = z_0, \ \dot{z}(0) = z_1,$$
(2.9)

for all $\phi \in V$, where $\langle \cdot, \cdot \rangle_{V^*, V}$ is the usual duality pairing (see [21]) and

$$\begin{aligned}
\sigma_1(\psi, \phi) &= \langle A_1 \psi, \phi \rangle_{V^*, V} = \langle EID^2 \psi, D^2 \phi \rangle_{L_2(0, \ell)}, \\
\sigma_2(\psi, \phi) &= \langle A_2 \psi, \phi \rangle_{V^*, V} = \langle c_D ID^2 \psi, D^2 \phi \rangle_{L_2(0, \ell)}, \\
g(t) &= D^2 E_p d_{31} [h + T][H_1 - H_2] u(t).
\end{aligned}$$
(2.10)

We can readily argue that σ_1, σ_2 are V-elliptic, symmetric, and continuous sesquilinear forms on $V \times V$ so $A_i \in \mathcal{L}(V, V^*)$. Let $\mathcal{V} = V \times V$ and $\mathcal{H} = V \times H$ so that $\mathcal{H} \simeq \mathcal{H}^*$, $\mathcal{V}^* = V \times V^*$, and $\mathcal{V} \hookrightarrow \mathcal{H} \hookrightarrow \mathcal{V}^*$. To establish that our model is well-posed, we may appeal to the following theorem from [4, 5].

Theorem 2.1 *If the sesquilinear forms σ_1 and σ_2 are continuous and V-elliptic with σ_1 symmetric, and $g \in L_2((0, T), V^*)$, then for each $y_0 = (z_0, z_1) \in \mathcal{V}$, the initial value problem (2.9) has a unique solution $y = (z, z_t) \in L_2((0, T), \mathcal{V})$. Moreover, this solution depends continuously on g and y_0 in the sense that the mapping $(y_0, g) \rightarrow y$ is continuous from $\mathcal{V} \times L_2((0, T), V^*)$ to $L_2((0, T), \mathcal{V})$.*

A complete discussion for the mathematical foundations of such formulations can be found in [4, 5, 21] with details for specific examples in [1, 2, 4, 10, 13, 18, 19].

3 The Control Problem and Approximation Schemes

In this section, we formulate a typical linear quadratic regulator (LQR) problem for the infinite dimensional system (2.8). We then formulate a class of approximation problems and summarize one set of convergence results for feedback gains, controlled trajectories, and controls that are available for such problems.

The first order form of (2.8) can be written by defining $y = (z, \dot{z})^T$ in $\mathcal{V}$. Then

$$\dot{y}(t) = \mathcal{A}y(t) + \mathcal{B}u(t) \quad in \; \mathcal{V}^*$$
$$y(0) = y_0 = (z_0, z_1)^T,$$

(3.1)

where

$$\mathcal{A} = \begin{bmatrix} 0 & I \\ -A_1 & -A_2 \end{bmatrix}, \quad \mathcal{B} = \begin{bmatrix} 0 \\ B \end{bmatrix}.$$

In this setting, the standard LQR control problem takes the form

$$\min_{u \in L_2((0,\infty),U)} J(y_0, u) = \int_0^\infty (\langle \mathcal{C}y(t), \mathcal{C}y(t) \rangle_Z + \langle Ru(t), u(t) \rangle_U) dt \quad (3.2)$$

subject to (3.1), where $\mathcal{C} \in \mathcal{L}(\mathcal{H}, Z)$, Z is the output or observation space, and $R \in \mathcal{L}(U)$ is self-adjoint, positive definite. The following well-posedness results for the feedback form of the optimal solution to the LQR problem are extensions of well-known results in the case that $\mathcal{B}$ is bounded and can be found, for example, in [3, 4].

Theorem 3.1 *If $(\mathcal{A}, \mathcal{B})$ is stabilizable, $(\mathcal{A}, \mathcal{C})$ is detectable, then there exists a unique optimal solution to the control problem, $\bar{u} \in L_2((0, \infty), U)$. This optimal control can be written in the feedback form*

$$\bar{u}(t) = -R^{-1}\mathcal{B}^* \Pi_\infty S(t) y_0 \qquad (3.3)$$

where $\Pi_\infty \in \mathcal{L}(\mathcal{V}^*, \mathcal{V})$ is the unique nonnegative self-adjoint solution of the algebraic Riccati equation

$$\mathcal{A}^* \Pi + \Pi \mathcal{A} - \Pi \mathcal{B} R^{-1} \mathcal{B}^* \Pi + \mathcal{C}^* \mathcal{C} - 0 \qquad (3.4)$$

and $S(t) = e^{(\mathcal{A} - \mathcal{B}R^{-1}\mathcal{B}^*\Pi_\infty)t}$ is the exponentially stable semigroup generated by $\mathcal{A} - \mathcal{B}R^{-1}\mathcal{B}^* \Pi_\infty$.

The optimal control, $\bar{u}$, can be written as the $\mathcal{H}$ inner product of a gain K and the state y. To determine the precise form of the feedback control, one can write the Riccati operator Π_∞ as

$$\Pi_\infty = \begin{bmatrix} \Pi_{11} & \Pi_{12} \\ \Pi_{21} & \Pi_{22} \end{bmatrix},$$

where $\Pi_{11} \in \mathcal{L}(V, V)$, $\Pi_{12} \in \mathcal{L}(V^*, V)$, $\Pi_{21} \in \mathcal{L}(V, V)$, and $\Pi_{22} \in \mathcal{L}(V^*, V)$. Then, if we choose the control weighting operator $R = I$, (3.3) implies that $\bar{u}$ can be written

$$\begin{aligned} \bar{u}(t) &= -[0 \; B^*] \begin{bmatrix} \Pi_{11} & \Pi_{12} \\ \Pi_{21} & \Pi_{22} \end{bmatrix} \begin{bmatrix} z(t) \\ z_t(t) \end{bmatrix}, \\ &= -B^* (\Pi_{21} z(t) + \Pi_{22} z_t(t)). \end{aligned}$$

If we use the fact that

$$\langle Bu(t), \phi \rangle = \langle u(t), B^* \phi \rangle = c \int_0^\ell [H_1 - H_2] u(t) D^2 \phi \, dx \quad \text{for } \phi \in V$$

where $c = E_p d_{31}[h + T]$ and define $\hat{\chi}(x) = \int_0^x \int_0^s \chi_p(\xi) d\xi ds$ where χ_p is as defined in Section 2, then

$$\begin{aligned}
\bar{u}(t) &= c \langle \hat{\chi}, (\Pi_{21} z(t) + \Pi_{22} z_t(t)) \rangle_V \\
&= \langle k_1, z(t) \rangle_V + \langle k_2, z_t(t) \rangle_H, \\
&= \langle \tilde{k}_1, D^2 z(t) \rangle_H + \langle k_2, z_t(t) \rangle_H,
\end{aligned} \tag{3.5}$$

where $k_1 = c\Pi_{21}^* \hat{\chi}$, $\tilde{k}_1 = D^2 k_1$ is the bending gain and $k_2 = cD^4\Pi_{22}^* \hat{\chi}$ is the velocity gain.

The Approximation Framework

The objective of an approximation scheme is to produce a sequence of suboptimal controls, $\bar{u}^N(t)$, such that $\bar{u}^N(t)$ converges to $\bar{u}(t)$ in an appropriate sense and if $\bar{u}^N(t)$ is applied to the infinite dimensional system in (3.1), the system response is stable and nearly optimal for any initial condition. Any convergent approximation scheme should yield approximate LQR problems and controllers that satisfy these minimal requirements. One standard way to construct reasonable approximations is to project the infinite dimensional control problem onto a series of finite dimensional subspaces and solve the resulting LQR problems.

To this end, let P_H^N denote a sequence of orthogonal projections $P_H^N : H \to H^N$ where H^N is the associated sequence of finite dimensional subspaces with $H^N \subset V \subset H$. Also, let $\mathcal{H}^N = H^N \times H^N$ with associated projections $P_{\mathcal{H}}^N$. Define $\mathcal{A}^N : \mathcal{H}^N \to \mathcal{H}^N$ as the restriction of $\mathcal{A}$ to $\mathcal{H}^N$, $\mathcal{C}^N$ as the restriction of $\mathcal{C}$ to $\mathcal{H}^N$ and $\mathcal{B}^N = (0, \tilde{B}^N)^T$ where, for a given $B \in \mathcal{L}(U, V^*)$, we define $\tilde{B}^N \in \mathcal{L}(U, H^N)$ by

$$\langle \tilde{B}^N u, \phi \rangle = \langle u, B^* \phi \rangle \quad \text{for all } \phi \in H^N.$$

Then the associated sequence of finite dimensional LQR problems on H^N is given by

$$\min_{u \in L_2((0,\infty),U)} J^N(y_0^N, u) =$$
$$\int_0^\infty (\langle \mathcal{C}^N y^N(t), \mathcal{C}^N y^N(t) \rangle_Z + \langle Ru(t), u(t) \rangle_U) dt \tag{3.6}$$

subject to

$$\begin{aligned}
\dot{y}^N(t) &= \mathcal{A}^N y(t) + \mathcal{B}^N u(t), \quad t > 0 \\
y^N(0) &= y_0^N = P^N y_0.
\end{aligned} \tag{3.7}$$

By Theorem 3.1, if $(\mathcal{A}^N, \mathcal{B}^N)$ is stabilizable and $(\mathcal{A}^N, \mathcal{C}^N)$ is detectable, then there exists a nonnegative self-adjoint solution Π^N to the algebraic Riccati equation in $\mathcal{H}^N$

$$\mathcal{A}^{N*}\Pi^N + \Pi^N \mathcal{A}^N - \Pi^N \mathcal{B}^N R^{-1} \mathcal{B}^{N*}\Pi^N + \mathcal{C}^{N*}\mathcal{C}^N = 0. \qquad (3.8)$$

The corresponding optimal control, $\bar{u}^N$, for the finite dimensional system is given by

$$\bar{u}^N(t) = -R^{-1}\mathcal{B}^{N*}\Pi^N S^N(t)y_0^N \qquad (3.9)$$

where $S^N(t) = e^{(\mathcal{A}^N - \mathcal{B}^N R^{-1}\mathcal{B}^{N*}\Pi^N)t}$ is the C_0-semigroup generated by $\mathcal{A}^N - \mathcal{B}^N R^{-1}\mathcal{B}^{N*}\Pi^N$ on $\mathcal{H}^N$.

To establish convergence of the approximation scheme for the second order system defined on V and H, one might attempt to write the problem as a first order system on the product spaces $\mathcal{V}$ and $\mathcal{H}$ and apply an extension (to systems with unbounded input operator $\mathcal{B}$) of the fundamental convergence theorem from [6] for first order systems. However, one condition of that theorem would require $\mathcal{V}$ to be compactly embedded in $\mathcal{H}$. Banks and Ito point out in [3] that this condition is false in general for infinite dimensional spaces, and in particular for our example. Thus, they provide the following theorem which can be used to establish the desired minimal requirements mentioned above for an approximation scheme.

Theorem 3.2 *Suppose $(\mathcal{A}, \mathcal{B})$ is stabilizable, $(\mathcal{A}, \mathcal{C})$ is detectable, and the following hold:*

(C1) *For each $z \in V$, there exists an element $\tilde{z}^N$ in H^N such that*

$$|z - \tilde{z}^N|_V \leq \epsilon(N), \quad \text{where } \epsilon(N) \to 0 \text{ as } N \to \infty,$$

(C2) *The embedding of V into H is compact.*

Then the approximate system (3.7) - (3.9) satisfies

$$\Pi^N P^N \xi \to \Pi\xi \text{ in } \mathcal{V} \quad \text{for every } \xi \in \mathcal{V}^*,$$

$$S^N(t)P^N\xi \to S(t)\xi \text{ in } \mathcal{V} \quad \text{for every } \xi \in \mathcal{V}^*, t > 0,$$

and therefore, $\bar{u}^N(t) \to \bar{u}(t)$, uniformly in t on compact intervals.

4 An Approximation Scheme

The Galerkin approximation ideas of the previous section are rather standard for this problem. If $\{\beta_1^N, \beta_2^N, \ldots, \beta_N^N\}$ denotes a basis for H^N, then the state $z(t)$ can be approximated by

$$z^N(t) = \sum_{i=1}^{N} \omega_i^N(t)\beta_i^N. \qquad (4.1)$$

When this approximation to the state is substituted into the variational form (2.9) and ϕ is allowed to range over the set $\{\beta_1^N, \beta_2^N, \ldots, \beta_N^N\}$, one obtains the system of equations

$$M^N \omega_{tt}^N(t) + D^N \omega_t^N(t) + K^N \omega^N(t) = B_0^N u(t) \tag{4.2}$$

where the mass matrix M^N, the damping matrix D^N, the stiffness matrix K^N, and the actuator influence vector B_0^N are given by

$$
\begin{aligned}
M_{ij}^N &= \langle \rho \beta_i^N, \beta_j^N \rangle, \\
D_{ij}^N &= \langle c_D I D^2 \beta_i^N, D^2 \beta_j^N \rangle, \\
K_{ij}^N &= \langle EI D^2 \beta_i^N, D^2 \beta_j^N \rangle, \\
B_{0j}^N &= \langle E_p d_{31} (h + T) [H_1 - H_2], D^2 \beta_j^N \rangle, \\
\omega^N(t) &= [\omega_1^N(t), \ldots, \omega_N^N(t)]^T.
\end{aligned}
\tag{4.3}
$$

This finite dimensional system can be written in first order form by letting $\eta^N = (\omega^N, \omega_t^N)^T$:

$$\eta_t^N(t) = A^N \eta^N(t) + B^N u(t), \tag{4.4}$$

where

$$A^N = \begin{bmatrix} 0 & I \\ -(M^N)^{-1} K^N & -(M^N)^{-1} D^N \end{bmatrix},$$

$$B^N = \begin{bmatrix} 0 \\ -(M^N)^{-1} B_0^N \end{bmatrix}.$$

In the computations, cubic B-splines were used as a basis for H^N. Since $H^N \subset V$, the basis elements must satisfy the boundary conditions $\beta_i^N(0) = 0 = \beta_i^N(\ell)$, $D\beta_i^N(0) = 0 = D\beta_i^N(\ell)$. Thus, if the ith B-spline corresponding to the partition $\{\frac{i\ell}{N}\}_{i=0}^N$ of $[0, \ell]$ is denoted b_i^N (e.g., see [20]), a nonorthogonal basis for H^N can be written

$$\beta_i^N = \begin{cases} b_0^N - 2b_{-1}^N - 2b_1^N & i = 1 \\ b_i^N & i = 2, \ldots, N - 2 \\ b_N^N - 2b_{N-1}^N - 2b_{N+1}^N & i = N - 1. \end{cases} \tag{4.5}$$

In this case, H^N is of dimension $N - 1$ and the matrices and vectors in (4.3) are $(N - 1) \times (N - 1)$ and $N - 1$ dimensional, respectively. This approximation scheme for (2.9) satisfies the conditions for convergence given in the previous section, and so, the suboptimal controls and trajectories for the approximations to the linear system are guaranteed to converge to the optimal control and trajectory for the infinite dimensional linear system. More specifically, the approximate solutions (see (4.1))

$z^N(t) = \sum_{i=1}^{N-1} \omega_i^N(t)\beta_i^N = w^N(t,\cdot)$ in H^N satisfy (2.9) for all $\phi \in H^N$, an equivalent equation to (4.2) or, in first order form, (4.4). The optimal control in (3.9) can be equivalently written

$$
\begin{aligned}
\bar{u}^N(t) \;\; &= \;\; -\mathcal{K}^N \left[\begin{array}{c} z^N(t) \\ z_t^N(t) \end{array} \right] \\
&= \;\; -\int_0^l D^2 k_1^N(x) D^2 z^N(t)\,dx - \int_0^l k_2^N(x) z_t^N(t)\,dx \qquad (4.6) \\
&= \;\; -\int_0^l \tilde{k}_1^N(x) D^2 z^N(t)\,dx - \int_0^l k_2^N(x) z_t^N(t)\,dx
\end{aligned}
$$

where the approximate functional gains $\tilde{k}_1^N$, k_2^N are in $L_2[0,l]$. The theory for these problems cited in Theorem 3.2 and in [3] also guarantees that for N sufficiently large, k_1^N, k_2^N are reasonable approximations to the corresponding optimal functional gains (see (3.5)) for the infinite dimensional LQR problem governed by (2.9).

5   Numerical results

To investigate the behavior of the control scheme, the above approximation scheme was implemented and the resulting LQR problem solved for the case of an aluminum beam with parameters listed in the table below. The parameters for the piezoceramic patch were taken from [10]. Note that in practice, d_{31} can vary significantly based on the construction of the patch and thus, parameter estimation must be done to accurately determine the specific value. In this section, we shall discuss the resulting gains, $\tilde{k}_1^N$, k_2^N, for four examples regarding patch placement and size. Then we will present a summary of the simulation results.

Table 5.1: Beam and Piezoceramic Patch Parameters

l	$1\,m$
h	$0.005\,m$
ρ_b	$1.35\,\frac{kg}{m}$
ρ_p	$2.115\,\frac{kg}{m}$
EI_b	$73.96\,Nm^2$
EI_p	$125.4\,Nm^2$
E_p	$6.3 \times 10^{10}\,\frac{N}{m^2}$
$c_D I_b$	$.001\,\frac{kg}{m\,sec}$
$c_D I_p$	$.00125\,\frac{kg}{m\,sec}$
d_{31}	$6.7 \times 10^{-7}\,\frac{m}{volts}$
T	$.0005\,m$

Gains

To investigate properties of the functional gains for bending and velocity, four examples concerning patch pair placement and size were considered. In the first example, a centered patch pair of length equal to one third of the beam length was considered. A centered patch pair was also used in the second example, but of length equal to only one sixth of the beam length. In the third example, an off-centered patch pair of length one sixth the length of the beam was used. The fourth example had two patch pairs, symmetrically placed about the center of the beam, each of which were of length equal to one sixth the beam length. These examples are schematically represented in the figure below.

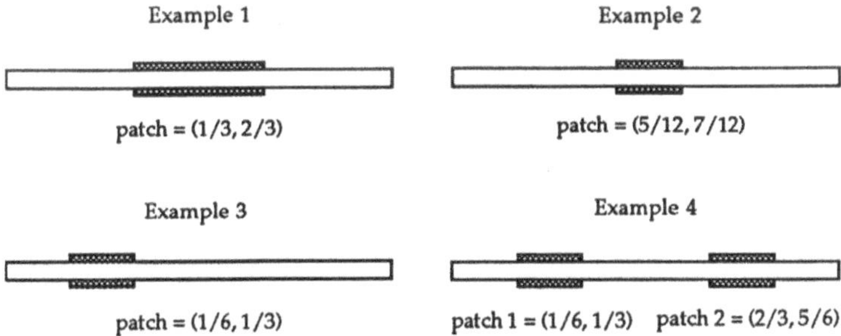

Figure 5.1: Patch Pair Placement and Size Examples

For each of these examples, the LQR problem was solved for $N = 12, 24, 36,$ and 48. In all cases, the gains showed the same pattern of converging to some nonzero function along the part of the beam to which the patches are bonded, and converging to a function oscillating about zero along the portion of the beam with no patches. Careful consideration of (3.5) and (4.6) leads one to expect that the optimal gains (at least $\tilde{k}_1$) might be discontinuous. In this case, use of continuous basis elements as in our examples here would, not surprisingly, produce approximate gains exhibiting Gibbs-like phenomena.

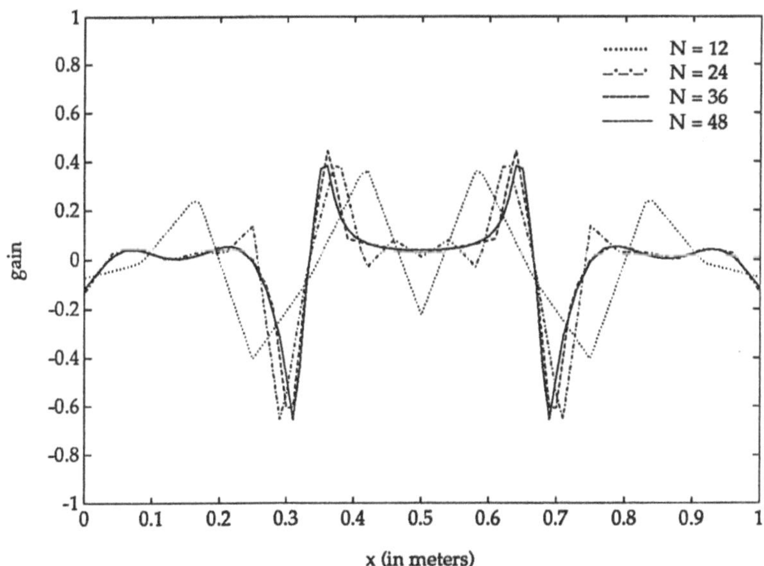

Figure 5.2: Bending Gain, $\tilde{k}_1^N$, Patch Example 1

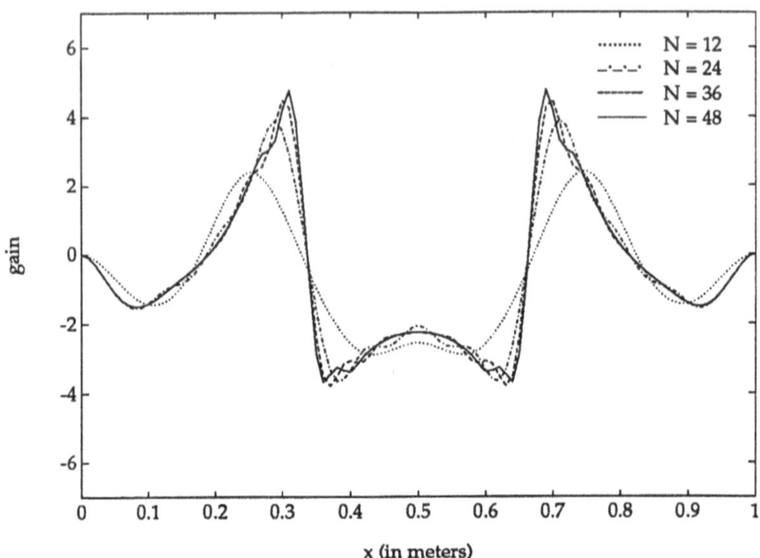

Figure 5.3: Velocity Gain, k_2^N, Patch Example 1

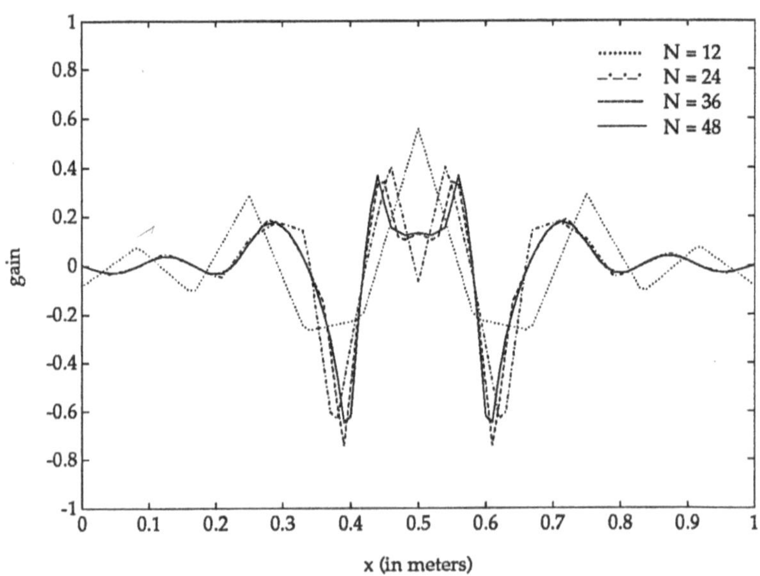

Figure 5.4: Bending Gain, $\tilde{k}_1^N$, Patch Example 2

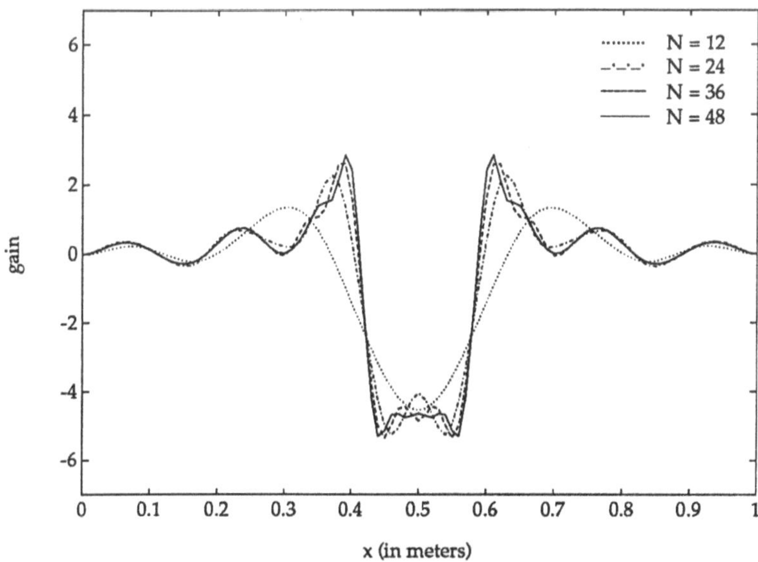

Figure 5.5: Velocity Gain, k_2^N, Patch Example 2

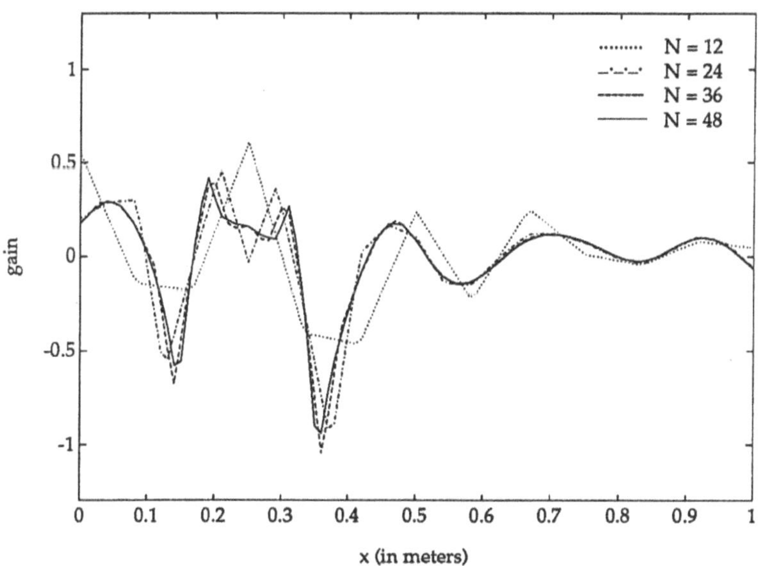

Figure 5.6: Bending Gain, $\tilde{k}_1^N$, Patch Example 3

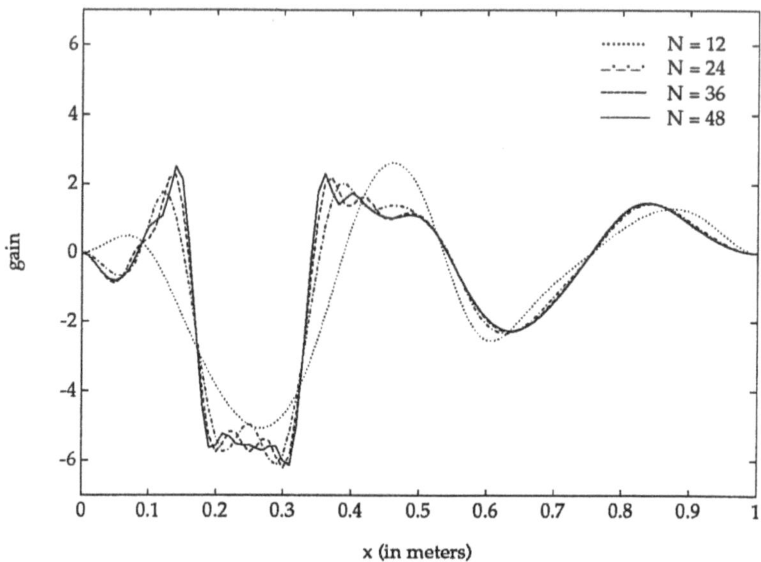

Figure 5.7: Velocity Gain, k_2^N, Patch Example 3

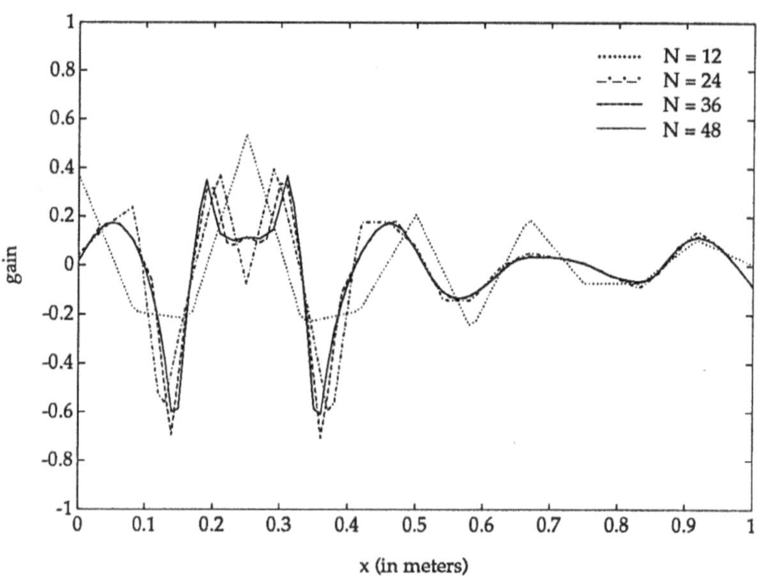

Figure 5.8: Bending Gain, Patch 1, $\tilde{k}_1^N$, Patch Example 4

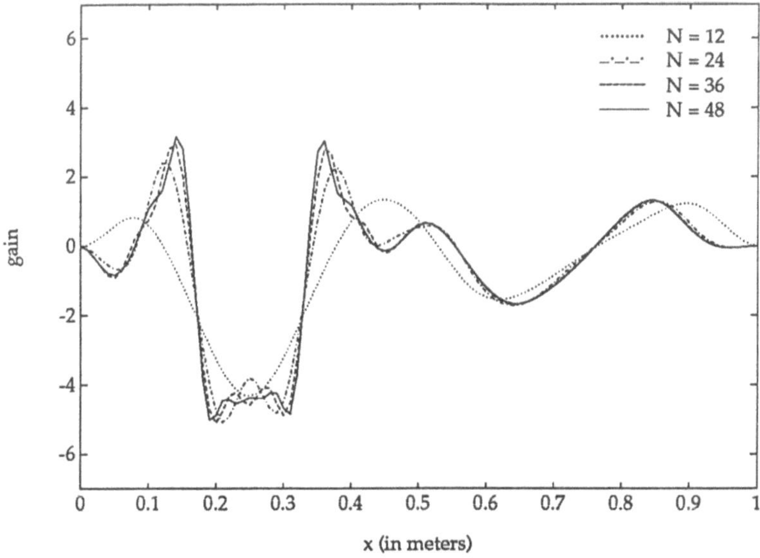

Figure 5.9: Velocity Gain, k_2^N, Patch 1, Patch Example 4

We point out that Example 4 has additional bending and velocity gains for the second patch. Those gains are simply Figures 5.8 and 5.9 reflected symmetrically about the center of the beam (i.e., $x = .5$), and are therefore not shown here.

Simulations

Three initial conditions were chosen to test the performance of the control scheme, one symmetric about the center of the beam, one antisymmetric, and one asymmetric about the center of the beam. The specific initial conditions are shown in Figure 5.10 below. For each initial condition, the control from each of the patch examples was applied, although not all cases are shown here. The examples are referred to as "Example a.b" where "a" refers to the initial condition and "b" refers to the patch placement, e.g., Example 1.2 refers to the symmetric initial condition with the centered patch pair one-sixth of the beam length. For each example, the open and closed loop displacement at a particular point along the beam is plotted for values of t between 0 and 1 second. The particular point chosen corresponds to that x-value for which the initial displacement is a maximum.

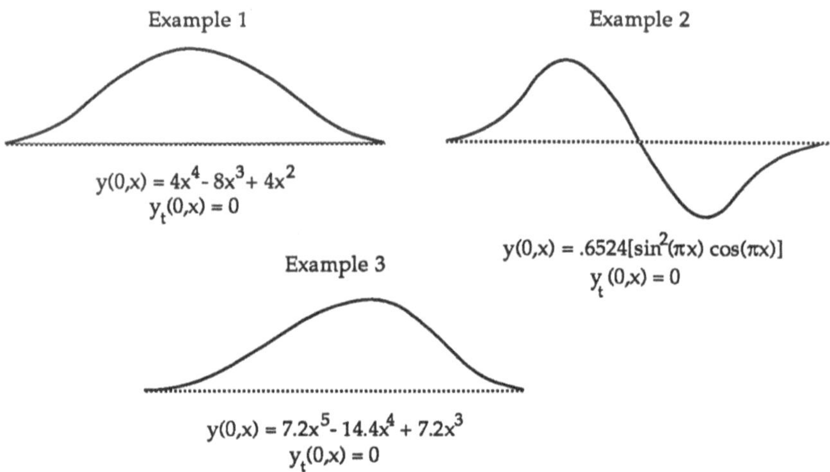

Figure 5.10: Initial Condition Examples

For the symmetric initial condition, the open loop displacement at the point $x = .5$ is qualitatively the same for all patch pair placements, i.e., the rate of decay over the given time interval is essentially the same. This behavior is shown for Example 1.1 in Figure 5.11 below.

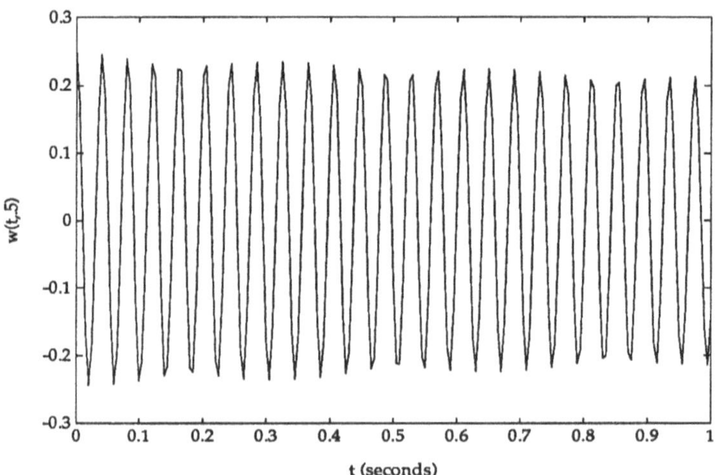

Figure 5.11: Example 1.1 Open Loop Simulation, $w(t,.5)$

In Figure 5.12, we see a rapid decay in the closed loop displacement at $x = .5$ for Example 1.1. The magnitude of $w(t, .5)$ is zero within .06 seconds. Example 1.2 exhibited similar behavior, although the rate of decay was slightly less with decay of $w(t, .5)$ to zero in .12 seconds and a greater negative displacement of .07 at .02 seconds.

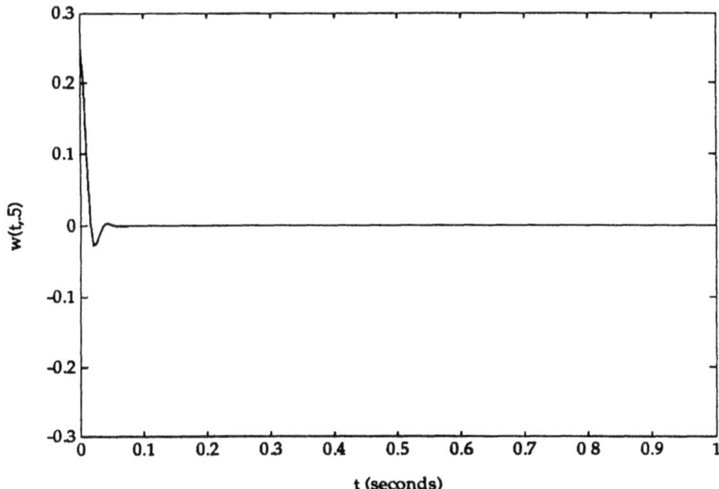

Figure 5.12: Example 1.1 Closed Loop Simulation, w(t,.5)

In Figure 5.13, the closed loop displacement at $x = .5$ corresponding to the symmetric initial condition is shown when the control is applied via the off-centered patch pair. Decay of $w(t, .5)$ to zero takes much longer, occurring at about .6 seconds with rapid oscillation of the beam before that time.

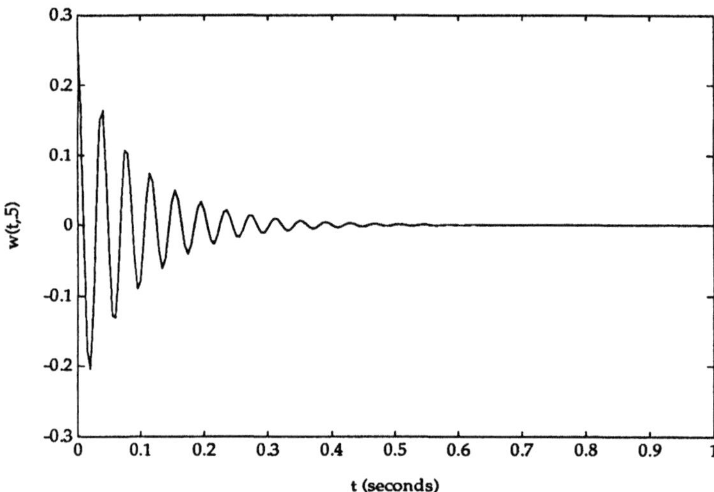

Figure 5.13: Example 1.3 Closed Loop Simulation, w(t,.5)

When two off-centered patch pairs are used, the same oscillatory displacement at $x = .5$ is observed except that the decay of $w(t, .5)$ to zero occurs in approximately .4 seconds.

In Figure 5.14, the open loop displacement at $x = .3125$ corresponding to the antisymmetric initial condition is shown when a centered patch pair is used. One would expect virtually no effect from this placement of the patches, and, in fact, the closed loop displacement is indistinguishable from the open loop displacement.

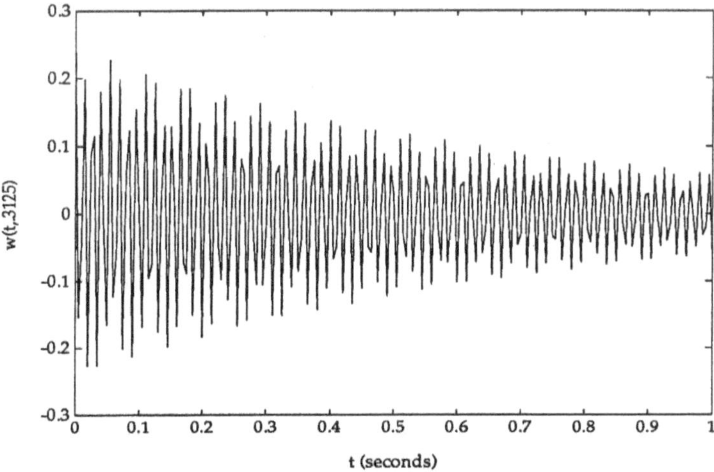

Figure 5.14: Example 2.2 Open Loop Simulation, w(t,.3125)

When the off-centered patch pair is used, the beam is driven to rest rapidly as is shown in Figure 5.15. As one would expect, the two patch pair actuator drives the beam to rest even more rapidly, within .02 seconds to be specific; this example is not shown here.

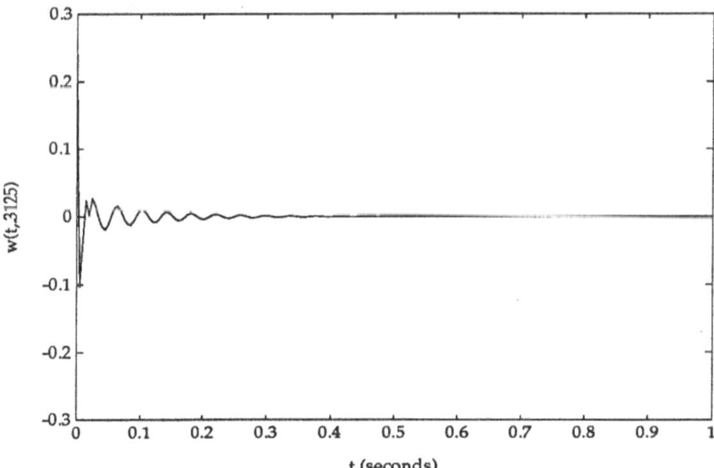

Figure 5.15: Example 2.3 Closed Loop Simulation, w(t,.3125)

The open loop displacement at $x = .6042$ corresponding to the asymmetric initial condition is shown below when the off-centered patch pair is used. The closed loop displacement shown in Figure 5.17 exhibits rapid decay to zero. It is interesting to note that the choice of the end of the beam to which the pair is affixed has little effect on the controlled beam displacement. The closed loop displacements for Examples 3.1, 3.2, and 3.4 are similar to Examples 1.1, 1.2 and 1.4 both qualitatively and in the magnitude of displacements and are not shown here.

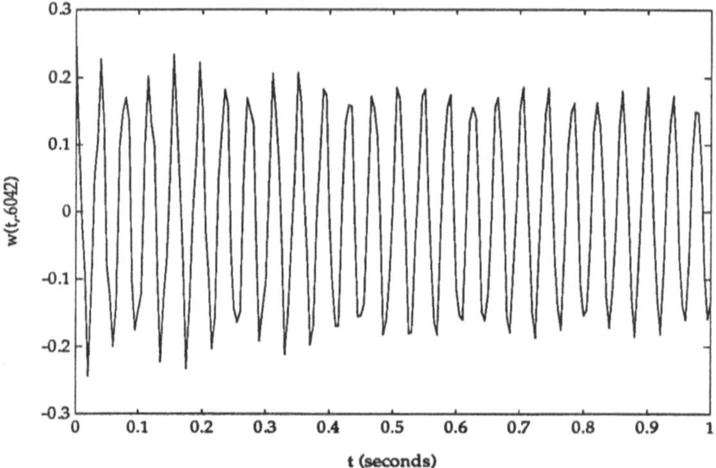

Figure 5.16: Example 3.3 Open Loop Simulation, w(t,.6042)

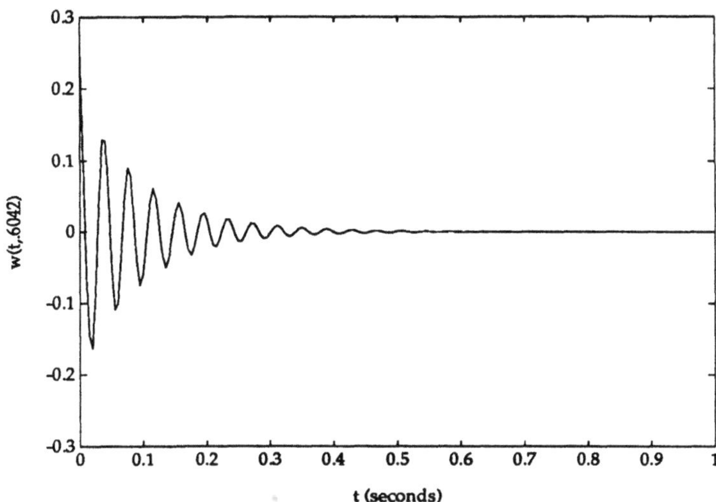

Figure 5.17: Example 3.3 Closed Loop Simulation, w(t,.6042)

As a final note, we observe that when one correctly includes the spatially varying aspect of structural parameters $(\rho, c_D I, EI)$ in the model, the influence of patch pair placement on open loop dynamics is substantial. The following two figures amply demonstrate this. Figure 5.18 depicts the antisymmetric initial condition with the centered patch pair of length equal to one-sixth of the beam length, while Figure 5.19 depicts the same initial condition with the off-centered patch pair. In these figures, we graph the displacement of the entire beam whereas in the previous figures, we graphed the displacement of only one point along the beam. The beam displacement is graphed at 10 time samples (each 1/10th of a second apart) starting at $t = 0$.

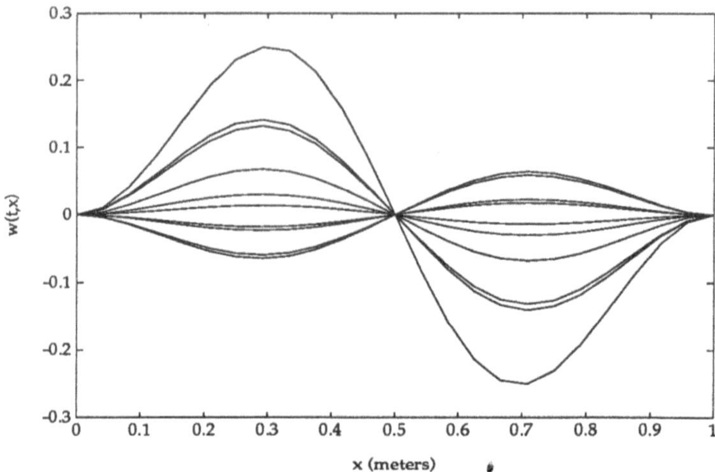

Figure 5.18: Example 2.2 Open Loop Simulation

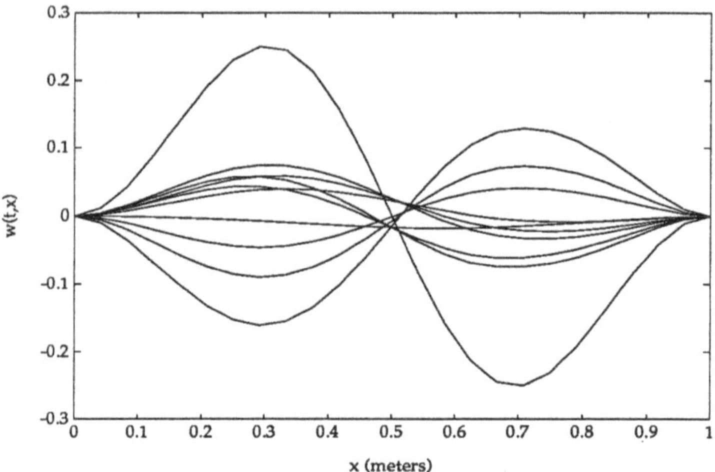

Figure 5.19: Example 2.3 Open Loop Simulation

6 Conclusions

In conclusion, we have presented results concerning modeling, control, approximation, and computation for structures with piezoceramic actuators. Further theoretical and computational studies are underway to more completely characterize the continuity properties of the gains. The computational results presented give general ideas as to strategies for piezoceramic actuator design with regard to placement and size for certain types of initial conditions. The importance of patch pair location on open loop dynamics was demonstrated.

References

[1] H.T. BANKS, W. FANG, R.J. SILCOX and R.C. SMITH, "Approximation Methods for Control of Acoustic/Structure Models with Piezoceramic Actuators," *J. Intelligent Material Systems and Structures*, v. 4, 1993, pp. 98-116.

[2] H.T. BANKS, S. GATES, G. ROSEN and Y. WANG, "The Identification of a Distributed Parameter Model for a Flexible Structure," *SIAM J. Contr. and Opt.*, v. 26, 1988, pp. 743-762.

[3] H.T. BANKS and K. ITO, "Approximation in LQR Problems for Infinite Dimensional Systems with Unbounded Input Operators," SIAM Conf. on Control, San Francisco, May 1990, preprint.

[4] H.T. BANKS, K. ITO, and Y. WANG, "Computational Methods for Identification and Feedback Control in Structures with Piezoceramic Actuators and Sensors," in *Recent Advances in Adaptive and Sensor Materials and their Applications*, (C.A. Rogers and R.C. Rogers, ed.), Technomic Publ., 1992, pp. 111-119.

[5] H.T. BANKS, K. ITO, and Y. WANG, "Well-posedness for Damped Second Order Systems with Unbounded Input Operators," to appear.

[6] H.T. BANKS and K. KUNISCH, The linear regulator problem for parabolic systems, *SIAM J. Contr. and Opt.*, v. 22, 1984, pp. 684-698.

[7] H.T. BANKS and R.C. SMITH, "Feedback Control of Noise in a 2-D Nonlinear Structural Acoustics Model," preprint.

[8] H.T. BANKS and R.C. SMITH, "Models for Control in Smart Material Structures," *Proc. 10th Annual Joint Summer Research Conf. in the Math. Sciences* (Mt. Holyoke College, July 11-17, 1992), SIAM, Philadelphia, to appear.

[9] H.T. BANKS and R.C. SMITH, "Parameter Estimation in a 2-D Acoustic/Structure Interaction Model with Fully Nonlinear Coupling Conditions," preprint.

[10] H.T. BANKS, R. SILCOX, R. SMITH, "The Modeling and Control of Acoustic/Structure Interaction Problems via Piezoceramic Actuators: 2-D Numerical Examples," Center for Research in Scientific Computation Technical Report CRSC-TR92-1, N. C. State University, April 1992; J. Vibrations and Acoustics, submitted.

[11] H.T. BANKS, R.C. SMITH and Y. WANG, "Modeling Aspects for Piezoelectric Patch Activation of Shells, Plates and Beams," Center for Research in Scientific Computation Technical Report CRSC-TR92-12, N. C. State University, November 1992.

[12] H.T. BANKS, R.C. SMITH and Y. WANG, "The Modeling of Piezoceramic Patch Interactions with Shells, Plates and Beams," *Quarterly of Applied Mathematics*, to appear.

[13] H.T. BANKS, Y. WANG, D.J. INMAN, and J.C. SLATER, "Variable Coefficient Distributed Parameter System Models for Structures with Piezoceramic Actuators and Sensors," Center for Research in Scientific Computation Technical Report CRSC-TR92-9, N. C. State University, September 1992; *Proc. 31st IEEE Conf. on Decision and Control*, Tucson, Dec. 1992, pp. 1803-1808.

[14] E.F. CRAWLEY, J. de LUIS, N.W. HAGOOD and E.H. ANDERSON, "Development of Piezoelectric Technology for Applications in Control of Intelligent Structures," Applications in Control of Intelligent Structures, American Controls Conference, Atlanta, June 1988, pp. 1890-1896.

[15] J.J. DOSCH, D.J. INMAN and E. GARCIA, "A Self-Sensing Piezoelectric Actuator for Collocated Control," *Journal of Intelligent Material Systems and Structures*, v. 3, 1992, pp. 166-185.

[16] J.L. FANSAN and T.K. CAUGHEY, "Positive Position Feedback Control for Large Space Structures," AIAA Paper No. 87-0902, Proceedings of the 28th AIAA/ ASME/ASCE/AHS Structures, Structural Dynamics and Materials Conference, AIAA Dynamics Specialists Conference, Part IIb, Monterey, CA, April 9-10, 1987, pp. 588-598.

[17] H. HAGOOD, W. CHUNG and A. von FLOTOW, "Modelling of Piezoelectric Actuator Dynamics for Active Structural Control," *Journal of Intelligent Material Systems and Structures*, v. 1, 1990, pp. 327-354.

[18] B.B. KING, *Modeling and Control of Multiple Component Structures*, Ph.D dissertation, Clemson University, December 1991.

[19] B.B. KING, "Modeling and Control of a Multiple Component Structure," Center for Research in Scientific Computation Technical Report CRSC-TR92-3, N. C. State University, June 1992; *J. Mathematical Systems, Estimation, and Control*, to appear.

[20] M.H. SCHULTZ, *Spline Analysis*, Prentice-Hall, Englewood Cliffs, 1973.

[21] J. WLOKA, *Partial Differential Equations*, Cambridge University Press, New York, 1987.

MODELING AND APPROXIMATION OF A COUPLED 3-D STRUCTURAL ACOUSTICS PROBLEM

H.T. Banks[*] and R.C. Smith[†]

*Center for Research in Scientific Computation
North Carolina State University
Raleigh, NC 27695

†Institute for Computer Applications in Science and Engineering
NASA Langley Research Center
Hampton, VA 23681

1 Introduction

A growing area of research in the structural acoustics community concerns the problem of reducing structure-borne noise levels within an acoustic cavity. A specific example of this is motivated by the development of a new class of turboprob and turbofan engines which are very fuel efficient but also very noisy. The low frequency high amplitude acoustic fields produced by these engines cause vibrations in the fuselage which in turn generate unwanted interior noise. The passive control techniques which were initially considered were in general undesirable since the increased weight offset the advantages gained through the use of the new engines and lighter airframe materials. This then led to the study of active control techniques for this problem both in a frequency domain setting and from a time domain approach (PDE approach).

In this work we develop the numerical methods needed to extend previous 2-D time domain results [1, 3] to a 3-D geometry in which the parameter estimation and control techniques can be experimentally tested. The domain Ω of interest consists of a cylinder of length ℓ and radius a as pictured in Figure 1.1. At one end of the cylinder is a clamped flexible plate of thickness h which is assumed to have Kelvin-Voigt damping. The other end of the cylinder is closed and hardwall boundary conditions are assumed on all walls of the cylinder other than the flexible plate. The choice of this geometry and configuration results from the experimental setup which consists of a concrete pipe with a clamped aluminum plate at one end.

---

*†The research of H.T.B. was supported in part by the Air Force Office of Scientific Research under grant AFOSR-90-0091. This research was also supported by the National Aeronautics and Space Administration under NASA Contract Numbers NAS1-18605 and NAS1-19480 while H.T.B. was a visiting scientist and R.C.S. was in residence at the Institute for Computer Applications in Science and Engineering (ICASE), NASA Langley Research Center, Hampton, VA 23681.

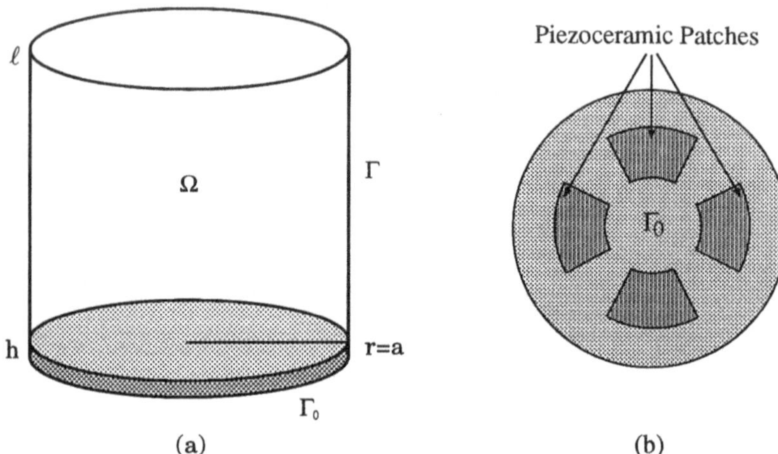

Figure 1.1: (a) The cylindrical acoustic cavity; (b) The circular plate with piezoceramic patches.

Control is implemented in the model via piezoceramic patches on the plate (see Figure 1.1) which produce pure bending moments and/or extensional forces when a voltage is applied [4]. Because of the coupling between the interior acoustic field and the structural vibrations, this provides a means of controlling the interior acoustic pressure levels through the control of the bounding structure dynamics. The benefits of using the piezoceramic patches as actuators are augmented by the fact that they are inexpensive, lightweight and space efficient.

For acoustic waves having small amplitude, the cavity dynamics can be modeled by the undamped wave equation

$$\phi_{tt} = c^2 \Delta\phi \qquad , \quad (r, \theta, z) \in \Omega \ , t > 0 \ ,$$

$$\nabla\phi \cdot \hat{n} = 0 \qquad , \quad (r, \theta, z) \in \Gamma \ , t > 0$$

where c is the speed of sound in the cavity, ϕ is the velocity potential, $\hat{n}$ is the outer normal to the boundary Γ, z is the axial coordinate, and the Laplacian in cylindrical coordinates is given by

$$\Delta\phi = \frac{\partial^2\phi}{\partial r^2} + \frac{1}{r}\frac{\partial\phi}{\partial r} + \frac{1}{r^2}\frac{\partial^2\phi}{\partial\theta^2} + \frac{\partial^2\phi}{\partial z^2} \ .$$

The boundary conditions on Γ (all walls other than the plate) model the hard wall conditions of the experimental concrete cylinder. Damping was omitted due to the relatively small dimensions of the cylinder ($a = 9\,in$,

$\ell = 42\,in$). Finally, we point out that the acoustic pressure p is related to the potential through the relationship $p = \rho_f \phi_t$ where ρ_f is the equilibrium density of the atmosphere.

The motion of the plate is influenced by a forcing function f which models the exterior noise source, the moments generated by the patches, and the backpressure due to pressure oscillations within the cavity. In formulating the patch contributions, we have assumed that s piezoceramic patches have been bonded to the plate as depicted in Figure 1.1. As detailed in [5], the strong form of the equations of motion for a fixed-edge, damped circular plate that is subjected to these moments and forces is

$$\rho_p h w_{tt} + \frac{\partial^2 \mathcal{M}_r}{\partial r^2} + \frac{2}{r}\frac{\partial \mathcal{M}_r}{\partial r} - \frac{1}{r}\frac{\partial \mathcal{M}_\theta}{\partial r} + \frac{2}{r}\frac{\partial^2 \mathcal{M}_{r\theta}}{\partial r \partial\theta} + \frac{2}{r^2}\frac{\partial \mathcal{M}_{r\theta}}{\partial\theta} + \frac{1}{r^2}\frac{\partial^2 \mathcal{M}_\theta}{\partial\theta^2}$$

$$= -\rho_f \phi_t(t, r, \theta, w(t, r, \theta)) + f(t, r, \theta) ,$$

$$w(t, a, \theta) = \frac{\partial w}{\partial r}(t, a, \theta) = 0$$

where w is the transverse displacement and ρ_p is the density of the plate. The general moments are given by

$$
\begin{aligned}
\mathcal{M}_r &= M_r - (M_r)_{pe} \\
\mathcal{M}_\theta &= M_\theta - (M_\theta)_{pe} \\
\mathcal{M}_{r\theta} &= M_{r\theta}
\end{aligned}
$$

where

$$M_r = D\left(\frac{\partial^2 w}{\partial r^2} + \frac{\nu}{r}\frac{\partial w}{\partial r} + \frac{\nu}{r^2}\frac{\partial^2 w}{\partial\theta^2}\right) + c_D\left(\frac{\partial^3 w}{\partial r^2 \partial t} + \frac{\nu}{r}\frac{\partial^2 w}{\partial r \partial t} + \frac{\nu}{r^2}\frac{\partial^3 w}{\partial\theta^2 \partial t}\right)$$

$$M_\theta = D\left(\frac{1}{r}\frac{\partial w}{\partial r} + \frac{1}{r^2}\frac{\partial^2 w}{\partial\theta^2} + \nu\frac{\partial^2 w}{\partial r^2}\right) + c_D\left(\frac{1}{r}\frac{\partial^2 w}{\partial r \partial t} + \frac{1}{r^2}\frac{\partial^3 w}{\partial\theta^2 \partial t} + \nu\frac{\partial^3 w}{\partial r^2 \partial t}\right) \quad (1.1)$$

$$M_{r\theta} = D(1-\nu)\left(\frac{1}{r}\frac{\partial^2 w}{\partial r \partial\theta} - \frac{1}{r^2}\frac{\partial w}{\partial\theta}\right) + c_D(1-\nu)\left(\frac{1}{r}\frac{\partial^3 w}{\partial r \partial\theta \partial t} - \frac{1}{r^2}\frac{\partial^2 w}{\partial\theta \partial t}\right)$$

are the internal plate moments (including discontinuous changes in D and c_D due to the bonding of the patches to the plate – see [4]), and

$$(M_r)_{pe} = (M_\theta)_{pe} = \sum_{i=1}^{s} \mathcal{K}_i u_i(t)[H(r-r_{i1}) - H(r-r_{i2})][H(\theta-\theta_{i1}) - H(\theta-\theta_{i2})]$$

are the applied patch moments. With E denoting the Young's modulus, the parameters $D = \frac{Eh^3}{12(1-\nu^2)}$, c_D and ν represent the flexural rigidity, damping coefficient and Poisson's ratio for the plate/patch structure. Here H

denotes the Heaviside function, $u_i(t)$ is the voltage into the ith patch, and $\mathcal{K}_i$ is a parameter which depends on the geometry, piezoceramic material properties, and piezoelectric strain constant (see [4] for details). The piezoceramic material parameters $\mathcal{K}_i, i = 1, \cdots, s$ as well as the plate parameters ρ_p, D, c_D and ν are ultimately considered to be unknown and will be approximated using parameter estimation techniques.

The final coupling equation is the continuity of velocity condition

$$\frac{\partial \phi}{\partial z}(t, r, \theta, w(t, r, \theta)) = -w_t(t, r, \theta) \quad , \quad (r, \theta) \in \Gamma_0 \ , \ t > 0$$

which results from the assumption that the plate is impenetrable to air. We note that both the backpressure and velocity (momentum) coupling conditions are in general nonlinear due to the fact that they occur on the surface of the plate.

Under the assumption of small displacements which is inherent in the Love-Kirchhoff plate theory, these nonlinear coupling terms can be replaced by their linear approximations. This yields the approximate system model

$$\phi_{tt} = c^2 \Delta\phi \quad , \quad (r, \theta, z) \in \Omega \ , \ t > 0 \ ,$$

$$\nabla\phi \cdot \hat{n} = 0 \quad , \quad (r, \theta, z) \in \Gamma \ , \ t > 0 \ ,$$

$$\frac{\partial \phi}{\partial z}(t, r, \theta, 0) = -w_t(t, r, \theta) \quad , \quad (r, \theta) \in \Gamma_0 \ , \ t > 0 \ ,$$

$$\rho_p h w_{tt} + \frac{\partial^2 M_r}{\partial r^2} + \frac{2}{r}\frac{\partial M_r}{\partial r} - \frac{1}{r}\frac{\partial M_\theta}{\partial r} + \frac{2}{r}\frac{\partial^2 M_{r\theta}}{\partial r \partial \theta} + \frac{2}{r^2}\frac{\partial M_{r\theta}}{\partial \theta} + \frac{1}{r^2}\frac{\partial^2 M_\theta}{\partial \theta^2}$$

$$= \frac{\partial^2 (M_r)_{pe}}{\partial r^2} + \frac{2}{r}\frac{\partial (M_r)_{pe}}{\partial r} - \frac{1}{r}\frac{\partial (M_\theta)_{pe}}{\partial r} + \frac{1}{r^2}\frac{\partial^2 (M_\theta)_{pe}}{\partial \theta^2} \qquad (1.2)$$

$$-\rho_f \phi_t(t, r, \theta, 0) + f(t, r, \theta) \ ,$$

$$w(t, a, \theta) = \frac{\partial w}{\partial r}(t, a, \theta) = 0 \ ,$$

$$\phi(0, r, \theta, z) = \phi_0(r, \theta, z) \quad , \quad w(0, r, \theta) = w_0(r, \theta) \ ,$$

$$\phi_t(0, r, \theta, z) = \phi_1(r, \theta, z) \quad , \quad w_t(0, r, \theta) = w_1(r, \theta) \ .$$

In the experimental setup, the dimensions of the cylinder and plate are $a = .2286\,m\ (9\,in)$, $\ell = 1.0668\,m\ (42\,in)$ and $h = .00127\,m\ (.05\,in)$. In order to test the numerical schemes, the "book" values $\rho_p = 2700\,kg/m^3$, $\nu = .33$, $E = 7.1 \times 10^{10}\,N/m^2$, $\rho_f = 1.21\,kg/m^3$ and $c = 343\,m/sec$ were used for the physical parameters. This then yielded the flexural rigidity $D = 13.6007\,N \cdot m$. Finally, we assumed that the Kelvin-Voigt damping parameter had the value $c_D = .00011222\,N \cdot m \cdot sec$.

Sections 2 and 3 contain the development of an infinite dimensional formulation and approximation framework which is amenable to the estimation of physical parameters as well as the application of LQR optimal control techniques. From the standpoint of performing forward simulations and estimating parameters, it is desirable to have a numerical scheme which is accurate, robust, efficient and easily implemented. From a control perspective, it is desirable to have a scheme which uniformly preserves stability margins as the dimension of the approximating system increases (see [2]). Finally, all of these criteria must be satisfied in the presence of the coordinate singularity at the origin (the careful handling of the singularity is especially important when using spectral basis functions [7, 8, 9, 11]).

With the discretization techniques having been defined in Section 3, examples demonstrating the approximation of the natural frequencies of a clamped circular plate and a wave in a cylindrical domain are given in Sections 4 and 5. The purpose of these examples is twofold; they provide a means of testing the accuracy and efficiency of the approximation techniques (since the approximated frequencies can be compared to analytic values), and they allow us to determine the frequencies of the component plate and wave which can then be compared to the coupled system dynamics which are presented in Section 6. Finally, eigenvalue results are presented in Section 7 which demonstrate that the approximation scheme uniformly preserves the stability margins. These results along with those in earlier sections demonstrate that the numerical method satisfies the previously mentioned approximation and control criteria.

2 Variational Form of the System

In the system model (1.2), the plate and acoustic equations are in strong form which leads to difficulties in the control problem since it involves the differentiation of the Heaviside function and the Dirac delta which then yields an unbounded control input term. Moreover, due to the presence of the piezoceramic patches and their differing material properties, it is assumed that ρ_p, D, c_D and ν for the combined structure are piecewise constant in nature (see [4, 6]). Hence these parameters will be expanded in terms of a Heaviside basis with the edges of the patches defining the support of the basis functions. This also leads to problems in the strong form, however, since it necessitates the differentiation of discontinuous material parameters. To avoid these difficulties, it is advantageous to formulate the problem in a weak or variational form.

In order to pose the problem in a manner which is conducive to approximation, parameter estimation and control, the state is taken to be $z = (\phi, w)$ in the Hilbert space $H = \bar{L}^2(\Omega) \times L^2(\Gamma_0)$. The choice of the space $\bar{L}^2(\Omega)$, the quotient space of L^2 over the constants, is motivated by the fact that the potentials are determined only up to a constant.

To provide a class of functions (test functions) which are considered when defining a variational form of the problem, we also define the Hilbert space $V = \bar{H}^1(\Omega) \times H_0^2(\Gamma_0)$ where $\bar{H}^1(\Omega)$ is the quotient space of H^1 over the constant functions and $H_0^2(\Gamma_0) = \{\psi \in H^2(\Gamma_0) : \psi = \psi_r = 0 \text{ at } r = a\}$.

A complete discussion concerning the derivation of a variational formulation of the circular plate equations from energy principles as well as the formulation of a weak form of the coupled system equations can be found in [5]. For our purposes here, we simply note that integration in combination with the use of Green's theorem yields the second-order variational system

$$\int_\Omega \frac{\rho_f}{c^2} \phi_{tt} \bar{\xi} d\omega + \int_{\Gamma_0} \rho_p h w_{tt} \bar{\eta} d\gamma + \int_\Omega \rho_f \nabla \phi \cdot \overline{\nabla \xi} d\omega$$

$$+ \int_{\Gamma_0} M_r \overline{\eta_{rr}} d\gamma + \int_{\Gamma_0} \frac{1}{r} M_\theta \overline{\eta_r} d\gamma + \int_{\Gamma_0} \frac{1}{r^2} M_{r\theta} \overline{\eta_{\theta\theta}} d\gamma$$

$$+ 2\int_{\Gamma_0} \frac{1}{r} M_{r\theta} \overline{\eta_{r\theta}} d\gamma - 2\int_{\Gamma_0} \frac{1}{r^2} M_{r\theta} \overline{\eta_\theta} d\gamma + \int_{\Gamma_0} \rho_f \left(\phi_t \bar{\eta} - w_t \bar{\xi} \right) d\gamma \qquad (2.1)$$

$$= \int_{\Gamma_0} \sum_{i=1}^{s} \mathcal{K}_i u_i(t) \left[H_{i1}(r) - H_{i2}(r) \right] \left[H_{i1}(\theta) - H_{i2}(\theta) \right] \overline{\nabla^2 \eta} d\gamma$$

$$+ \int_{\Gamma_0} f \bar{\eta} d\gamma$$

for all $(\xi, \eta) \in V$ (note that $d\omega = r\,dr\,d\theta\,dz$ and $d\gamma = r\,dr\,d\theta$). The internal plate moments M_r, M_θ and $M_{r\theta}$ are defined in (1.1). Also, in order to simplify the above expression, we have adopted the Heaviside notation $H_j(r) \equiv H(r - r_j)$ and $H_{ij}(\theta) \equiv H(\theta - \theta_{ij})$ for $i = 1, \cdots, s$, $j = 1, 2$.

We point out that in this variational form, the derivatives have been transferred from the plate and patch moments onto the test functions. This eliminates the problem of having to approximate the derivatives of the Heaviside function and the Dirac delta which is the case in the strong form of the equations.

The system can then be written in first-order form by defining the product spaces $\mathcal{V} = V \times V$ and $\mathcal{H} = V \times H$ and taking the state to be $Z = (\phi, w, \phi_t, w_t)$. Note that the state now contains a multiple of the pressure since $p = \rho_f \phi_t$. Further details concerning the formulation of the first-order infinite dimensional system in weak form is given in [5] where the problem is posed in terms of sesquilinear forms and the bounded operators which they define.

3 Finite Dimensional Approximation

To approximate the solution to the coupled system, suitable expansions must be chosen for the state variables w and ϕ. The plate displacement is approximated by

$$w^{\mathcal{N}}(t, r, \theta) = \sum_{m=-M_p}^{M_p} \sum_{n=1}^{N_p^m} w_{mn}(t) e^{im\theta} r^{|\hat{m}|} B_n^m(r)$$

where $B_n^m(r)$ is the nth modified cubic spline satisfying $B_n^m(a) = \frac{dB_n^m(a)}{dr} = 0$ with the condition $\frac{dB_n^m(0)}{dr} = 0$ being enforced when $m = 0$ (this latter condition guarantees differentiability at the origin and implies that $N_p^m = N_p + 1$ when $m \neq 0$ and $N_p^m = N_p$ when $m = 0$, where N_p denotes the number of modified cubic splines). The total number of plate basis functions is $\mathcal{N} = (2M_p + 1)(N_p + 1) - 1$. As discussed in the report [5], the inclusion of the weighting term $r^{|\hat{m}|}$ with

$$\hat{m} = \begin{cases} 0 & , \quad m = 0 \\ 1 & , \quad m \neq 0 \end{cases}$$

is motivated by the asymptotic behavior of the Bessel functions (which make up the analytic plate solution) as $r \to 0$. It also serves to ensure the uniqueness of the solution at the origin. The Fourier coefficient in the weight is truncated to control the conditioning of the mass and stiffness matrices (see the examples in [5]).

A suitable Fourier-Galerkin expansion of the potential is

$$\phi^{\mathcal{M}}(t, r, \theta, z) = \sum_{p=0}^{P_w} \sum_{m=-M_w}^{M_w} \sum_{\substack{n=0 \\ p+|m|+n \neq 0}}^{N_w^{p,m}} \phi_{pmn}(t) e^{im\theta} r^{|\hat{m}|} P_n^{p,m}(r) P_p(z)$$

where

$$\hat{m} = \begin{cases} m & , \quad |m| = 0, \cdots, 5 \\ 5 & , \quad |m| = 6, \cdots, M_w \end{cases}$$

and

$$P_n^{p,m}(r) = \begin{cases} P_1(r) - 1/3 & , \quad p = m = 0, n = 1 \\ P_n(r) & , \quad \text{otherwise} . \end{cases}$$

Here $P_n(r)$ and $P_p(z)$ are the nth and p^{th} Legendre polynomials which have been mapped to the intervals $(0, a)$ and $(0, \ell)$, respectively. The term $P_1(r) - 1/3$ when $p = m = 0, n = 1$ results from the orthogonality properties of the Legendre polynomials and arises when enforcing the condition

$\int_\Omega \phi^\mathcal{M}(t, r, \theta, z)d\omega = 0$ so as to guarantee that the functions are suitable as a basis for the quotient space. The inclusion of the weight $r^{|\hat{m}|}$ again incorporates the decay of the analytic solution near the origin while ensuring its uniqueness at that point. Finally, we note that the limit $N_w^{p,m}$ is given by $N_w^{p,m} = N_w + 1$ when $p + |m| \neq 0$ and $N_w^{p,m} = N_w$ when $p = m = 0$, which implies that $\mathcal{M} = (2M_w + 1)(N_w + 1)(P_w + 1) - 1$ basis functions are used in the wave expansion.

The $\mathcal{N}$ and $\mathcal{M}$ dimensional approximating plate and cavity subspaces are taken to be $H_p^\mathcal{N} = \mathrm{span}\{B_i^\mathcal{N}\}_{i=1}^\mathcal{N}$ and $H_c^\mathcal{M} = \mathrm{span}\{B_i^\mathcal{M}\}_{i=1}^\mathcal{M}$, respectively, where $B_i^\mathcal{N}$ and $B_i^\mathcal{M}$ are the ith plate and cavity bases described above. Defining $\mathcal{P} = \mathcal{N} + \mathcal{M}$, the approximating state space is $H^\mathcal{P} = H_c^\mathcal{M} \times H_p^\mathcal{N}$ and the product space for the first order system is $\mathcal{H}^\mathcal{P} = H^\mathcal{P} \times H^\mathcal{P}$.

As shown in [5], the restriction of the infinite dimensional system (2.1) to $\mathcal{H}^\mathcal{P} \times \mathcal{H}^\mathcal{P}$ yields the matrix system

$$M^\mathcal{P} \dot{y}^\mathcal{P}(t) = \tilde{A}^\mathcal{P} y^\mathcal{P}(t) + \tilde{B}^\mathcal{P} u(t) + \tilde{F}^\mathcal{P}(t)$$
$$M^\mathcal{P} y^\mathcal{P}(0) = \tilde{y}_0^\mathcal{P}$$

(3.1)

where

$$y^\mathcal{P}(t) = \begin{pmatrix} \vartheta^\mathcal{P}(t) \\ \dot{\vartheta}^\mathcal{P}(t) \end{pmatrix}.$$

Here $\vartheta^\mathcal{P}(t) = [\phi_1^\mathcal{P}(t), \phi_2^\mathcal{P}(t), \cdots, \phi_\mathcal{M}^\mathcal{P}(t), w_1^\mathcal{P}(t), w_2^\mathcal{P}(t), \cdots, w_\mathcal{N}^\mathcal{P}(t)]^T$ denotes the approximate state vector coefficients while $u(t) = [u_1(t), \cdots, u_s(t)]^T$ contains the s control variables. The system matrices and vectors are

$$M^\mathcal{P} = \begin{bmatrix} M_1^\mathcal{P} & 0 \\ 0 & M_2^\mathcal{P} \end{bmatrix} \quad , \quad \tilde{A}^\mathcal{P} = \begin{bmatrix} 0 & M_1^\mathcal{P} \\ -A_1^\mathcal{P} & -A_2^\mathcal{P} \end{bmatrix}$$

$$\tilde{B}^\mathcal{P} = \begin{bmatrix} 0 \\ \tilde{B}^\mathcal{P} \end{bmatrix} \quad , \quad \tilde{F}^\mathcal{P}(t) = \begin{bmatrix} 0 \\ \tilde{F}^\mathcal{P}(t) \end{bmatrix} \quad , \quad \tilde{y}_0^\mathcal{P} = \begin{bmatrix} g_1^\mathcal{P} \\ g_2^\mathcal{P} \end{bmatrix}$$

with

$$M_1^\mathcal{P} = \mathrm{diag}[M_{11}^\mathcal{P}, M_{12}^\mathcal{P}] ,$$
$$M_2^\mathcal{P} = \mathrm{diag}[M_{21}^\mathcal{P}, M_{22}^\mathcal{P}] , \quad A_2^\mathcal{P} = \begin{bmatrix} 0 & A_{31}^\mathcal{P} \\ A_{32}^\mathcal{P} & A_{22}^\mathcal{P} \end{bmatrix}$$
$$A_1^\mathcal{P} = \mathrm{diag}[A_{11}^\mathcal{P}, A_{12}^\mathcal{P}] ,$$

and

$$\tilde{B}^\mathcal{P} = \begin{bmatrix} 0, \tilde{B}_2^\mathcal{P} \end{bmatrix}^T \quad , \quad \tilde{F}^\mathcal{P}(t) = \begin{bmatrix} 0, \tilde{F}_2^\mathcal{P}(t) \end{bmatrix}^T .$$

The matrices $M_{21}^\mathcal{P}$ and $A_{11}^\mathcal{P}$ are the mass and stiffness matrices which arise when solving the uncoupled wave equation with Neumann boundary conditions while M_{22}^P, A_{12}^P and A_{22}^P are the mass, stiffness and damping matrices which arise when solving the damped plate equation with fixed boundary conditions. The matrices M_{11}^P and M_{12}^P result from the choice of V inner product (see [5]). The contributions from the coupling terms are contained in the matrices A_{31}^P and A_{32}^P while the control, forcing and initial terms are contained in B_2^P, $\tilde{F}_2^P(t)$ and $\tilde{y}_0^P$, respectively. A more detail description of the various component matrices can be found in [5].

4 Uncoupled Plate Dynamics

To test the accuracy and efficiency of the plate expansion as well as to determine the natural frequencies of the uncoupled plate, we consider the equation of motion for an undamped circular plate with constant stiffness and density:

$$\rho_p h \frac{\partial^2 w}{\partial t^2} + D\nabla^4 w = 0 \ .$$

In order to determine the analytic frequencies, free vibrations are assumed, and the displacements are taken to be of the form

$$w = W\cos(\omega t) \ .$$

Here ω is the circular frequency (with units of radians/sec) and W contains the remaining spatial contributions. The substitution of this expression into the equation of motion yields the eigenvalue problem

$$\nabla^4 W = \gamma^4 W \tag{4.1}$$

where

$$\gamma^4 = \frac{\rho_p h \omega^2}{D} \ .$$

After enforcing the boundary conditions, the frequencies ω for a fixed plate are determined by solving the nonlinear problem

$$I_n(\lambda)\frac{dJ_n(\lambda)}{dr} - J_n(\lambda)\frac{dI_n(\lambda)}{dr} = 0$$

where J_n is the nth Bessel function of the first kind, $I_n(z) = i^{-n}J_n(iz)$ is the modified Bessel function of the first kind, and λ is related to γ by $\lambda = \gamma a$.

By noting the relationship $f = \frac{1}{2\pi}\omega$ where f is the frequency expressed in hertz, the natural frequencies of the fixed circular plate can be written as

$$f = \frac{1}{2\pi}\left(\frac{\lambda}{a}\right)^2\sqrt{\frac{D}{\rho_p h}} \ .$$

The experimental dimensions $a = .2286\,m$ $(9\,in)$, $\rho_p = 2700\,kg/m^3$, $h = .00127\,m$ $(.05\,in)$, and parameter choices $E = 7.1 \times 10^{10}\,N/m^2$ and

$\nu = .33$ are used which then yields the flexural rigidity $D = 13.6007\ N \cdot m$. For these values, several frequencies deriving from the Bessel solutions λ^2 as reported on page 8 of [10] are given in Table 4.1. In the tables, n and s refer to the number of nodal diameters and circles (not including the boundary).

To compare the results reported in Table 4.1 with those obtained via the Fourier-Galerkin scheme, it is noted that under approximation, the infinite dimensional eigenvalue problem (4.1) yields the matrix eigenvalue problem

$$A_{12}^{\mathcal{P}}\vartheta^{\mathcal{P}} = \gamma^4 M_{22}^{\mathcal{P}}\vartheta^{\mathcal{P}}$$

where $M_{22}^{\mathcal{P}}$ and $A_{12}^{\mathcal{P}}$ are the mass and stiffness matrices for the undamped and uncoupled plate. Note that these matrices are components in the full system (3.1) and hence this example provides a test of their accuracy. The frequencies obtained by solving this generalized matrix eigenvalue problem are reported in Table 4.2. In this case, the basis limits are $M_p = 6$, $N_p = 24$. By comparing the Bessel and Galerkin results in Tables 4.1 and 4.2, respectively, it can be seen that the Fourier-Galerkin scheme does a very good job of approximating the plate dynamics and hence determining the natural frequencies.

s	$n = 0$	$n = 1$	$n = 2$	$n = 3$	$n = 4$	$n = 5$	$n = 6$
0	61.96	128.95	211.56	309.58	422.56	550.38	692.75
1	241.23	368.90	513.02	673.33	849.82	1042.07	1249.92
2	540.46	728.34	932.93	1154.26	1392.14	1646.34	1916.70
3	959.46	1207.39	1472.15	1753.95	2052.63	2367.90	
4	1498.20	1806.12	2131.29	2473.02			
5	2156.69	2524.45	2909.31	3311.57			
6	2934.91	3362.52	3807.60	4269.79			

Table 4.1: Natural frequencies deriving from the Bessel expansions (in hertz).

s	$n = 0$	$n = 1$	$n = 2$	$n = 3$	$n = 4$	$n = 5$	$n = 6$
0	61.96	128.95	211.55	309.52	422.56	550.38	692.75
1	241.23	368.96	513.04	673.40	849.83	1042.08	1249.93
2	540.46	728.35	932.98	1154.31	1392.18	1646.40	1916.79
3	959.50	1207.41	1472.31	1754.16	2052.83	2368.17	
4	1498.37	1806.28	2131.37	2473.57			
5	2157.22	2525.17	2910.49	3313.05			
6	2936.35	3364.46	3810.17	4273.26			

Table 4.2: Natural frequencies obtained via the Fourier-Galerkin scheme with $M_p = 6, N_p = 24$ basis functions (in hertz).

5 Uncoupled Wave Dynamics

To test the wave discretization and to find the natural frequencies for the wave equation in a cylindrical domain, we consider the problem

$$\phi_{tt} = c^2 \Delta\phi \quad , \quad (r,\theta,z) \in \Omega \ , t > 0 \ ,$$

$$\nabla\phi \cdot \hat{n} = 0 \quad , \quad (r,\theta,z) \in \Gamma \ , t > 0 \ .$$

Through the separation of variables $\phi(t,r,\theta,z) = T(t)\Phi(r,\theta,z)$, one arrives at Helmholz's equation

$$\Delta\Phi + \gamma^2\Phi = 0 \quad , \ (r,\theta,z) \in \Omega \ ,$$

$$\nabla\Phi \cdot \hat{n} = 0 \quad \quad , \ (r,\theta,z) \in \Gamma$$

$$(5.1)$$

and the relation $T'' + \omega^2 T = 0$, $t > 0$. The separation constant here is $\gamma = \frac{\omega}{c}$ where ω is the circular frequency with units of radians/sec. To find Φ, we separate variables once more. Letting $\Phi(r,\theta,z) = R(r)\Theta(\theta)Z(z)$, the expansion of Helmholz's equation yields the expression

$$\frac{1}{rR}\frac{d}{dr}\left(r\frac{dR}{dr}\right) + \frac{1}{r^2\Theta}\frac{d^2\Theta}{d\theta^2} + \frac{1}{Z}\frac{\partial^2 Z}{dz^2} + \gamma^2 = 0 \ .$$

This implies that R, Θ and Z must satisfy the differential equations

$$\frac{1}{rR}\frac{d}{dr}\left(r\frac{dR}{dr}\right) - \frac{n^2}{r^2} - k^2 + \gamma^2 = 0$$

$$\frac{dR(a)}{dr} = 0 \ ,$$

$$(5.2)$$

$$\frac{d^2\Theta}{d\theta^2} + n^2\Theta = 0$$

$$\Theta(0) = \Theta(2\pi)$$

$$(5.3)$$

and

$$\frac{d^2 Z}{dz^2} + k^2 Z = 0$$

$$\frac{dZ(0)}{dz} = \frac{dZ(\ell)}{dz} = 0 \ ,$$

$$(5.4)$$

respectively.

    The general solution to (5.4) is $Z(z) = A_1\cos(kz) + B_1\sin(kz)$, and by enforcing the boundary conditions, we find that $B_1 = 0$ and the frequencies are $k_p = \frac{p\pi}{\ell}$, $p = 0,1,2,\cdots$. Similarly, the general solution to (5.3) is periodic and is given by $\Theta(\theta) = A_2\cos(n\theta) + B_2\sin(n\theta)$. Finally, by enforcing differentiability constraints, we find the general solution of (5.2) to be $R(r) = A_3 J_n(\Lambda r)$ where $\Lambda^2 = -k^2 + \gamma^2$ and J_n is the nth Bessel function of the first kind. The eigenvalues of (5.2) are then determined by applying the boundary condition and solving for the zeros of the nonlinear equation

$$\frac{dJ_n(\lambda)}{dr} = 0 \tag{5.5}$$

where $\lambda = \Lambda a$. This then yields the set of eigenvalues

$$\gamma_{nsp}^2 = \left(\frac{p\pi}{\ell}\right)^2 + \left(\frac{\lambda_{ns}}{a}\right)^2 \tag{5.6}$$

where again, λ_{ns} solves (5.5).

To determine the natural frequencies for the problem, we recall that $f = \frac{1}{2\pi}\omega$ where f has units of hertz. By combining this with (5.6) and the fact that $\gamma = \frac{\omega}{c}$, the natural frequencies can be expressed as

$$f_{nsp} = \frac{1}{2\pi}c\sqrt{\left(\frac{p\pi}{\ell}\right)^2 + \left(\frac{\lambda_{ns}}{a}\right)^2}$$

where $p = 0, 1, 2, \cdots$, $n = 0, 1, 2, \cdots$ and $s = 1, 2, \cdots$. For the experimental parameter choices $a = .2286\,m$ (9 in) , $\ell = 1.0668\,m\,(42\,in)$ and $c = 343\,m/sec$, several frequencies deriving from this expression are reported in Tables 5.1-5.3. The results in Table 5.1 are the radial and tangential frequencies ($p = 0$) while those in Table 5.2 are the axial values ($\lambda_{ns} = 0$). Finally, Table 5.3 contains the cross frequencies obtained when $n = 0$.

To compare these results with those obtained via the Fourier-Galerkin expansion, it is noted that under approximation, Helmholz's equation (5.1) yields the matrix eigenvalue problem

$$A_{11}^P \vartheta^P = \gamma^2 M_{21}^P \vartheta^P \tag{5.7}$$

(note that the mass and stiffness matrices M_{21}^P and A_{11}^P are components in the system (3.1)). The frequencies obtained by solving (5.7) with the basis limits $M_w = 1$, $N_w = 15$ and $P_w = 15$ are reported in Tables 5.4 - 5.6. (the choice of the Fourier number $M_w = 1$ was dictated by limitations encountered when solving the generalized eigenvalue problem).

By comparing the analytic results in Tables 5.1 - 5.3 with those in Tables 5.4 - 5.6, it can be seen that the Fourier-Galerkin method is performing well for this problem. Although omitted here, numerical results obtained with $M_w = 1, N_w = 12$ and $P_w = 12$ demonstrate that the method is also converging as expected. Thus the wave approximation in the system (3.1) appears to be both accurate and efficient.

s	$n = 0$	$n = 1$	$n = 2$	$n = 3$	$n = 4$
0	0.0000	439.6823	729.3492	1003.2551	1269.8537
1	915.0178	1273.1492	1601.4303	1914.0461	2216.6560
2	1675.3396	2038.4858	2380.7369	2709.4240	3028.4635

Table 5.1: Radial and tangential frequencies $f_{ns0} = \frac{\lambda_{ns}c}{2\pi a}$.

	$p = 1$	$p = 2$	$p = 3$	$p = 4$	$p = 5$	$p = 6$
k_p	160.7612	321.5223	482.2835	643.0446	803.8058	964.5669

Table 5.2: Axial frequencies $f_{00p} = \frac{pc}{2\ell}$.

s	$p = 1$	$p = 2$	$p = 3$	$p = 4$	$p = 5$	$p = 6$
0	160.7612	321.5223	482.2835	643.0446	803.8058	964.5669
1	929.0326	969.8629	1034.3379	1118.3756	1217.9332	1329.5288
2	1683.0350	1705.9130	1743.3760	1794.5108	1858.1890	1933.1715

Table 5.3: Cross frequencies $f_{0sp} = \frac{1}{2\pi}c\sqrt{\left(\frac{p\pi}{\ell}\right)^2 + \left(\frac{\lambda_{0s}}{a}\right)^2}$.

s	$n = 0$	$n = 1$
0		439.6784
1	915.0192	1273.1594
2	1675.3364	2038.4897

Table 5.4: Radial and tangential frequencies obtained with $M_w = 1$, $N_w = 15$ and $P_w = 15$.

	$p = 1$	$p = 2$	$p = 3$	$p = 4$	$p = 5$	$p = 6$
k_p	160.7612	321.5223	482.2835	643.0446	803.8058	964.5735

Table 5.5: Axial frequencies obtained with $M_w = 1$, $N_w = 15$ and $P_w = 15$.

s	$p = 1$	$p = 2$	$p = 3$	$p = 4$	$p = 5$	$p = 6$
0	160.7612	321.5223	482.2835	643.0446	803.8058	964.5735
1	929.0341	969.8643	1034.3391	1118.3767	1217.9343	1329.5346
2	1683.0318	1705.9099	1743.3730	1794.5078	1858.1861	1933.1720

Table 5.6: Cross frequencies obtained with $M_w = 1$, $N_w = 15$ and $P_w = 15$.

6   Dynamics of the Coupled System

In order to determine the dynamics of the coupled system, the plate was subjected to a numerical impulse at its center; that is, the forcing function f was taken to be

$$f(t, r, \theta) = \delta(t)\delta(r, \theta) \ .$$

So as to approximate the dynamics of the experimental apparatus as closely as possible, the dimensions were taken to be $\ell = 1.0668\,m$ $(42\,in)$, $a = .2286\,m$ $(9\,in)$, and $h = .00127\,m$ $(.05\,in)$ while the choices $\nu = .33$, $\rho_p = 2700\,kg/m^3$, $D = 13.6007\,N \cdot m$, $c_D = .00011222\,N \cdot m \cdot sec$, and $\rho_f = 1.21\,kg/m^3$, $c = 343\,m/sec$ were made for the physical parameters.

The time history of the system response throughout the interval $[0, 1/6]$ was recorded at several points on the plate and in the cavity and from this, the system frequencies were determined and compared to those of the uncoupled plate and cavity. This temporal interval was chosen since it was sufficiently long so as to demonstrate the system dynamics but short enough so that the higher frequency responses were not completely lost.

Because the impulse was applied at the center of the plate, the response was uniform in θ which implied that the Fourier limits could be taken to be $M_p = M_w = 0$. The plate and pressure responses obtained with the basis choices $M_p = 0, N_p = 12$ and $M_w = 0, N_w = 9, P_w = 9$ at the plate and cavity points $(r, \theta) = (0, 0)$ and $(r, \theta, z) = (0, 0, .05)$ are plotted in Figure 6.1 with corresponding frequency plots in Figure 6.2 (the range of frequencies being examined are resolved with the above limit choices). The system frequencies measured on the plate and in the cavity are summarized in Table 6.1.

In order to compare these harmonics to the natural frequencies of the individual cavity and undamped plate, the values corresponding to the symmetric plate and cavity modes were taken from Tables 4.1 and 5.1 - 5.3, and summarized in Tables 6.2 and 6.3. From Figure 6.2 and Table 6.1, it can be seen that although the system frequencies agree quite closely with those in Tables 6.2 and 6.3, there are slight differences due to the fact that the system (1.2) involves not only coupling between the plate and cavity (due to the backpressure and continuity of velocity) but also includes damping in the plate. Specifically, the system frequencies measured on the plate tend to be slightly less than those of the undamped plate while those measured in the cavity are slightly higher than the natural frequencies of a wave within a cavity having this geometry. These tendencies are consistent with the results obtained in a 2-D analogue of this problem [3] and reinforce the notion that although the system frequencies are close to those of the component plate and cavity, care must be taken when describing the dynamics of the coupled system in terms of the properties of the undamped and uncoupled plate and cavity.

MODELING AND APPROXIMATION OF AN ACOUSTICS PROBLEM

The effects of the coupling can be noted in the observation that a response of frequency 164.06 hertz is excited in the plate as a result of the pressure oscillations in the cavity (similarly, responses at 59.77 and 240.23 hertz are excited in the cavity and are due to the vibrations of the plate). The decay in the magnitude of the pressure oscillations also indicates the effect of the coupling since the only damping in the system is the Kelvin-Voigt damping in the plate.

From Figure 6.2, it can be seen that the highest energy pressure responses are at 164.06, 930.08 and 970.53 hertz. The first corresponds to the first axial mode for a cavity of this size while the response at 930.08 hertz corresponds to the first cross harmonic. Buried within this peak is a response at 915.13 hertz which corresponds to the first symmetric radial/tangential mode. This response can be detected if fewer sample points are used, but this strategy leads to aliasing problems and those results are not graphed here. Finally, the strong response at 970.53 hertz corresponds to the second cross frequency. These frequency results can then be used when determining strategies for the parameter estimation and control problems.

Harmonics Measured at $(0,0)$	59.77	164.06	240.23	535.55
Harmonics Measured at $(0,0,.05)$	59.77	164.06	240.23	324.78
	483.82	644.53	806.25	
	915.13	930.08	970.53	

Table 6.1: System frequencies at the plate and cavity points $(0,0)$ and $(0,0,.05)$ obtained with $M_p = 0, N_p = 12, M_w = 0, N_w = 12$ and $P_w = 9$ basis functions.

s	0	1	2	3
$n = 0$	61.96	241.23	540.46	959.46

Table 6.2: Analytic frequencies of the symmetric plate modes,
$$f = \frac{1}{2\pi} \left(\frac{\lambda}{a}\right)^2 \sqrt{\frac{D}{\rho_p h}}.$$

s	$p=0$	$p=1$	$p=2$	$p=3$	$p=4$	$p=5$	$p=6$
0		160.76	321.52	482.28	643.04	803.81	964.57
1	915.02	929.03	969.86				

Table 6.3: Analytic frequencies of the symmetric wave modes,
$$f_{0sp} = \frac{1}{2\pi} c \sqrt{\left(\frac{p\pi}{\ell}\right)^2 + \left(\frac{\lambda_{0s}}{a}\right)^2}.$$

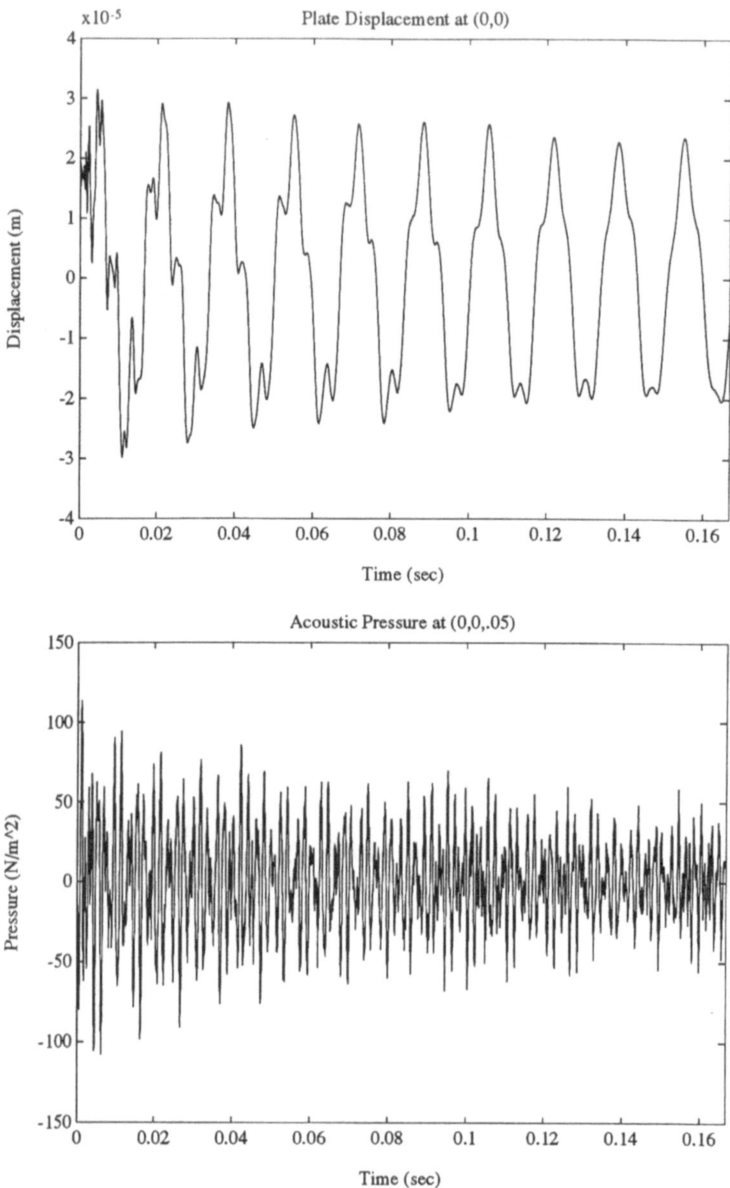

Figure 6.1: The plate and pressure responses to a centered impact.

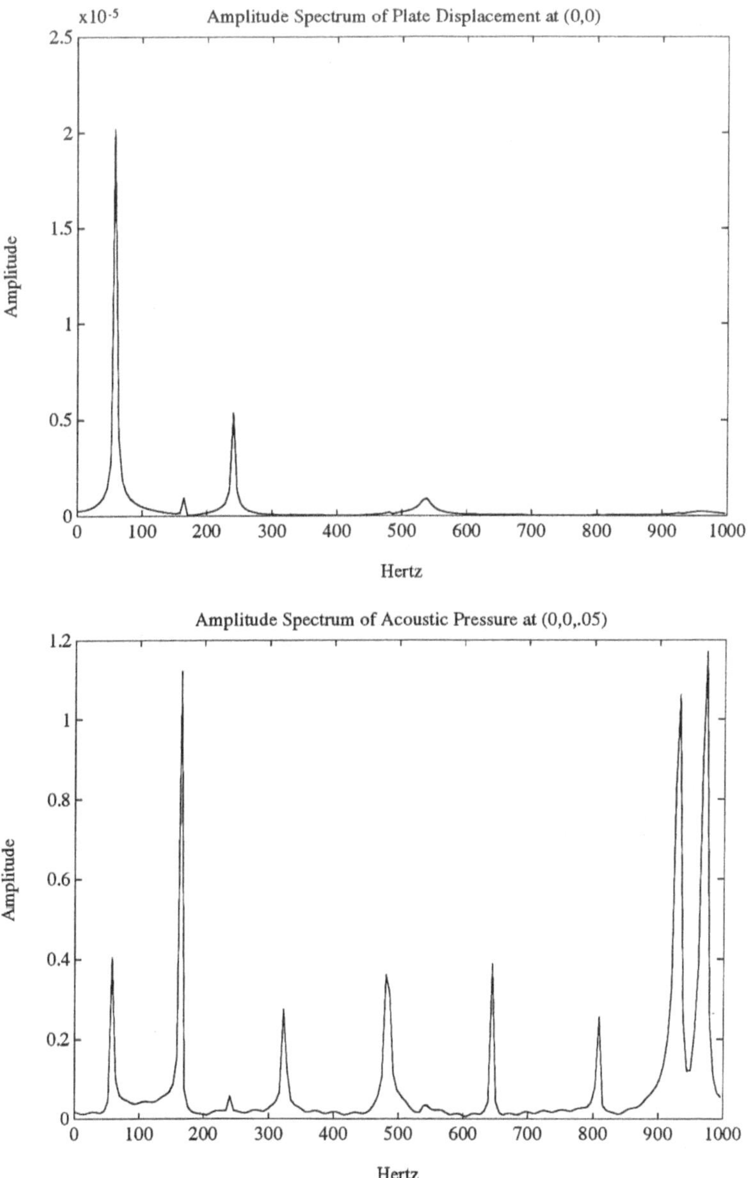

Figure 6.2: The plate and pressure frequency responses to a centered impact.

7 Stability Margin for the Coupled System

In order to check for the uniform preservation of exponential stability for the open loop approximating system with this discretization scheme, the maximum eigenvalues for the system matrix $A^{\mathcal{P}} = (M^{\mathcal{P}})^{-1}\tilde{A}^{\mathcal{P}}$ in (3.1) are given in Tables 7.1 and 7.2. The results in these tables correspond to the physical parameter choices used in the last example and were obtained with the choices $M_p = 0$ and $M_p = 1$ for the Fourier limits (similar results were obtained with larger choices for the the Fourier limits). These results along with a more extensive set of examples reported in [5] demonstrate that the exponential stability margins are uniformly maintained in the open loop approximations of this system using this Fourier-Galerkin scheme. As discussed in [2], this is an important consideration when approximating weakly damped systems with hyperbolic components in a control setting.

M_p	N_p	M_w	N_w	P_w	$size(A^{\mathcal{P}})$	$\max\left\{Re\lambda, \lambda \in \sigma\left(A^{\mathcal{P}}\right)\right\}$
0	5	0	2	2	26×26	$-.5853 - 3$
0	5	0	4	4	58×58	$-.2064 - 2$
0	10	0	6	6	116×116	$-.2911 - 2$
0	20	0	8	8	200×200	$-.3034 - 2$

Table 7.1: Margin between the open loop eigenvalues and the imaginary axis with $M_p = 0$.

M_p	N_p	M_w	N_w	P_w	$size(A^{\mathcal{P}})$	$\max\left\{Re\lambda, \lambda \in \sigma\left(A^{\mathcal{P}}\right)\right\}$
1	5	1	2	2	86×86	$-.5853 - 3$
1	5	1	4	4	182×182	$-.2064 - 2$
1	10	1	6	6	356×356	$-.2911 - 2$
1	20	1	8	8	608×608	$-.3034 - 2$

Table 7.2: Margin between the open loop eigenvalues and the imaginary axis with $M_p = 1$.

8 Conclusion

In this work, the 2-D structural acoustics model developed in [1] and [3] has been extended to a 3-D cylindrical domain with a flexible plate at one end. This geometry was chosen since it models the experimental apparatus which is being constructed to test the control methodology which was developed in [1]. The final system of equations is considered in variational form since this avoids the problems associated with the differentiation of the discontinuous plate material parameters as well as the Heaviside function and Dirac delta which arise in the control input term.

A Fourier-Galerkin scheme utilizing modified cubic splines in the radial direction was chosen for the plate expansion. The choice of the splines resulted from the fact that they fit the smoothness criteria, were easily adapted to the clamped boundary conditions, and were suitable for approximations involving the discontinuous plate parameters. As noted in the plate example (as well as in several examples in [5]), this expansion provides accurate results when approximating the dynamics of an undamped plate which are then translated into accurate approximations of the plate motion in the fully coupled system.

A Fourier-Galerkin expansion was also chosen for discretizing the wave equation in the cylindrical domain, but in this case, translated Legendre functions were used in the radial and axial directions. The results in [2] concerning the decay of stability margins under approximation with finite elements and finite differences in weakly damped hyperbolic systems were one motivation for this choice and indeed, the results in the last section indicate that a uniform margin of stability is being maintained with this approximation scheme. Because natural boundary conditions exist on all walls of the cavity, no basis modifications were necessary for boundary conditions, and due to the orthogonality properties of the Legendre polynomials, the functions were easily adapted so as to be suitable as a basis for the quotient space. Finally, for relatively smooth forcing functions, this provided a wave solution which was exponentially accurate.

The use of these expansions for approximating the state variables w and ϕ in the weak form of the system equations has yielded a scheme which accurately and efficiently approximates the dynamics of the coupled system. Hence it should be suitable for forward simulations, parameter estimation, and the calculation of gains so as to reduce interior noise through LQR optimal control techniques.

Acknowledgements

The authors would like to thank the staff of the Department of Mathematical Sciences at Montana State University for the assistance that they provided during this conference. We also express our sincere appreciation to H.C. Lester and R.J. Silcox of the Acoustics Division, NASA Langley Research Center, for numerous consultations concerning the modeling of this problem.

References

[1] H.T. BANKS, W. FANG, R.J. SILCOX and R.C. SMITH, "Approximation Methods for Control of Acoustic/Structure Models with Piezoceramic Actuators," *Journal of Intelligent Material Systems and Structures*, 4(1), 1993, pp. 98-116.

[2] H.T. BANKS, K. ITO and C. WANG, "Exponentially Stable Approximations of Weakly Damped Wave Equations," *International Series in Numerical Mathematics*, Birkhäuser, 100, 1991, pp. 1-33.

[3] H.T. BANKS, R.J. SILCOX and R.C. SMITH, "The Modeling and Control of Acoustic/Structure Interaction Problems Via Piezoceramic Actuators: 2-D Numerical Examples," submitted to the *ASME Journal of Vibration and Acoustics.*

[4] H.T. BANKS, R.C. SMITH and Y. WANG, "Modeling Aspects for Piezoceramic Patch Activation of Shells, Plates and Beams," Center for Research in Scientific Computation Technical Report, CRSC-TR92-12, North Carolina State University, November 1992.

[5] H.T. BANKS and R.C. SMITH, "The Modeling and Approximation of a Structural Acoustics Problem in a Hard-Walled Cylindrical Domain," Center for Research in Scientific Computation Technical Report, work in progress.

[6] H.T. BANKS, Y. WANG, D.J. INMAN and J.C. SLATER, "Variable Coefficient Distributed Parameter System Models for Structures with Piezoceramic Actuators and Sensors," *Proceedings of the 31st Conference on Decision and Control*, Tucson, AZ, December 16-18, 1992, pp. 1803-1808.

[7] S. BOUAOUDIA and P.S. MARCUS, "Fast and Accurate Spectral Treatment of Coordinate Singularities," *Journal of Computational Physics*, 96, 1991, pp. 217-223.

[8] C. CANUTO, M.Y. HUSSAINI, A. QUARTERONI and T.A. ZANG, *Spectral Methods in Fluid Dynamics*, Springer-Verlag, New York, 1988.

[9] D. GOTTLIEB and S.A. ORSZAG, *Numerical Analysis of Spectral Methods: Theory and Applications*, SIAM, Philadelphia, 1977.

[10] A.W. LEISSA, *Vibration of Plates*, NASA SP-160, Washington, D.C., 1969.

[11] T.A. ZANG, C.L. STREETT and M.Y. HUSSAINI, "Spectral Methods for CFD," ICASE Report 89-13, 1989.

PARAMETER IDENTIFICATION IN THE FREQUENCY DOMAIN

H.T. Banks* and Yun Wang*

Center for Research in Scientific Computation
North Carolina State University
Raleigh, NC 27695-8205

1 Introduction

In studying vibrations of flexible structures, estimation of system parameters using observations in the time domain gave poor results when the observations contained several vibration modes. In response to this difficulty in using time domain optimization techniques, we attempted to carry out identification in the frequency domain. The underlying idea for this procedure involves taking the discrete Fourier transform (DFT) of the data and defining the cost function by using this transformed data and transforms of the model solution. In this paper we outline the theoretical foundations for general frequency domain parameter estimation techniques for second order systems described in terms of sesquilinear forms and operators in a Hilbert space. To illustrate the ideas and techniques, we apply them to the problem of estimating damping parameters in Timoshenko beams.

2 The Abstract Problem

Let V and H be complex Hilbert spaces satisfying $V \hookrightarrow H = H^* \hookrightarrow V^*$ (see [16] for the construction of this so-called Gelfand triple), where we denote their topological duals by V^* and H^*, respectively. Let Q be the admissible parameter metric space with metric d. We consider the parameter dependent second order abstract inhomogeneous initial value problem in V^*

$$\begin{aligned}
&\ddot{u}(t) + B(q)\dot{u}(t) + A(q)u(t) = f(t) \\
&u(0) = u_0 \\
&\dot{u}(0) = u_1,
\end{aligned} \qquad (2.1)$$

where $A(q)$ and $B(q)$ are parameter dependent differential operators and $q \in Q$. The corresponding variational formulation is given by

*Research supported in part by the Air Force Office of Scientific Research under grant AFOSR-90-0091.

$$< \ddot{u}(t), \psi >_{V^*,V} + \sigma_1(u(t), \psi) + \sigma_2(\dot{u}(t), \psi) = < f(t), \psi >_{V^*,V}$$
$$u(0) = u_0 \qquad\qquad\qquad\qquad \text{for } \psi \in V$$
$$\dot{u}(0) = u_1, \tag{2.2}$$

with $< \cdot, \cdot >_{V^*,V}$ denoting the duality product [16]. We assume that the sesquilinear forms $\sigma_1(q)$ and $\sigma_2(q)$, where $\sigma_i(q) : V \times V \to \mathbb{C}$, satisfy the following conditions:

(A1) *Boundedness.* There exist $c_i > 0$, $i = 1, 2$ such that for $q \in Q$

$$|\sigma_i(q)(\phi, \psi)| \leq c_i |\phi|_V \cdot |\psi|_V \qquad \text{for } \phi, \psi \in V;$$

(A2) *V-Coercivity.* There exist $k_i > 0$ and $\lambda_i > 0$, $i = 1, 2$ such that for $q \in Q$

$$Re\, \sigma_i(q)(\phi, \phi) \geq k_i |\phi|_V^2 - \lambda_i |\phi|_H^2, \qquad \phi \in V;$$

(A3) *Continuity.* For $q, \tilde{q} \in Q$ and $i = 1, 2$

$$|\sigma_i(q)(\phi, \psi) - \sigma_i(\tilde{q})(\phi, \psi)| \leq d_i(q, \tilde{q}) |\phi|_V |\psi|_V, \quad \phi, \psi \in V,$$

where $d_i(q, \tilde{q}) \to 0$ as $d(q, \tilde{q}) \to 0$.

If (A1) holds, then σ_1, σ_2 define operators $A(q)$, $B(q) \in \mathcal{L}(V, V^*)$ by

$$\sigma_1(q)(\phi, \psi) = < A(q)\phi, \psi >_{V^*,V}$$
$$\sigma_2(q)(\phi, \psi) = < B(q)\phi, \psi >_{V^*,V} \qquad \text{for } \phi, \psi \in V.$$

In this manner, we have the equivalence of 2.2 and 2.1. The conditions (A1)-(A3) are sufficient to establish well posedness and continuous dependence results for 2.1 and 2.2.

Theorem 2.1 *If the sesquilinear forms σ_1 and σ_2 satisfy conditions (A1)-(A3) with σ_1 symmetric and $f \in L_2\big((0,T), V^*\big)$, then, for each $w_0 = (u_0, u_1) \in \mathcal{H} = V \times H$, the initial value problem 2.2 has a unique solution $w(t) = \big(u(t), \dot{u}(t)\big) \in L_2\big((0,T), V \times V\big)$. Moreover, this solution depends continuously on f and w_0 in the sense that the mapping $\{w_0, f\} \to w = (u, \dot{u})$ is continuous from $\mathcal{H} \times L_2\big((0,T), V^*\big)$ to $L_2\big((0,T), V \times V\big)$.*

We have in Theorem 2.1 stated the well posedness of the system 2.2 in a weak variational setting. We can take an alternative (but, as we shall see, equivalent) approach using the theory of semigroups [11], [12].

We can rewrite the second order system 2.1 as a first order system for $w(t) = \left(u(t), \dot{u}(t)\right)^T$ on a product space. We define the product space $\mathcal{V} = V \times V$ in addition to $\mathcal{H} = V \times H$ above and observe that $\mathcal{V}^* = V \times V^*$ in the Gelfand triple $\mathcal{V} \hookrightarrow \mathcal{H} \hookrightarrow \mathcal{V}^*$. The first order system can be written as

$$\dot{w}(t) = \mathcal{A}\, w(t) + F(t) \qquad \text{in } \mathcal{V}^*$$
$$w(0) = w_0, \qquad\qquad (2.3)$$

where $F(t) = \left(0, f(t)\right)^T \in \mathcal{V}^*, w_0 = \left(u_0, u_1\right)^T \in \mathcal{H}$ and

$$\mathcal{A} = \begin{pmatrix} 0 & I \\ -A & -B \end{pmatrix} \in \mathcal{L}(\mathcal{V}, \mathcal{V}^*). \qquad (2.4)$$

With the assumptions on σ_1 and σ_2 given in Theorem 2.1, the operator $\mathcal{A}$ is the infinitesimal generator of an analytic semigroup $T(t)$ on $\mathcal{V}^*$ (see [4] or [2]). Then, by definition, mild solutions of 2.3 in $\mathcal{V}^*$ are given by

$$w(t; q) = T(t)\, w_0 + \int_0^t T(t - s)\, F(s)\, ds\,. \qquad (2.5)$$

Theorem 2.2 *Suppose $w_0 = (u_0, u_1)^T \in \mathcal{H} = V \times H$, $f \in L_2\left((0, T), V^*\right)$, and sesquilinear forms σ_1 and σ_2 are given as in Theorem 2.1. Then 2.2 has a unique solution in $L_2\left((0, T), V \times V\right)$ and it is given by the mild solution 2.5.*

For the proofs of both Theorem 2.1 and Theorem 2.2 see [4].

For computational efforts in control and estimation of these systems, it is an important result to note that the weak formulation and the semigroup formulation yield the same solutions. In actuality, this equivalence can be given under weaker assumptions on σ_2 than (A2). If one relaxes the assumption on σ_2 to H-semiellipticity, one can show that the operator $\mathcal{A}$ of 2.4 defines a C_0-semigroup on $\mathcal{H}$ which can be extended to a space $\mathcal{Y}$, a proper subset of $\mathcal{V}^*$ which contains elements of the form $(0, v^*)$, $v^* \in V^*$. Then 2.5 can still be used to define mild solutions and the equivalence of solutions from Theorem 2.1 with mild solutions can be established (see [4] for details).

3   The Optimization Problem

We formulate the estimation problem as a least squares fit to observations. We seek $\bar{q} \in Q$ which minimizes

$$J(u, z; q) = \left| \tilde{C}_2\left(\tilde{C}_1 \left\{ u(t_i, \tilde{x}; q) \right\} - \left\{ z(t_i) \right\} \right) \right|^2. \qquad (3.1)$$

In 3.1, $u(t_i; q)$ is the solution to 2.2 (or the first component of the state vector $w(t; q)$ in 2.5) evaluated at t_i, $z(t_i)$ are pointwise time and pointwise space measurements. The operator $\tilde{C}_1$ may be in the form of the identity, time differentiation d/dt, or time differentiation twice d^2/dt^2, each followed by pointwise evaluation in time and space (at $x = \tilde{x}$). The operator $\tilde{C}_2$ may be the identity (corresponding to time domain identification procedures) or the Fourier transform (corresponding to identification in the frequency domain). In this paper we only treat the operator $\tilde{C}_2$ in form of the Fourier transform. If the measurements are taken with fixed sampling time, i.e., $\Delta t = t_i - t_{i-1} = t_{i+1} - t_i$ for all i and with a total of $\bar{N}$ samples at the fixed space position $x = \tilde{x}$ (this is often the case in experiments), then the Fourier series coefficients for $0 \leq i \leq \bar{N}$ are given by

$$\tilde{C}_2 \left\{ \tilde{C}_1 \{ u(t_i, \tilde{x}; q) \} \right\}_k = U(k; q) \tag{3.2}$$

$$= \frac{1}{\bar{N}} \sum_{i=0}^{\bar{N}-1} \tilde{C}_1 \{ u(t_i, \tilde{x}; q) \} \, e^{-jk(2\pi/\bar{N})i},$$

$$\tilde{C}_2 \{ z(t_i) \}_k = Z(k) = \frac{1}{\bar{N}} \sum_{i=0}^{\bar{N}-1} z(t_i) \, e^{-jk(2\pi/\bar{N})i}, \tag{3.3}$$

where $t_i = i \cdot \Delta t$ and $k = 0, 1, \ldots, \bar{N} - 1$. In 3.2, we use the generic symbol $U(k; q)$ to represent the Fourier coefficients of the transform of $\tilde{C}_1 \{ u(\cdot, \tilde{x}; q) \}$ for all three forms of $\tilde{C}_1$. With Δt as the sampling time, the kth value of the frequency corresponding to the kth coefficient is given by

$$\mathrm{f}_k = \frac{1}{\Delta t \cdot \bar{N}} \, k \tag{3.4}$$

and the corresponding magnitudes are given by $|U(k; q)|$ and $|Z(k)|$.

We assume that there are a finite and distinct number $(< \bar{N})$ of "spikes" among the $Z(k)$. Since each spike can be described by its frequency (corresponding to its index), magnitude and width of the spike, we will make some modifications to the cost function 3.1 when $\tilde{C}_2$ is the Fourier transform. Let $\bar{N}_M$ be the number of spikes among the $Z(k)$. We shall assume

(A4) The number of spikes of the solution $U(k; q)$ is the same as $\bar{N}_M$.

After reindexing coefficients of spikes among the $Z(k)$ and $U(k; q)$ and denoting the indices by k_ℓ^z, k_ℓ^u for $\ell = 1, \ldots, \bar{N}_M$ with $0 \leq k_\ell^z, k_\ell^u \leq \bar{N} - 1$, the frequency domain cost function can be more appropriately expressed by

$$\hat{J}(q) = \hat{J}(u, z; q) \qquad (3.5)$$

$$= \sum_{\ell=1}^{\bar{N}_M} \left(\epsilon_1 \left| \mathfrak{f}_{k_\ell^u}(q) - \mathfrak{f}_{k_\ell^z} \right|^2 \right.$$

$$\left. + \epsilon_2 \sum_{j=-n_\ell}^{N_\ell} ||U(k_\ell^u + j; q)| - |Z(k_\ell^z + j)||^2 \right),$$

where ϵ_1, ϵ_2 are weight constants, and n_ℓ, N_ℓ are certain lower and upper limits associated with the width (or the support) of the ℓth spike. The first part of the cost function 3.5 is related to the frequencies and the second part is related to the magnitude and the width of each spike. The limits n_ℓ and N_ℓ depend on the ℓth spike and are chosen so that n_ℓ and N_ℓ are the last i and last j respectively, for which the following conditions are satisfied:

$$|Z(k_\ell^k - i)| \geq 20\%|Z(k_\ell^z)|$$

for $i = 1, 2, \ldots, n_\ell$, and

$$|Z(k_\ell^z + j)| \geq 20\%|Z(k_\ell^z)|$$

for $j = 1, 2, \ldots, N_\ell$. The motivation behind our choice related to the width of the spike is that in traditional modal analysis, the width at approximately 30% of the peak value of the spike is used to estimate the damping ratio for the ℓth mode. Hence taking a conservative approach and using the width at 20% of the peak value should guarantee the inclusion of substantial damping information in the Fourier coefficients.

If the parameter $\bar{q}$ minimizes 3.5, then we shall take $\bar{q}$ as the estimate of the parameter which best describes the system in the frequency domain, i.e. the least squares fit of the model to data in the frequency domain sense. Hereafter, we interpret 3.1 as 3.5 whenever $\tilde{C}_2$ is the Fourier transform.

4 Approximation Technique

The minimization in our parameter estimation problem involves an infinite dimensional state space governed by 2.1. For computational purpose, finite dimensional approximations are necessary. To make these approximations, we first select a sequence of finite dimensional spaces H^N which are subspaces of H. We define orthogonal projections $P_H^N : H \rightarrow H^N$, $P_V^N : V \rightarrow H^N$, and choose the finite dimensional spaces $\mathcal{H}^N = H^N \times H^N$. We denote the orthogonal projections of $\mathcal{H} = V \times H$ onto $\mathcal{H}^N$ by $P_{\mathcal{H}}^N$. For convenience, we will hereafter restrict our consideration to the case where $f \in C^1([0,T]; V)$. Then the approximating estimation problems with finite dimensional state spaces can be stated as finding $q \in Q$ which minimizes

$$J^N(u^N, z; q) = \left| \tilde{C}_2 \left(\tilde{C}_1 \{u^N(t_i, \tilde{x}; q)\} - \{z(t_i)\} \right) \right|^2, \qquad (4.1)$$

or

$$\hat{J}(q) = \hat{J}^N(u^N, z; q) \qquad (4.2)$$

$$= \sum_{\ell=1}^{\bar{N}_M} \left(\epsilon_1 \left| f_{k_\ell^u{}^N}(q) - f_{k_\ell^z} \right|^2 \right.$$

$$\left. + \epsilon_2 \sum_{j=-n_\ell}^{N_\ell} \left| |U^N(k_\ell^u{}^N + j; q)| - |Z(k_\ell^z + j)| \right|^2 \right).$$

In 4.1 and 4.2, u^N is an approximate solution which satisfies a readily-solved finite dimensional system approximating 2.1 given by

$$\ddot{u}^N(t) + B^N(q)\dot{u}^N(t) + A^N(q)u^N(t) = P_H^N f(t)$$
$$u^N(0) = P_V^N u_0 \qquad (4.3)$$
$$\dot{u}^N(0) = P_H^N u_1,$$

or equivalently the first coordinate of the state vector

$$w^N(t; q) = T^N(t) P_{\mathcal{H}}^N w_0 + \int_0^t T^N(t-s) P_{\mathcal{H}}^N F(s)\, ds. \qquad (4.4)$$

Here A^N and B^N are Galerkin approximations to A and B, respectively, and $T^N(t)$ is the obvious corresponding approximation to $T(t)$, the semigroup of 2.5 generated by the operators of 2.4. Moreover, $U^N(k; q)$, for $k = 0, 1, \ldots, \bar{N} - 1$, in 4.2 is given by

$$U^N(k; q) = \frac{1}{\bar{N}} \sum_{i=0}^{\bar{N}-1} \tilde{C}_1 \{u^N(t_i, \tilde{x}; q)\}\, e^{-jk(2\pi/\bar{N})i}, \qquad (4.5)$$

and $f_{k_\ell^u{}^N}$ is defined in the same manner as $f_{k_\ell^z}$.

In 4.2, we have assumed that the number of "spikes" present in the approximate solution is the same as $\bar{N}_M$, the number of "spikes" in the data z. If one chooses N such that $N \geq 2\bar{N}_M$, then this assumption, which is (A4) for the approximation problems, is guaranteed.

Solving the estimation problems with finite dimensional state spaces, we obtain a sequence of estimates $\{\bar{q}^N\}$. To obtain parameter estimate convergence and continuous dependence (with respect to the observations $\{z(t_i)\}$) results, when $\tilde{C}_2$ is the Fourier transform operator, it has been shown in [1], [5] that it suffices, under the assumption that Q is a compact set, to argue: for arbitrary $\{q^N\} \subset Q$ with $q^N \to q$, we have

$$\tilde{C}_2\,\tilde{C}_1\, u^N(t;q^N) \to \tilde{C}_2\,\tilde{C}_1\, u(t;q)$$

for each t.

We first observe that the "solutions" $U^N(k;q^N)$ corresponding to the approximating systems provide approximate solutions to the original system "solutions" $U(k;q)$.

Theorem 4.1 *Suppose $\{q^N\} \subset Q$ is an arbitrary sequence with $q^N \to \bar{q}$ as $N \to \infty$. Let $U^N(k;q^N)$ denote the Fourier series coefficients for $\tilde{C}_1\{u^N(t;q^N)\}$ where u^N is the solution to the initial value problem 4.3 corresponding to q^N and let $U(k;\bar{q})$ denote the Fourier series coefficients for $\tilde{C}_1\{u(t;\bar{q})\}$ where u is the solution to the initial value problem 2.1 corresponding to $\bar{q}$. If $\tilde{C}_1\{u^N(t;q^N)\} \to \tilde{C}_1\{u(t;\bar{q})\}$ in V norm, and pointwise evaluation is continuous in the V norm, then*

$$\sum_{\ell=1}^{\bar{N}_M}\left(\ \epsilon_1\ \left|f_{k_\ell^u{}^N}(q^N) - f_{k_\ell^u}(\bar{q})\right|^2\right.$$

$$\left. +\ \epsilon_2 \sum_{j=-n_\ell}^{N_\ell}\ \left|\,|U^N(k_\ell^u{}^N + j;q^N)| - |U(k_\ell^u + j;\bar{q})|\,\right|^2\right) \to 0$$

as $N \to \infty$.

Now we are ready to state the main theorem for parameter estimation in the frequency domain formulation.

Theorem 4.2 *Assume that the parameter space Q is a compact subset of Euclidean space. Then each of the approximating estimation problems for 4.2 has a solution $\bar{q}^N$. Moreover, the sequence $\{\bar{q}^N\} \subset Q$ admits a convergent subsequence $\{\bar{q}^{N_j}\}$ with $\bar{q}^{N_j} \to \bar{q} \in Q$ as $j \to \infty$. If for each $q \in Q$, $U(k;q)$ is defined as in Theorem 4.1, then $\bar{q}$ is a solution to the original optimization problem for 3.5.*

For the proof of both these theorems, see [7].

Continuous dependence of parameter estimates on observations (an analogue of the "method stability" of [1], [5]) can be established for frequency domain estimation problems using the ideas above with the arguments given in [1], [5] and [7].

Next we consider conditions under which $\tilde{C}_1\{u^N(t;q^N)\}$ would converge to $\tilde{C}_1\{u(t;q)\}$ in V-norm. For the following theorems, some assumptions on the finite dimensional spaces H^N are required. We assume

(A5) $H^N \subset V \subset H$.

(A6) For each $z \in V$, there exists $\hat{z} \in H^N$ such that $|z - \hat{z}^N|_V \to 0$ as $N \to \infty$.

Theorem 4.3 *Suppose both $\sigma_1(q)$ and $\sigma_2(q)$ in 2.2 satisfy the conditions (A1)-(A3) and that conditions (A5)-(A6) hold. Let q^N be arbitrary such that $q^N \to q$ in Q. Then*

$$T^N(t;q^N)P_{\mathcal{H}}^N \xi \to T(t;q)\xi, \qquad \xi \in \mathcal{H}, \ t > 0$$

and

$$\mathcal{A}^N(q^N)T^N(t;q^N)P_{\mathcal{H}}^N \xi \to \mathcal{A}(q)T(t;q)\xi, \qquad \xi \in \mathcal{H}, \ t > 0$$

in V norm, with the convergence being uniform in t on compact subintervals. Here $T^N(t;q)$ and $T(t;q)$ are the analytic semigroups generated by $\mathcal{A}^N(q)$ and $\mathcal{A}(q)$, respectively.

For a proof of this theorem, see [3], [6].

Corollary 4.1 *Let $\mathcal{A}^N(q)$ and $\mathcal{A}(q)$ be the infinitesimal generators of the analytic semigroups $T^N(t;q)$ and $T(t;q)$, respectively, and $f \in C^1([0,T];V)$. Then for $\tilde{C}_1$ in one of the forms: identity, d/dt or d^2/dt^2 we have*

$$\tilde{C}_1\{u^N(t;q^N)\} \to \tilde{C}_1\{u(t;q)\}$$

in V-norm, where $u^N(t;q^N)$ and $u(t;q)$ are the first coordinate of 4.4 and 2.5 respectively.

This corollary follows from Theorem 4.3. For the detailed proof when $\tilde{C}_1 = d^2/dt^2$ see [6].

Before concluding this section, some comments on the assumption (A4) are appropriate. All our parameter estimation investigations have shown (and simple analysis of 2nd order damped scalar systems suggest) that the frequencies of the vibration of a beam are primarily determined by parameters representing stiffness, mass density of the beam and mass of the tip body, whereas the magnitude of each excited mode is determined by the damping parameters (as well as the excitation force, of course). Stiffness and mass can be measured and calculated quite accurately through the experiments and these values can be used in the process of parameter identification (ID). That is, those measured quantities can be used as initial values to begin optimization. Using 4.2 as a cost function, we have carried out our parameter ID using the following three steps to ensure that (A4) was satisfied. First, we fix damping parameters with values from our knowledge of previous experience with experimental configurations similar to the one being studied. We use the measured stiffness and mass as initial values and optimize on those parameters. Then we fix the stiffness and mass parameters at the optimal values resulting from the first step and optimize on the damping parameters. Finally, we proceed to carry out an optimization on all parameters using the optimal values of the parameters from the previous steps as an initial guess. Our numerical efforts with such a procedure here (and in previously reported findings) have proved most satisfactory.

The basic mathematical model that we have considered in connection with the efforts discussed in this paper is the Timoshenko equations for a cantilevered beam with tip body. We shall describe the model in some detail in the next section.

5 Example

As an example, we apply the techniques outlined above to the flexural vibrations of elastic beams represented by models that include the effects of rotary inertia and shear deformation. We consider a cantilevered beam with tip body, internal or material damping, and an applied transverse force. The partial differential equation together with boundary conditions based on the Timoshenko theory in terms of the bending moment $M(t, x)$ and the shear force $S(t, x)$ are given by (see [8], [9], [13], [14] and [15])

$$\rho \frac{\partial^2 u}{\partial t^2}(t, x) - \frac{\partial S}{\partial x}(t, x) + \gamma \frac{\partial u}{\partial t}(t, x) = \tilde{f}(t, x),$$

$$\rho r^2 \frac{\partial^2 \alpha}{\partial t^2}(t, x) - \frac{\partial M}{\partial x}(t, x) - S(t, x) = 0, \qquad 0 < x < \ell, \, t > 0,$$

$$mc \frac{\partial^2 u}{\partial t^2}(t, \ell) + (J_o + mc^2)\frac{\partial^2 \alpha}{\partial t^2}(t, \ell) + M(t, \ell) = 0, \qquad t > 0, \qquad (5.1)$$

$$m \frac{\partial^2 u}{\partial t^2}(t, \ell) + mc \frac{\partial^2 \alpha}{\partial t^2}(t, \ell) + S(t, \ell) = 0, \qquad t > 0,$$

$$u(t, 0) = \alpha(t, 0) = 0, \qquad t > 0.$$

Here ρ is the linear mass density, $u(t, x)$ is the transverse displacement, $\alpha(t, x)$ is the rotation of the beam cross section, $\tilde{f}(t, x)$ is the external applied transverse forces, $r^2 = I/A$ where I is the moment of inertia of the cross sectional area A and ℓ is the length of the beam. In 5.1, viscous (air) damping has been taken into account with γ as damping coefficient. We have assumed that the tip body has mass m and moment of inertia J_o about its center of mass which is assumed to be located at a distance c from the tip of the beam along the beam's axis or centerline. The bending moment and shear force with Kelvin-Voigt damping are given by

$$M(t, x) = EI \frac{\partial \alpha}{\partial x}(t, x) + c_D I \frac{\partial^2 \alpha}{\partial t \partial x}(t, x), \tag{5.2}$$

$$S(t, x) = \kappa A G \, \beta(t, x) + \kappa A c_s \frac{\partial \beta}{\partial t}(t, x), \tag{5.3}$$

where G is the shear modulus, κ is a correction factor, the shear distortion $\beta(t, x)$ is defined by $\partial u(t, x)/\partial x - \alpha(t, x)$, $c_D I$ is the bending damping coefficient, and c_s represents resistance related to shear strain rate. The possible parameters of the system to be considered are $q = (\rho, A, G, EI, m, c, J_o, \gamma, c_D I, c_s) \in Q \subset \mathbb{R}^{10}$. In view of the physical meaning of each parameter, the admissible parameter set will be taken to be a compact subset of $\mathbb{R}^{10}$ with $c, \gamma \geq 0$ and each of $\rho, A, G, EI, m, J_o, c_D I, c_s$ bounded below by some positive constants.

The Hilbert spaces H and V are defined by $H = \mathbb{R}^2 \times H^0(0, \ell) \times H^0(0, \ell)$ with inner product for $\Phi = (\nu_1, \nu_2, \phi_1, \phi_2)$, $\Psi = (\xi_1, \xi_2, \psi_1, \psi_2) \in H$

$$< \Phi, \Psi >_H \; = \; < \Lambda(\nu_1, \nu_2), (\xi_1, \xi_2) >_{\mathbb{R}^2} \tag{5.4}$$
$$+ < \rho \phi_1, \psi_1 > + < \rho r^2 \phi_2, \psi_2 >$$

where Λ is given by

$$\Lambda = \begin{pmatrix} m & mc \\ mc & mc^2 + J_o \end{pmatrix}, \tag{5.5}$$

and $V = \{(\nu_1, \nu_2, \phi_1, \phi_2) \in H \,|\, \phi_i \in H_L^1(0, \ell), \nu_i = \phi_i(\ell), i = 1, 2\}$ with inner product for $\Phi, \Psi \in V$

$$< \Phi, \Psi >_V = < (D\phi_1 - \phi_2), (D\psi_1 - \psi_2) > + < D\phi_2, D\psi_2 > . \tag{5.6}$$

Here we use the notation $D = \partial/\partial x$ and $H_L^1(0, \ell) = \{\phi \in H^1(0, \ell) | \phi(0) = 0\}$. A normalized variational form of 5.1 has the same form as 2.2

$$< \ddot{z}(t), \Phi >_{V^*, V} + \sigma_1(q)(z(t), \Phi) + \sigma_2(q)(\dot{z}(t), \Phi) = < f(t), \Phi >_H$$
$$z(0) = 0 \qquad\qquad\qquad\qquad\qquad\qquad\qquad\qquad \forall \Phi \in V$$
$$\dot{z}(0) = 0,$$

with

$$z(t) = \big(u(t,1), \alpha(t,1), u(t,\cdot), \alpha(t,\cdot)\big)$$
$$\Phi = \big(\phi_1(1), \phi_2(1), \phi_1, \phi_2\big)$$
$$f(t) = \big(0, 0, \rho^{-1}\tilde{f}(t,\cdot), 0\big),$$

and

$$\sigma_1(q)(z(t), \Phi) = \; < \kappa AG\,(Du - \alpha)\,,(D\phi_1 - \phi_2) > \tag{5.7}$$
$$+< EI\,D\alpha\,, D\phi_2 >,$$
$$\sigma_2(q)(\dot{z}(t), \Phi) = \; < \kappa Ac_s\,(D\dot{u} - \dot{\alpha})\,,(D\phi_1 - \phi_2) > \tag{5.8}$$
$$+< c_D I\,D\dot{\alpha}\,, D\phi_2 > + < \gamma \dot{u}\,, \phi_1 >.$$

The first order abstract form of 5.1 for $w(t) = (z(t), \dot{z}(t))^T$ is

$$\dot{w}(t) = \mathcal{A}(q)w(t) + F(t),$$

where

$$\mathcal{A}(q) = \begin{pmatrix} 0 & I \\ -A(q) & -B(q) \end{pmatrix}$$

with

$$A(q)z(t) = \Big(\Lambda^{-1}\big(-\kappa AG(\alpha(1) - Du(1)), EI\,D\alpha(1)\big), \tag{5.9}$$
$$\rho^{-1}D\big(\kappa AG(\alpha - Du)\big),$$
$$(\rho r^2)^{-1}\big(D(EI\,D\alpha) - \kappa AG(\alpha - Du)\big)\Big),$$

$$B(q)\dot{z}(t) = \Big(\Lambda^{-1}\big(-\kappa Ac_s\,(\dot{\alpha}(1) - D\dot{u}(1)), c_D I\,D\dot{\alpha}(1)\big), \tag{5.10}$$
$$\rho^{-1}D\big(\kappa Ac_s(\dot{\alpha} - D\dot{u})\big) + \rho^{-1}\gamma\dot{u},$$
$$(\rho r^2)^{-1}\big(D(c_D I\,D\dot{\alpha}) - \kappa Ac_s(\dot{\alpha} - D\dot{u})\big)\Big).$$

The domain of $\mathcal{A}(q)$ is defined by

$$dom\big(\mathcal{A}(q)\big) = \{\chi = (\Phi, \Psi) \in V \times H \,|\, \Psi \in V, EI\,D\phi_2 + c_D ID\psi_2 \in H^1(0,1),$$
$$\kappa AG(\phi_2 - D\phi_1) + \kappa Ac_s(\psi_2 - D\psi_1) \in H^1(0,1)\}.$$

With the chosen parameter space, both $\sigma_1(q)$ and $\sigma_2(q)$ satisfy (A1)-(A3) hence $\mathcal{A}(q)$ generates an analytic semigroup.

A Galerkin method can be applied in developing an approximation scheme (see [3], [15] for details) with cubic splines chosen to generate a set of basis elements (e.g. see [5]). It can be shown that this choice results in a finite dimensional space H^N which satisfies both (A5) and (A6).

A sample numerical result is depicted in Figure 5.1. The solid line is the experimental data and the dashed line is the response for the model with the optimal parameters. For detailed descriptions of the experiments and parameter identification procedures used, see [15] and [7]. The significant difference in magnitude between the experimental data and model response in the 6th mode is caused by the type of damping, Kelvin-Voigt, we have chosen. A main characteristic of Kelvin-Voigt type damping is the tendency for overdamping of the high frequency modes.

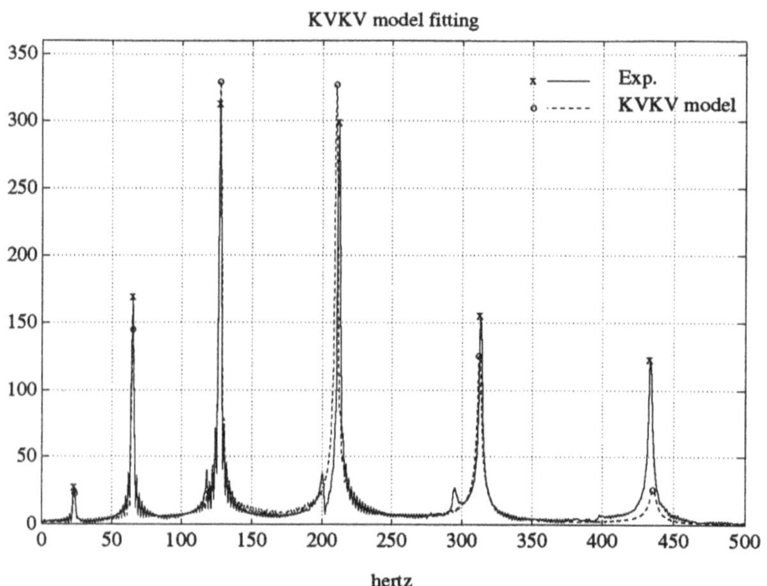

Figure 5.1: KVKV model response in frequency domain.

Here we depict only one example. We have successfully applied the frequency domain ID techniques outlined in this paper to many other problems involving systems such as Euler-Bernoulli beams, beams with piezoceramic patches as sensors and actuators, etc.. These ideas can also be applied to parameter identification problems involving plate vibrations and a number of other problems, as long as the system eigenvalues are distinct.

References

[1] H.T. BANKS, "On a Variational Approach to Some Parameter Estimation Problems," in *Distributed Parameter Systems, Springer Lecture Notes in Control and Information Sciences*, Vol. 75 (1985), pp. 1-23.

[2] H.T. BANKS and J.A. BURNS, "Introduction to Control of Distributed Parameter Systems," *Birkhaüser, Boston*, 1993.

[3] H.T. BANKS, and K. ITO, "A Unified Framework for Approximation in Inverse Problems for Distributed Parameter Systems," *Control-Theory and Advanced Technology*, **4** (1988), pp. 73-90.

[4] H.T. BANKS, K. ITO and Y. WANG, "Well-posedness for Damped Second Order Systems with Unbounded Input Operators," *to appear*.

[5] H.T. BANKS and K. KUNISCH, "Estimation Techniques for Distributed Parameter Systems," *Birkhaüser, Boston*, 1989.

[6] H.T. BANKS and D.A. REBNORD, "Analytic Semigroups: Applications to Inverse Problems for Flexible Structures," *CAMS Report No. 90-3, University of Southern California*, January, 1990; in *Differential Equations with Applications (ed. by J. Goldstein, et. al)*, *Marcel Dekker*, 1991, pp. 21-35.

[7] H.T. BANKS, Y. WANG and D.J. INMAN, "Bending and Shear Damping in Beams: Frequency Domain Estimation Techniques," *Journal of Vibration and Acoustics, to appear*.

[8] R. DAVIS, R.D. HENSHELL and G.B. WARBURTON, "A Timoshenko Beam Element," *Journal of Sound and Vibration* **22** (4) (1972), pp. 475-487.

[9] T.C. HUANG, "The Effect of Rotatory Inertia and of Shear Deformation on the Frequency and Normal Mode Equations of Uniform Beams with Simple End Conditions," *ASME Transactions, Journal of Applied Mechanics*, Vol. 28, Dec. 1961, pp. 579-583.

[10] D.A. REBNORD, "Parameter Estimation for Two-Dimensional Grid Structures," *Ph.D. Thesis, Brown University*, 1989.

[11] R.E. SHOWALTER, "Hilbert Space Methods for Partial Differential Equations," *Pitman, London*, 1977.

[12] H. TANABE, "Equations of Evolution," *Pitman, London*, 1979.

[13] S. TIMOSHENKO, "Theory of Elasticity," *McGraw-Hill, New York*, 1934.

[14] S. TIMOSHENKO, D.H. YOUNG and W. WEAVER, "Vibration Problems in Engineering," 4th ed., *Wiley, New York*, 1974.

[15] Y. WANG, "Damping Modeling and Parameter Estimation in Timoshenko Beams," *Ph.D. Thesis, Brown University*, 1990.

[16] J. WLOKA, "Partial Differential Equations," *Cambridge University Press, New York*, 1987.

ON MODEL IDENTIFICATION OF GAUSSIAN RECIPROCAL PROCESSES FROM THE EIGENSTRUCTURE OF THEIR COVARIANCES

Carlos F. Borges

Code Ma/Bc
Naval Postgraduate School
Monterey, CA 93943

Ruggero Frezza

DEI, Univ. di Padova
via Gradenigo 6/A
35131 PADOVA - ITALY

1　Models of Gaussian reciprocal processes

A stochastic process $x(k)$ defined on $[1, N]$ is *reciprocal* if for any subinterval $[l, m]$ of $[1, N]$ the process in the interior of $[l, m]$ is conditionally independent of the process in $[1, N] - [l, m]$ given $x(l)$ and $x(m)$. For a more rigorous definition see [10]. Reciprocal processes generalize Markov processes since a Markov process is reciprocal while the converse, in general, is not true, see [10] for an example of a process which is reciprocal and not Markov.

Recently, it was shown [11,12,5] that, under suitable assumptions, a discrete time Gaussian reciprocal process $x(k)$ satisfies a nearest neighbor model like the following

$$m_0(k)x(k) - m_-(k)x(k-1) - m_+(k)x(k+1) = \xi(k) \qquad (1.1)$$

where $\xi(k)$ is a zero mean Gaussian process with covariance

$$
\begin{aligned}
E[\xi(k)\xi(k)] &= m_0(k) \\
E[\xi(k)\xi(k+1)] &= -m_-(k+1) = -m_+(k) \qquad (1.2) \\
E[\xi(k)\xi(k+l)] &= 0 \quad \text{for } l > 1
\end{aligned}
$$

In matrix form the model (1.1) can be written as

$$\Lambda \mathbf{x} = \xi \qquad (1.3)$$

where

$$\mathbf{x}^T = \begin{bmatrix} x(1) & x(2) & \cdots & x(N) \end{bmatrix}, \qquad (1.4)$$

$$\xi^T = \begin{bmatrix} \xi(1) & \xi(2) & \cdots & \xi(N) \end{bmatrix}, \qquad (1.5)$$

and Λ is the following Jacobi matrix

$$\Lambda = \begin{bmatrix} m_0(1) & -m_+(1) & & & \\ -m_+(1) & m_0(2) & -m_+(2) & & \\ & -m_+(2) & & \ddots & \\ & & \ddots & & -m_+(n-1) \\ & & & -m_+(n-1) & m_0(n) \end{bmatrix}$$

The covariance structure (1.2) of the noise process corresponds to

$$E[\xi \xi^T] = \Lambda. \tag{1.6}$$

From (1.3) and (1.6) we see that

$$E[\mathbf{x}\xi^T] = I \tag{1.7}$$

and, therefore, that

$$\Lambda \mathbf{R} = I \tag{1.8}$$

where $\mathbf{R} = E[\mathbf{x}\mathbf{x}^T]$. Thus, the matrix Λ characterizing the model of the reciprocal process x and the covariance $\mathbf{R}$ of $\mathbf{x}$ have a related eigenstructure. If $(\lambda_k, \mathbf{u}_k)$ are the eigenpairs of Λ then $(1/\lambda_k, \mathbf{u}_k)$ are the eigenpairs of $\mathbf{R}$.

This leads us to consider the possibility of identifying the reciprocal model (1.1) of such a process starting from the eigenfunctions of its covariance. Clearly, this is equivalent to reconstructing the matrix Λ from its eigenstructure and this is a well known problem in the literature see, for example,[8], [9] and references therein.

We propose an algorithm net yet studied in the literature on reconstructing Jacobi matrices. We reconstruct Λ from its two extremal eigenpairs $(\lambda_1, \mathbf{u}_1)$ and $(\lambda_n, \mathbf{u}_n)$. We show that the algorithm is well posed. The extremal eigenpairs also have the advantage that they can be easily computed from the covariance $\mathbf{R}$ using Krylov sequence methods like the Lanczos algorithm or power and inverse iteration [7].

The algorithm is straightforward and can be generalized to other classes of matrices. For example, we show how to reconstruct symmetric arrow matrices. We conjecture that this inverse problem is well-posed for all unreduced symmetric acyclic matrices which represent an interesting generalization of the reciprocal models.

The algorithm also serves to identify Markov models. In [12], it was shown that a Markov process $x(k)$ satisfying the model

$$\Omega \mathbf{x} = B\mathbf{w} \tag{1.9}$$

where

$$\Omega = \begin{bmatrix} 1 & & & & \\ -a(1) & 1 & & & \\ & -a(2) & \ddots & & \\ & & & \ddots & \\ & & & -a(n-1) & 1 \end{bmatrix} \qquad (1.10)$$

$$B = \mathrm{diag}(b(k)) \qquad (1.11)$$

and

$$\mathbf{w} = \begin{bmatrix} w(0) & w(1) & \cdots & w(N-1) \end{bmatrix} \qquad (1.12)$$

where $w(k)$ are Gaussian, zero mean, independent random variables with unitary variance, also satisfies a reciprocal model like (1.3) where

$$\Lambda = \Omega^T Q^{-1} \Omega \qquad (1.13)$$

with $Q = BB^T$ and

$$\xi = \Omega^T Q^{-1} \mathbf{w}. \qquad (1.14)$$

Therefore, Ω and B can be obtained from Λ by performing a Cholesky factorization.

In practice, the covariance $\mathbf{R}$ will be corrupted by noise; we will know the covariance $\mathbf{R}_y$ of the observations

$$\mathbf{y} = C\mathbf{x} + \mathbf{v} \qquad (1.15)$$

where $C = \mathrm{diag}_{N \times N}(c)$. If $v(k)$ are independent Gaussian random variables identically distributed with zero mean and variance v, then $\mathbf{R}_y$ is related to $\mathbf{R}$ by

$$\mathbf{R}_y = CRC^T + vI. \qquad (1.16)$$

One can show that as long as $v \neq 0$ the covariance $\mathbf{R}_y$ does not have a tridiagonal inverse. We can refine the algorithm so that given $\mathbf{R}_y$ it estimates v and Λ such that

$$\Lambda(\mathbf{R}_y - vI) = \mathrm{I}. \qquad (1.17)$$

2 Reconstructing a Jacobi Matrix

Let T be an unreduced $n \times n$ real symmetric tridiagonal matrix

$$T = \begin{bmatrix} \alpha_1 & \beta_1 & & & & \\ \beta_1 & \alpha_2 & \beta_2 & & & \\ & \beta_2 & & \ddots & & \\ & & \ddots & & & \beta_{n-1} \\ & & & & \beta_{n-1} & \alpha_n \end{bmatrix} \tag{2.1}$$

with $\beta_i > 0$ for $i = 1, 2, ..., n-1$. Using notation introduced in [15] we say $T \in \mathrm{UST}_+(n)$.

We want to reconstruct T from two of its eigenpairs $(\lambda, \mathbf{u})$ and $(\mu, \mathbf{v})$. The eigenvector recurrence for symmetric tridiagonal matrices is

$$\beta_{i-1} u_{i-1} + \alpha_i u_i + \beta_i u_{i+1} = \lambda u_i \tag{2.2}$$

where u_i is the i'th element of $\mathbf{u}$, and we define $\beta_0 \equiv \beta_n \equiv 0$. Applying this relation to both eigenpairs gives

$$\beta_{i-1} u_{i-1} + \alpha_i u_i + \beta_i u_{i+1} = \lambda u_i \tag{2.3}$$
$$\beta_{i-1} v_{i-1} + \alpha_i v_i + \beta_i v_{i+1} = \mu v_i \tag{2.4}$$

and, clearly

$$\begin{bmatrix} u_i & u_{i+1} \\ v_i & v_{i+1} \end{bmatrix} \begin{bmatrix} \alpha_i \\ \beta_i \end{bmatrix} = \begin{bmatrix} \lambda u_i - \beta_{i-1} u_{i-1} \\ \mu v_i - \beta_{i-1} v_{i-1} \end{bmatrix}. \tag{2.5}$$

This is a *forward* recurrence for the elements of T (there is an analogous backward recurrence). Setting $i = 1$ in (2.5) and solving gives the initial condition

$$\alpha_1 = \frac{\lambda u_1 v_2 - \mu v_1 u_2}{u_1 v_2 - v_1 u_2}, \tag{2.6}$$

$$\beta_1 = \frac{\mu - \lambda}{u_1 v_2 - v_1 u_2} u_1 v_1. \tag{2.7}$$

The terminal condition can be found in a similar manner

$$\alpha_n = \frac{\mu v_n u_{n-1} - \lambda u_n v_{n-1}}{u_{n-1} v_n - v_{n-1} u_n}, \tag{2.8}$$

$$\beta_{n-1} = \frac{\lambda - \mu}{u_{n-1} v_n - v_{n-1} u_n} u_n v_n. \tag{2.9}$$

Applying Cramer's rule to (2.5) we get

$$\alpha_i = \frac{\lambda u_i v_{i+1} - \mu v_i u_{i+1} + \beta_{i-1}(v_{i-1}u_{i+1} - u_{i-1}v_{i+1})}{u_i v_{i+1} - v_i u_{i+1}} \qquad (2.10)$$

$$\beta_i = \frac{(\mu - \lambda)u_i v_i + \beta_{i-1}(u_{i-1}v_i - v_{i-1}u_i)}{u_i v_{i+1} - v_i u_{i+1}} \qquad (2.11)$$

Combining equations (2.11) and (2.7) it follows that

$$\beta_i = \frac{\mu - \lambda}{u_i v_{i+1} - v_i u_{i+1}} \sum_{k=1}^{i} u_k v_k \qquad (2.12)$$

for $i = 1, 2, ..., n - 1$. The backward formula

$$\beta_i = \frac{\lambda - \mu}{u_i v_{i+1} - v_i u_{i+1}} \sum_{k=i+1}^{n} u_k v_k. \qquad (2.13)$$

follows directly from (2.12) and the orthogonality of **u** and **v**.

The cost of reconstructing the β_i is $7n-8$ flops, and an additional $9n-10$ flops for the α_i given that the computation is properly implemented.

The reconstruction formulas require that the denominators of (2.11) and (2.10) are not zero, this is always the case if the two eigenpairs in question are the extremal ones. We introduce the following theorem from [14]

Theorem 2.1 *If $T \in \mathrm{UST}_+(n)$ then the number of sign changes between consecutive elements of the k'th eigenvector of T, denoted s_k, is k.*

We refer the reader to [14] for a proof but note that it follows from the Sturm sequence property for the characteristic polynomials of the principal minors. With this fact we prove the following theorem.

Theorem 2.2 *If $T \in \mathrm{UST}_+(n)$ and if $(\lambda, \mathbf{u})$ and $(\mu, \mathbf{v})$ are the largest and smallest eigenpairs of T, respectively, then $u_i v_{i+1} - v_i u_{i+1} \neq 0$ for $i = 1, 2, ..., n - 1$.*

Proof. Since v_i and v_{i+1} have opposite sign and u_i and u_{i+1} have the same sign it follows that both terms in $u_i v_{i+1} - v_i u_{i+1}$ have the same sign. Moreover, the strict interlacing property for unreduced symmetric tridiagonals (see [16] p. 300) guarantees that all terms are nonzero, and the theorem follows. ∎

Hence, using the two extremal eigenpairs of $T \in \mathrm{UST}_+$ we can always reconstruct the original matrix using formulas (2.10), (2.12), and (2.13). The restriction that $\beta_i > 0$ is an artificial one that simplifies the proofs and may be relaxed to simple unreducedness, $\beta_i \neq 1$. Finally, note that the denominator is computed without cancellation because of the sign pattern.

3   Reconstructing the Arrow Matrix

To demonstrate the generality of this approach we now show how to reconstruct the arrow matrix in a similar manner. The arrow is of some importance as it occurs in certain parallel divide and conquer schemes for solving eigenproblems [1] and also occurs in certain problems of physics [13].

The general form of an arrow matrix is

$$
T = \begin{bmatrix}
\alpha_1 & & & & \beta_1 \\
& \alpha_2 & & & \beta_2 \\
& & \ddots & & \vdots \\
& & & \alpha_{n-1} & \beta_{n-1} \\
\beta_1 & \beta_2 & & \beta_{n-1} & \gamma
\end{bmatrix}
\tag{3.1}
$$

If $\beta_i > 0$ for $i = 1, 2, ..., n-1$ then we shall say that $T \in \mathbf{USA}_+(n)$, where $\mathbf{USA}_+(n)$ is the set of unreduced symmetric arrow matrices with $\beta_i > 0$. Proceeding as before, we let $(\lambda, \mathbf{u})$ and $(\mu, \mathbf{v})$ be two eigenpairs of T. The eigenvector recurrence yields

$$
\begin{bmatrix}
u_i & u_n \\
v_i & v_n
\end{bmatrix}
\begin{bmatrix}
\alpha_i \\
\beta_i
\end{bmatrix}
=
\begin{bmatrix}
\lambda u_i \\
\mu v_i
\end{bmatrix}
\tag{3.2}
$$

for $i = 1, 2, ..., n-1$. Applying Cramer's rule gives

$$
\alpha_i = \frac{\lambda v_n u_i - \mu u_n v_i}{v_n u_i - u_n v_i}
\tag{3.3}
$$

$$
\beta_i = \frac{u_i v_i (\mu - \lambda)}{v_n u_i - u_n v_i}
\tag{3.4}
$$

Moreover, the eigenvector relation also gives

$$
\gamma = \lambda - \frac{1}{u_n} \sum_{i=1}^{n-1} \beta_i u_i
\tag{3.5}
$$

This gives a simple, parallel reconstruction algorithm. The only question is whether or not the determinants $v_n u_i - u_n v_i$ are all nonzero. To demonstrate that they are, under the correct conditions, we need to establish some facts about the eigenvectors of an unreduced arrow matrix. We begin by noting that

$$
T - \lambda I = \begin{bmatrix}
A - \lambda I & \mathbf{b} \\
\mathbf{b}^T & \gamma - \lambda
\end{bmatrix}
\tag{3.6}
$$

where $A = \text{diag}(\alpha_1, \alpha_2, ..., \alpha_{n-1})$, and $\mathbf{b} = [\beta_1, \beta_2, ..., \beta_{n-1}]^T$. Provided that the α_i are all distinct the Gauss factorization of T is

$$\left[\begin{array}{cc} A - \lambda I & \mathbf{b} \\ \mathbf{b}^T & \gamma - \lambda \end{array} \right] = \left[\begin{array}{cc} I & 0 \\ \mathbf{b}^T (A - \lambda I)^{-1} & 1 \end{array} \right] \left[\begin{array}{cc} A - \lambda I & \mathbf{b} \\ 0^T & -f(\lambda) \end{array} \right] \quad (3.7)$$

where f, the *spectral function*, is a rational Pick function and is given by

$$f(\lambda) = \lambda - \gamma + \sum_{i=1}^{n-1} \frac{\beta_i^2}{\alpha_i - \lambda} \quad (3.8)$$

It is obvious from equations (3.7) and (3.8) that the zeros of f are the eigenvalues of T and that the α_i interlace the eigenvalues. Henceforth, we shall consider only those elements of $\mathbf{USA}_+(n)$ with distinct α_i. The eigenvector associated with a given eigenvalue λ is

$$\mathbf{u}(\lambda) = \left[\begin{array}{c} (\lambda I - A)^{-1} \mathbf{b} \\ 1 \end{array} \right] \quad (3.9)$$

Combining this description of the eigenvectors with the fact that the α_i interlace the eigenvalues, it is easy to verify the following theorem

Theorem 3.1 *Let $T \in \mathbf{USA}_+(n)$ have distinct elements α_i, and order its eigenvalues, λ_i, so that $\lambda_1 > \lambda_2 > ... > \lambda_n$. Then, the eigenvectors, $\mathbf{u}(\lambda_k)$, from (3.9) satisfy*
1. $\mathbf{u}_i(\lambda_k) \neq 0$ for any $i = 1, 2, ..., n$.
2. $\mathbf{u}(\lambda_k)$ has precisely $k - 1$ elements that are less than zero, and $n - k + 1$ elements that are greater than zero.

Proof. The proof follows directly from formula (3.9), the interlacing property, and the positivity of the β_i. ∎

This simplifies the reconstruction formulas since, if we normalize the eigenvectors so that their last elements are one, they become

$$\alpha_i = \frac{\lambda u_i - \mu v_i}{u_i - v_i} \quad (3.10)$$

$$\beta_i = \frac{(\mu - \lambda) u_i v_i}{u_i - v_i} \quad (3.11)$$

$$\gamma = \mu - \sum_{i=1}^{n-1} \beta_i v_i \quad (3.12)$$

With these formulas, we can reconstruct the arrow matrix in $8n - 7$ flops. To establish well-posedness we must verify that none of the denominators in these formulas are zero. If we use the extremal eigenpairs, then this follows directly from fact 2 of Theorem 3.1. However, for the symmetric arrow matrix a more general result is possible.

Theorem 3.2 *Let $T \in \mathrm{USA}_+(n)$ have distinct α_i. If λ and μ are distinct eigenvalues of T then $\mathbf{u}_i(\lambda) \neq \mathbf{u}_i(\mu)$.*

Proof. From (3.9) we have

$$\mathbf{u}_i(\lambda) = \frac{\beta_i}{\lambda - \alpha_i} \neq \frac{\beta_i}{\mu - \alpha_i} = \mathbf{u}_i(\mu) \tag{3.13}$$

∎

The reconstruction algorithm has another very important property – if the two extremal eigenpairs are used, then the β_i can be found, up to the scaling factor $\lambda_n - \lambda_1$, without cancellation. This follows from the fact that if the corresponding eigenvectors are eigenvectors corresponding to the two extremal eigenvalues of T, and if they are normalized so that their last elements are both ones, then all of their remaining elements must have opposite signs. This is fortuitous since it means that the differences that appear in the denominator do not involve cancellation. Moreover, if T is indefinite there are no cancellations whatsoever in computing the β_i. Conversely, if T is definite there are no cancellations in computing the α_i. If T is semi-definite then there is no cancellation at all (including the computation of γ). The computation of γ involves one cancellation if the matrix is indefinite, and none if it is definite, or semi-definite, provided we choose the correct eigenvector for its computation. In any case, whenever there is cancellation in this algorithm, it is benign.

References

[1] P. ARBENZ, *Divide and conquer algorithms for the bandsymmetric eigenvalue problem*, in Parallel Computing '91, D. J. Evans, G. R. Joubert, and H. Liddell, eds., Elsevier Science Publishers B. V., Amsterdam, 1992, pp. 151–158.

[2] R. ASH AND M. GARDNER, *Topics in Stochastic Processes*, Series on Probability and Mathematical Statistics, Academic Press, 1975.

[3] D. BOLEY AND G. GOLUB, *A modified method for reconstructing periodic Jacobi matrices*, Math. Comp., 42 (1984), pp. 143–150.

[4] C. BORGES, R. FREZZA, AND W. GRAGG, *Some inverse eigenproblems for Jacobi and arrow matrices*, J. Numer. Linear Algebra Appl., (1992). In review.

[5] R. FREZZA, *Modeling of Higher Order and Mixed Order Gaussian Reciprocal Processes with Application to the Smoothing Problem*, PhD thesis, University of California, Davis, 1990.

[6] I. GEL'FAND AND B. LEVITAN, *On the determination of a differential equation from its spectral function*, AMS Translations, 2 (1955), pp. 253–304.

[7] G. GOLUB AND C. V. LOAN, *Matrix Computations*, The Johns Hopkins University Press, 1983.

[8] W. GRAGG AND W. HARROD, *The numerically stable reconstruction of jacobi matrices from spectral data*, Numer. Math., 44 (1984), pp. 317–335.

[9] W. HEGLAND AND J. MARTI, *Algorithms for the reconstruction of special Jacobi matrices from their eigenvalues*, SIAM J. Matrix Anal. Appl., 10 (1989), pp. 219–228.

[10] B. JAMISON, *Reciprocal processes: The stationary Gaussian case*, Ann. of Math. Stat., 41 (1970), pp. 1624–1630.

[11] A. KRENER, R. FREZZA, AND B. LEVY, *Gaussian reciprocal processes and self adjoint stochastic differential equations of second order*, Stochastics and Stochastics Reports, 34 (1991), pp 29–56.

[12] B. LEVY, R. FREZZA, AND A. KRENER, *Modeling and estimation of discrete time Gaussian reciprocal processes*, IEEE Trans. Automat. Contr., AC–35 (1990), pp. 1013–1023.

[13] D. P. O'LEARY AND G. W. STEWART, *Computing the eigenvalues and eigenvectors of arrowhead matrices*, J. Comp. Phys., 90 (1990), pp. 497–505.

[14] B. PARLETT, *The Symmetric Eigenvalue Problem*, Prentice–Hall, Englewood Cliffs, NJ, 1980.

[15] B. PARLETT AND W.-D. WU, *Eigenvector matrices of symmetric tridiagonals*, Numer. Math., 44 (1984), pp. 103–110.

[16] J. WILKINSON, *The Algebraic Eigenvalue Problem*, Clarendon Press, Oxford, 1965.

AN INVERSE PROBLEM IN THERMAL IMAGING

Kurt Bryan *

Institute for Computer Applications in Science and Engineering
NASA Langley Research Center
Hampton, VA 23681

1 Introduction

Thermal imaging is a technique which has proven quite useful for the non-destructive evaluation of materials, especially locating cracks or disbonds in structures. In thermal imaging, one applies a specified heat flux to the boundary of an object and records the resulting surface temperature as a function of time. From this information one hopes to determine the internal thermal diffusivity of the object, perhaps to locate flaws–cracks, bubbles, corrosion, etc. Some recent work on this subject is detailed in [2], [3], [4] and [7].

In this paper the problem of detecting and identifying an unknown internal void in a planar domain using thermal methods is examined. The void could represent a defect in the material, or it could be a feature which is supposed to be present, e.g., a conduit, whose location or geometry is to be assessed. We will examine the case in which the thermal stimulus, an applied heat flux at the boundary of the sample, is a periodic point heat source. In this case one can separate the temporal and spatial variables, which leads to an inverse problem for an elliptic equation. This is solved with an optimization approach and uses a boundary integral equation formulation to approximate the heat conduction problem.

2 Mathematical Model

The sample (without void) to be tested will be denoted by Ω, a bounded region in $\mathbb{R}^2$ with piecewise C^2 boundary. The internal void will be denoted by D, where $D \subset\subset \Omega$ with C^2 boundary. The situation is illustrated in Figure 1. Since the heating is periodic, the temperature $T(t, x)$ can be written as

$$T(t, x) = \mathrm{Re}\{e^{i\omega t} T(x)\} \tag{2.1}$$

*This research was carried out while the author was in residence at the Institute for Computer Applications in Science and Engineering (ICASE), NASA Langley Research Center, Hampton, VA, 23681, which is operated under National Aeronautics and Space Administration contracts NAS1-18605 and NAS1-19480.

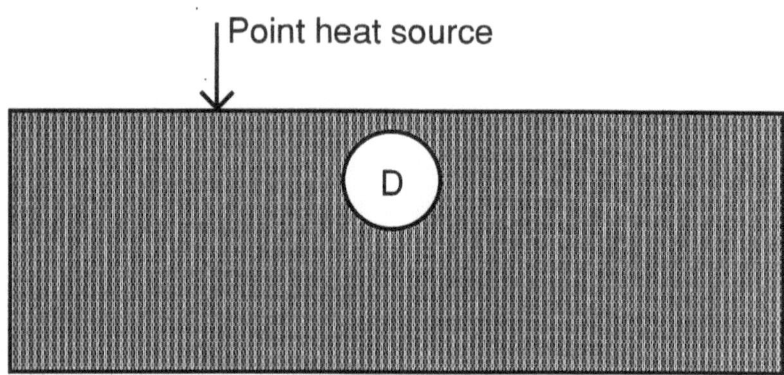

Figure 1: Sample geometry and heating.

where t denotes time, $x = (x_1, x_2)$ is the spatial variable and the periodic heating has frequency $\frac{\omega}{2\pi}$. The spatial part of the solution, $T(x)$, is assumed to satisfy

$$\begin{aligned}
\alpha \nabla^2 T - i\omega T &= 0 \quad \text{in } \Omega \\
k \frac{\partial T}{\partial n} &= g \quad \text{on } \partial\Omega \\
k \frac{\partial T}{\partial n} &= 0 \quad \text{on } \partial D
\end{aligned} \qquad (2.2)$$

where α is the sample thermal diffusivity, k the sample thermal conductivity and g is the heat flux at the boundary. In our case $g = \delta_P$ where P denotes the point at which the input heat flux is applied. The problem under consideration can be stated as follows: Given measurements of the solution T to equation (2.2) on the boundary of the sample, identify the void D. Note that the function T is complex-valued.

3 Assumptions and Identification Results

In order to obtain uniqueness and continuous dependence results, a few restrictions on the class of voids and their parameterization are needed. First, let us use $\tilde{C}^2[0, 1]$ to denote the space of C^2 functions on $[0, 1]$ where the endpoints 0 and 1 are identified with each other. This space can be normed by $\|\phi\| = \sup_{t \in [0,1]} |D^\beta \phi|$, $|\beta| \leq 2$. We assume that D depends on finitely many parameters, $D = D(q)$ with $q \in Q \subset\subset \mathbb{R}^m$ and where:

(a) $q_1 = q_2$ implies $D(q_1) = D(q_2)$ (unique parameterization) .

(b) $D(q) \subset \Omega' \subset\subset \Omega$ for $q \in Q$ ($D(q)$ stays away from $\partial\Omega$).

(c) The curves $\partial D(q)$ are parameterized as $x(q,t) = (x_1(q,t), x_2(q,t))$ for $q \in Q$, $0 \leq t < 1$, with $x_i(q,t)$ a $\tilde{C}^2$ function of t for each $q \in Q$ and $\frac{dS}{dt} = \sqrt{(x_1')^2 + (x_2')^2}$ bounded away from zero. Also, the map $q \to x(q,t)$ is continuous from $\mathbb{R}^m$ to $\tilde{C}^2[0,1]$.

With these assumptions one can prove:

- (Uniqueness) If T_1 and T_2 denote the solutions to equation (2.2) with $D = D(q_1)$ and $D = D(q_2)$, respectively, and S is any portion of $\partial\Omega$ with positive measure, then $T_1 = T_2$ on S implies $D(q_1) = D(q_2)$ and so $q_1 = q_2$.

- (Continuous dependence) Let q_n be a sequence in Q and $T(q_n)$ the corresponding solution to equation (2.2) with $D = D(q_n)$. Suppose $T(q_n) \to T(q^*)$ in $C(\partial(\Omega \setminus D))$ for some $q^* \in Q$. Then $q_n \to q^*$.

See [5] for a more detailed discussion and proofs.

4 Optimization and Boundary Integral Approach

One approach to finding the unknown void is to use an optimization technique. Specifically, suppose T_i, $i = 1, \ldots, n$, are measurements at points x_i of the boundary temperature of the sample with true void $D(q^*)$. One can then seek an estimate of $D(q^*)$ by finding that region $D(q)$ which minimizes the quadratic functional

$$J(q) = \sum_{i=1}^{n} |T_q(x_i) - T_i|^2 \tag{4.1}$$

where $T_q(x)$ is the solution to equation (2.2) with $D = D(q)$.

One of the drawbacks to the optimization approach is that minimizing the functional (4.1) requires many repeated solutions to the heat conduction problem with varying regions D. It is thus highly advantageous to have a means of rapidly solving the heat conduction problem. Moreover, standard methods for minimizing $J(q)$ also require the derivatives of J with respect to q. For these reasons the method of boundary integral equations was chosen for solving equation (2.2). If the sample without the void D is denoted by Ω then the function $T(x)$ satisfies the integral equation

$$-\frac{1}{2}T(x) + \int_{\partial(\Omega \setminus D)} T(y) \frac{\partial G(x,y)}{\partial n_y} \, dS_y = \frac{1}{k} \int_{\partial(\Omega \setminus D)} G(x,y) g(y) \, dS_y \tag{4.2}$$

for each $x \in \partial(\Omega \setminus D)$ where $\frac{\partial}{n_y}$ is the normal derivative in the y variable and dS_y is surface measure. The function $G(x,y)$ is a fundamental solution or Green's function for the operator $\nabla^2 - \frac{i\omega}{\alpha}$. Such a function is given by

$$G(x, y) = -\frac{1}{2\pi} \left(\mathrm{ker}(\sqrt{\frac{\omega}{\alpha}}r) + i\,\mathrm{kei}(\sqrt{\frac{\omega}{\alpha}}r) \right) \tag{4.3}$$

where $r = |x - y|$ and ker() and kei() are the Kelvin functions. For a derivation of the boundary integral equation see [6], chapter 3.

There are a variety of methods for solving the equation (4.2). We have chosen Nyström's method; see [1] for more details. The boundary integral equation offers several advantages for the current identification problem. They are:

- It reduces the dimension of the problem by one; the two-dimensional heat equation is reduced to a one-dimensional integral equation.

- It only solves for the temperature $T(x)$ where its value is needed, on the surface of the sample, with a corresponding increase in speed.

- Allows simultaneous computation of $\frac{\partial T}{\partial \bar{q}}$; these derivatives satisfy the integral equation (4.2) but with a different right hand side, thus much of the work involved in solving (4.2) can be re-used to solve for these derivatives.

Moreover, one can prove that as the Nyström computation is refined by adding more nodes, the solution to the optimization problem for the finite dimensional system (obtained by replacing $T_q(x)$ in (4.1) by $T_q^k(x)$, the k-node Nyström rule) converges to the solution of the optimization problem for the infinite dimensional system, i.e., the minimum of $J(q)$ over Q.

5 Strategy

Optimization approaches have a drawback, namely, the possibility of getting caught in a local minima which is not a global minima. The following example illustrates this point. The sample is taken to be a rectangular aluminum block with length 1.27 cm and height 0.32 cm. The point heat source has a power of one watt and is at a frequency of 3.0 Hz. The "true" void D^* is circular, centered at x_1, x_2 coordinates (0.9 cm, 0.24 cm) where the lower left corner of the sample is (0,0). The void radius is 0.06 cm. The heat source was applied near D^*, at x_1 coordinate 0.9 cm as illustrated in Figure 2. Equation 4.2 was solved using Nyström's method to yield the temperature $T(x)$ solving equation 2.2 on the boundary of the sample; the solution was obtained at 40 points on the top surface, denoted by T_i, $i = 1, \ldots, 40$. The "prospective" void, D, also a circle, was chosen to have the same x_2 coordinate and radius as D^* but the x_1 coordinate was varied

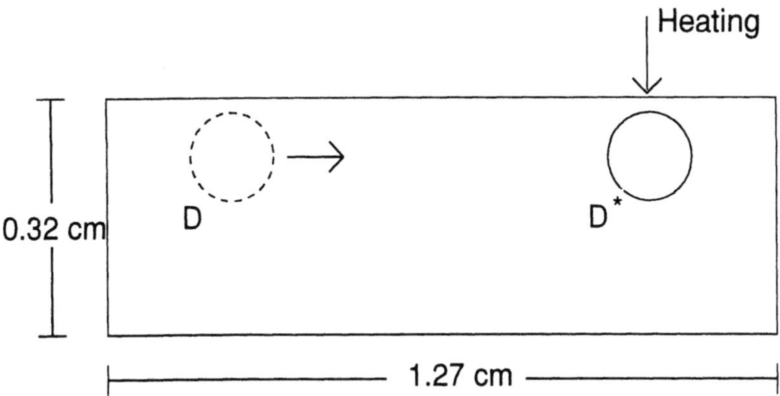

Figure 2: Set up for computation of least-square functional.

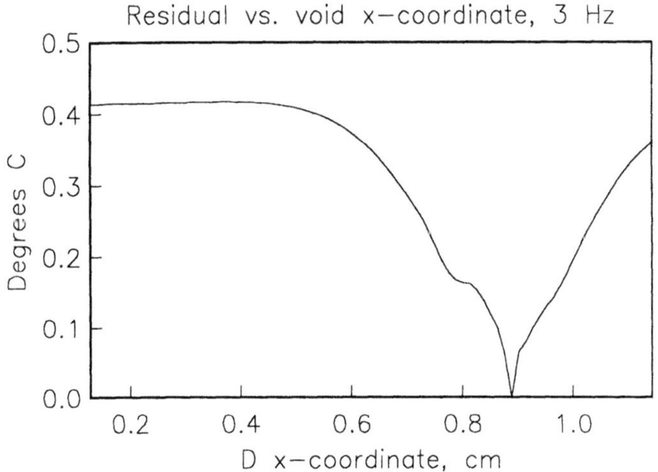

Figure 3: Least-squares functional.

from 0.15 cm to 1.15 cm. For each x_1 coordinate the functional $J(q)$ was computed and the x_1 coordinate of D versus $J(q)$ graphed to illustrate the nature of the least-squares functional, as shown in Figure 3.

As can be seen in Figure 3, the residual J is zero when the x_1 coordinates of D and D^* coincide. However, any optimization technique requires an initial guess at the parameters. In the present case if one chose an initial guess for the x_1 coordinate of D which was far from the correct coordinate (e.g., between 0.1 and 0.5), then the optimization routine would likely be unable to adjust the initial guess to find the correct x_1 coordinate. The residual curve is nearly flat in this region; in fact, it slopes slightly away from the correct value. Thus, a poor initial guess would probably not converge to the correct parameters for the void, particularly in the presence of noise. Moreover, if the heating is applied far from the true void then computational experiments show that the least-squares functional typically has many local minima which are not global minima; it is essential that the heat source be close enough to "illuminate" the void. For this reason, the strategy of applying a single heat source, but in multiple locations over the length of the sample, was adopted. By examining the temperature response at the sample surface as the heat source location changes, the void x_1 coordinate can be more accurately located. Computational experiments indicate that temperature response should peak as the heat source moves over a subsurface void. The graphs in Figure 4 illustrate this phenomena for

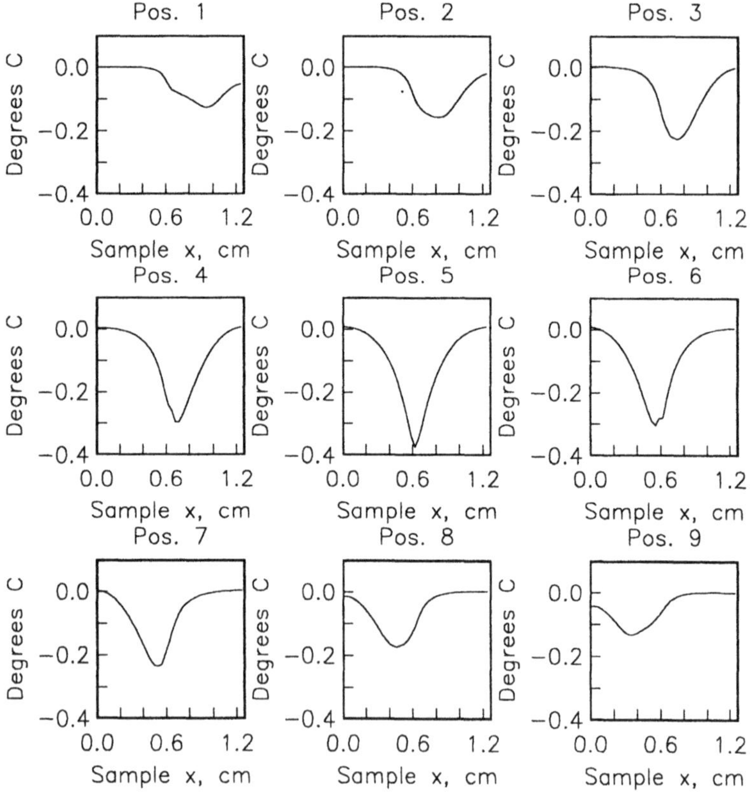

Figure 4: Thermal response for varying point heat source locations.

the imaginary component of the temperature, T. The simulated sample is as in Figure 2, with a circular void centered at x_1 coordinate 0.635 cm, x_2 coordinate 0.16 cm and with radius 0.12 cm. The heating is at 3 Hz, the heat source applied in 9 different equispaced locations, from right to left. Position 1 is 3/4 of the way along the length of the sample, position 9 is 1/4 of the way along the length and position 5 half way, directly over the void. Note that the response peaks as the heat source passes over the void.

6 Experimental Results

Figure 5 depicts the experimental configuration used to collect thermal data for testing the numerical method. The laser heating source was periodic with a cycle of 0.5 seconds on, 3.5 seconds off. An infrared camera, sensitive in the 8-12 micron range, measures the sample's thermal response at a rate of 15 frames per second. By doing a Fourier transform, one can recover the sample thermal response for frequencies from zero to 7.5 Hz, corresponding to the solution to equation (2.2). An aluminum block 1.27 cm in length, 0.32 cm in height and 0.20 cm deep was used as the sample. The infrared response was averaged over the z-direction of the top face to provide an approximation to a two-dimensional model. The void is cylindrical with radius 0.12 cm, x coordinate 0.635 cm and y coordinate 0.16 cm. As in the previous computational example, the heating source is applied at 9 equispaced points on the top surface.

As an example we use the temperature response at 0.94 Hz. The averaged top surface imaginary temperature response for each heating position is shown in Figure 6. It is clear that the peak response occurs for source number 5. The top surface data for source 5 is then used to recover an estimate of the void by performing the minimization of equation (4.1),

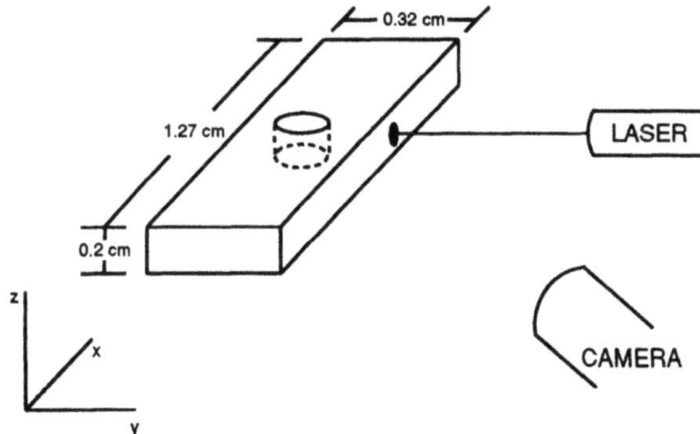

Figure 5: Experimental configuration.

using a Levenberg-Marquardt algorithm as outlined in [8] There is one twist: since the experimental data contains no information about the scale of the response, all computational and experimental data is re-scaled to a common scale, in this case, a root-mean-square value of 1.0 across the

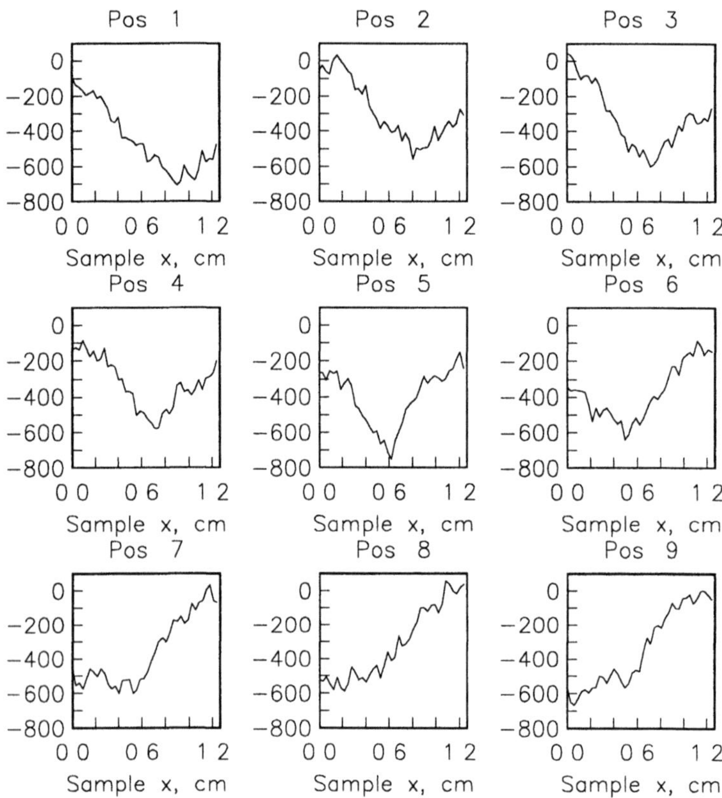

Figure 6: Experimental response for varying point heat source locations.

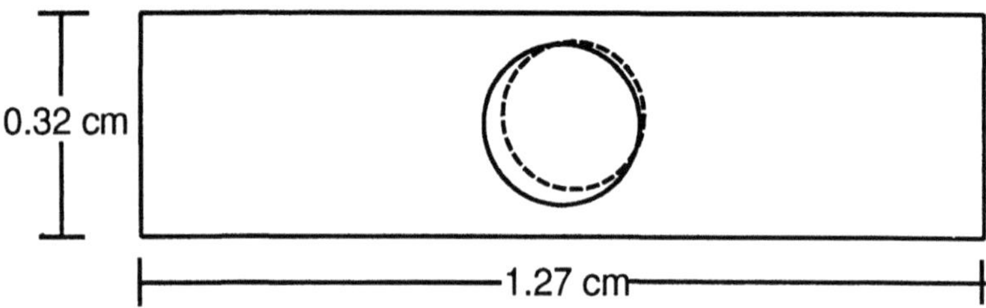

Figure 7: Actual and recovered voids.

sample top surface. This means that the optimization is attempting to fit the shape of the temperature response without regard to its magnitude. The actual void (solid outline) and recovered estimate (dotted outline) of the void are shown in Figure 7. The estimated void has radius 0.11 cm and center at (0.65 cm, 0.20 cm).

7 Conclusion

An algorithm based on a least-squares/optimization approach using a boundary integral method to solve the model heat conduction problem has been demonstrated for locating voids in a sample. As is common with optimization approaches, a reasonable initial guess at the void is needed. Much more experimental data remains to be analyzed, including an analysis of the method's resolution for voids of varying size, depth and shape. This may also include the development of a full three-dimensional boundary integral model for the heat conduction problem.

8 Acknowledgements

The experimental data was provided by W.P. Winfree of the Nondestructive Evaluation Sciences Branch, NASA Langley Research Center.

References

[1] Atkinson, K.E., *A survey of numerical methods for the solution of fredholm integral equations of the second kind*, SIAM, Philadelphia, PA, 1976.

[2] Banks, H.T. and F Kojima, "Approximations techniques for domain identification in two-dimensional parabolic systems under boundary observations," Proc. 20th IEEE CDC Conference, Los Angels, Dec. 9-11, (1987), pp.14411-1416.

[3] Banks, H.T. and F. Kojima, "Boundary shape identification problems in two-dimensional domains related to thermal testing of materials," Quart. Appl. Math., Vol. 47 (1989), pp. 273-293.

[4] Banks, H.T., F. Kojima and W.P. Winfree, "Boundary estimation problems arising in thermal tomography," Inverse Problems 6 (1990), pp. 897-922.

[5] Bryan, K., "A boundary integral method for an inverse problem in thermal imaging," ICASE report 92-38, submitted to the Journal of Mathematical Systems, Estimation and Control.

[6] Folland, Gerald B., *Introduction to partial differential equations.* Princeton, NJ: Princeton University Press, 1976.

[7] Kojima, F., "Identification of microscopic flaws arising in thermal tomography by domain decomposition method," Proc. Computation and Control II, MSU (1990), Birkhäuser.

[8] More', J. "The Levenberg-Marquardt algorithm: implementation and theory," Numerical Analysis (Edited by Watson, G.A.), pp. 105–116. *Lecture Notes in Math.* 630. Springer Verlag, 1977.

OPTIMAL FIXED-FINITE-DIMENSIONAL COMPENSATOR FOR BURGERS' EQUATION WITH UNBOUNDED INPUT/OUTPUT OPERATORS

John A. Burns * and Hamadi Marrekchi [†]

Interdisciplinary Center for Applied Mathematics
Department of Mathematics
Virginia Polytechnic Institute and State University,
Blacksburg, Virginia 24061

1 Introduction

In recent years considerable attention has been devoted to the problem of using feedback to control fluid dynamic systems. This problem is complex and particularly difficult when one is faced with phenomena such as shocks. Moreover, these systems are governed by nonlinear partial differential equations so that the natural state of the system is infinite dimensional. If one assumes that "full state feedback" is necessary to design practical controllers, then one would conclude that feedback control of fluid dynamic system is "not practical". However, it is well known that even in finite dimensional control systems one rarely has the ability to accurately sense all states, so that some form of dynamic compensation must be used.

This idea clearly extends to infinite dimensional problems and there is a growing literature on observers/compensators for distributed parameter systems. In this paper we consider a boundary control problem governed by Burgers' equation. We selected this problem because Burgers' equation is an infinite dimensional model that captures some phenomena (e.g., shocks) often observed in fluid flows and because it is simple enough to provide real insight into the problem. The goal is to show that it is possible to use modern control theory to produce practical finite dimensional dynamic compensators for boundary control of nonlinear partial differential equations of the type that occur naturally in fluid dynamics.

We shall present a short summary of one approach (the optimal projection method due to Bernstein and Hyland) and show how this approach can be used in conjunction with standard numerical schemes to produce a

*Supported in part by the Air Force Office of Scientific Research under Grant F-49620-92-J-0078, the National Science Foundation under Grant INT-89-2249 and by the National Aeronautics and Space Administration under Contract Nos. NAS1-18605 and NAS1-19480 while the author was a visiting scientist at the Institute for Computer Applications in Science and Engineering (ICASE), NASA Langley Research Center, Hampton, VA 23681-0001.

[†]Supported in part by the Air Force Office of Scientific Research under Grant F-49620-92-J-0078.

realizable low order controller. The optimal projection method is one of many approaches to this problem. However, we shall concentrate on this method because a very nice theory has already been developed (for bounded input and output operators) and we are more interested in illustrating (to non-experts) that recent results in distributed parameter control theory can be used to design practical feedback laws, than in discussing the "best" approach to the problem. It will be clear from our presentation that we are writing for those that are not necessarely "control experts". The extension of the general theoretical results to unbounded input and output operators will appear in a forthcoming paper. However, for the compensators presented here, we do not need the most general theory since we use the finite dimensional version of the optimal projection method.

As noted above it is almost impossible to observe the whole state. Controls and sensors are limited to a few points or segments of the boundary, so it is necessary to construct an appropriate observer (estimator) of the state and design a feedback control law (called a compensator) based on the information available from the observed (estimated) state variable. Boundary control and observation often leads to unbounded input and output operators. Stabilization by dynamic feedback or compensation has been considered by Curtain [5], Fujii [8], and Nambu [12] for classes of parabolic as well as hyperbolic systems, including control and observation at the boundary. All of these approaches produce stabilization schemes that either have the same finite order as that of a high-order approximate model, or alternatively, open-loop model reduction or closed-loop control reduction techniques are applied to achieve a lower-order compensator. An advance was made by Schumacher [15], when he gave a theory for designing finite-dimensional compensators for a large class of systems, including parabolic and delay systems. However, in his theory it was assumed that the control and observation operators are bounded. Curtain [4] presented an alternative compensator design which applied to the same class of systems, except that unbounded inputs and outputs were allowed. In [14], Pritchard and Salamon established a framework based on semigroup theory for treating the linear quadratic regulator problem for infinite-dimensional systems with unbounded input and output operators. Their approach is based on a weak formulation of the Riccati equations which characterize the optimal feedback law in an appropriate dual space.

Here we consider the problem of designing a fixed-finite-dimensional compensator for a class of distributed system governed by Burgers' equation, where the control and the observation are implemented at the boundary of the domain. The possibility of applying this approach to distributed parameter systems was first suggested by Johnson in [9] and Pearson [13]. The idea of fixing the order of the finite-dimensional compensator, while retaining the distributed parameter model was expanded and developed by

OPTIMAL FIXED-FINITE-DIMENSIONAL COMPENSATOR

Bernstein and Hyland in [1] and [2]. The method extends the full order LQG case to an "optimal fixed-finite-order compensator" characterized by four equations; two modified Riccati equations and two modified Lyapunov equations, coupled by an oblique projection whose rank is precisely equal to the order of the compensator. Bernstein and Hyland assumed that the control and observation operators were bounded and hence boundary control and observations were not covered by their theory.

We will present a Bernstein/Hyland type fixed-finite-dimensional compensator design, which does extend to unbounded input/output problems. In Section 2 we discuss the existence of a finite-dimensional compensator for parabolic distributed parameter systems with unbounded control and observation. In Section 3 we summarize the infinite-dimensional optimal projection theory from [1], and derive the corresponding equations and feedback gains which characterize the fixed-finite-order compensator. In Section 4 we present an example, construct the approximation schemes, and discuss the computational algorithm used for the optimal projection design synthesis. Finally, Section 5 contains numerical results and Section 6 is devoted to a few closing remarks.

2 A Theoretical Existence Result

We consider the following abstract Cauchy problem

$$
\begin{aligned}
\dot{z}(t) &= Az(t) + Bu(t), \quad z(0) = z_o \in H \qquad (2.1) \\
y(t) &= Cz(t) \qquad t \geq 0 \qquad (2.2)
\end{aligned}
$$

where H is a Hilbert space, $u(.) \in L^2(0\ T; \Re^m)$, $y(.) \in L^2(0\ T; \Re^\ell)$, and A is the infinitesimal generator of analytic semigroup $S(t)$ on H, generally unstable, with exponential growth rate

$$
w_0 = \lim_{t \to \infty} t^{-1} \log \|S(t)\|_{\mathcal{L}(H)} > 0 \qquad (2.3)
$$

so that

$$
\|S(t)\|_{\mathcal{L}(H)} \leq Me^{(w_0 + \epsilon)t} \quad \text{for all} \quad \epsilon > 0,\ t \geq 0 \qquad (2.4)
$$

for some constant $M = M(w_0, \epsilon) \geq 1$. Throughout the remainder of this paper we let $\hat{A}$ denote the translation $\hat{A} = -A + wI$, where w is fixed and $w > w_0$, so that $\hat{A}$ has well-defined fractional powers $(\hat{A})^\mu$ on H and $-\hat{A}$ is the generator of a strongly continuous analytic semigroup $\hat{S}(t)$ on H satisfying

$$
\|\hat{S}(t)\|_{\mathcal{L}(H)} \leq \hat{M}e^{-\hat{w}t}, \quad t \geq 0. \qquad (2.5)
$$

In order to allow for unbounded operators B and C, we assume that $B \in \mathcal{L}(\Re^m, V)$ and $C \in \mathcal{L}(W, \Re^\ell)$, where W and V are also Hilbert spaces such that

$$\mathcal{D}(A) \subseteq W \hookrightarrow H \hookrightarrow V \tag{2.6}$$

with continuous dense injections. More precisely, we assume that B^* is $[\hat{A}^*]^\gamma$-bounded, or equivalently,

$$[\hat{A}]^{-\gamma} B \in \mathcal{L}(\Re^m, H) \qquad \text{for } 0 \leq \gamma < 1. \tag{2.7}$$

Similarly, for the operator C we assume that

$$C[\hat{A}]^{-\gamma} \in \mathcal{L}(H, \Re^\ell) \qquad \text{for } 0 \leq \gamma < 1. \tag{2.8}$$

It is helpful to interpret (2.1-2.2) in mild form. In particular, the solution $z(t)$ is given by

$$z(t) = S(t)z_o + \int_0^t S(t-s)Bu(s)ds, \quad 0 \leq t \leq T \tag{2.9}$$

and the output by

$$y(t) = CS(t)z_0 + C \int_0^t S(t-s)Bu(s)ds. \tag{2.10}$$

We assume that $S(t)$ is also an analytic semigroup on W and that the following hypotheses are satisfied:

(H-1) There exists a constant $b(T) > 0$ such that for every $T > 0$, $\int_0^T S(T-s)Bu(s)ds \in W$ and

$$\left\| \int_0^T S(T-s)Bu(s)ds \right\|_W \leq b(T)\|u(\cdot)\|_{L^2(0\ T;\Re^m)} \tag{2.11}$$

for every $u(\cdot) \in L^2(0\ T; \Re^m)$.

(H-2) There exists a constant $c(T) > 0$ such that for every $T > 0$,

$$\int_0^T \|CS(t)x\|_{L^2(0\ T;\Re^\ell)}dt \leq c(T)\|x\|_V \tag{2.12}$$

for every $x \in W$.

We now give sufficient conditions which imply that the system (2.1-2.2) can be stabilized by a finite-dimensional compensator of the form

$$\begin{aligned}
\dot{w}(t) &= A_c w(t) - B_c y(t) \qquad w(0) = w_o \tag{2.13}\\
u(t) &= C_c w(t) \tag{2.14}
\end{aligned}$$

where $A_c \in \Re^{N_c \times N_c}$, $B_c \in \Re^{N_c \times m}$, and $C_c \in \Re^{\ell \times N_c}$ are suitably chosen matrices. We need the following well-posedness result for the connected system (2.1-2.2) and (2.13-2.14). This result and proof may be found in [4].

Proposition 2.1 *Let (H-1)-(H-2) be satisfied, then for all $z_o \in W$, $w_o \in \Re^{N_c}$ there exists a unique solution pair $z(t)$ and $w(t)$ of (2.1-2.2) and (2.13-2.14). This means that $z(t)$ is continuous in H and absolutely continuous in V, that (2.1) is satisfied for almost every $t \geq 0$ where $u(t)$ is given by (2.14), and that $w(t) \in \Re^{N_c}$ is continuously differentiable and satisfies (2.13) where $y(t)$ is given by (2.2).*

In addition to hypotheses (H-1) and (H-2), we assume:

(H-3) Stabilizability Condition (S.C.)
There exists an operator $F \in \mathcal{L}(H, \Re^m)$ such that $A_F = A + BF$ generates an analytic semigroup $S_F(t) = e^{(A+BF)t}$ and $S_F(t)$ is exponentially stable on H, i.e.,

$$\left\| S_F(t) \right\|_{\mathcal{L}(H)} \leq M_F e^{-w_F t}, \qquad \text{for } w_F > 0. \qquad (2.15)$$

(H-4) Detectability Condition (D.C.)
There exists an operator $G \in \mathcal{L}(\Re^\ell, H)$ such that $A_G = A + GC$ generates an analytic semigroup $S_G(t) = e^{(A+GC)t}$ and $S_G(t)$ is exponentially stable on H, i.e.,

$$\left\| S_G(t) \right\|_{\mathcal{L}(H)} \leq M_G e^{-w_G t}, \qquad \text{for } w_G > 0. \qquad (2.16)$$

(H-5) In addition to (H-3) and (H-4) there exists a finite-dimensional subspace $\aleph \subset W$, with $\dim \aleph \leq N_c$ such that

(i) $S_F(t)\aleph \subset \aleph$, for all $t \geq 0$,

(ii) Range $G \subset \aleph$,

(iii) $\aleph \subseteq \mathcal{D}(A_F)$.

Moreover, there exist linear maps $i \colon \Re^{N_c} \to \aleph$, $\pi \colon H \to \Re^{N_c}$ such that

$$\pi i = I_{N_c}, \qquad i\pi x = x \qquad \text{for } x \in \aleph. \qquad (2.17)$$

Note that (H-5) implies that $\pi A_F i$ is a well defined linear map on $\Re^{N_c}$. We will show that the system

$$\dot{w}(t) \;=\; \pi (A_F + GC)\, i\, w(t) - \pi G y(t), \qquad w(0) = w_o \qquad (2.18)$$
$$u(t) \;=\; F\, i\, w(t) \qquad\qquad\qquad\qquad\qquad\qquad\qquad (2.19)$$

defines a stabilizing compensator for the Cauchy problem (2.1-2.2). The following result is a slight extension of Theorem 2.5 in [7] for unbounded inputs and outputs.

Theorem 2.2 *If (H-1)-(H-5) are satisfied, then the closed-loop system defined by (2.1-2.2) and (2.18-2.19) is exponential stable.*

Proof: Note that without loss of generality we can assume that dim $\aleph = N_c$. By Proposition 2.1 it follows that the closed-loop system is a well-posed Cauchy problem. Let $z_o \in W$, $w_o \in \Re^{N_c}$ and $z(t)$, $w(t)$ be defined by (2.1-2.2) and (2.18-2.19), respectively. Since $z(t) \in W$, if $x(t)$ is defined by

$$x(t) \;=\; iw(t) - z(t) \qquad t \geq 0,$$

then $x(t)$ belongs to W and it is straightforward to show that

$$\dot{w}(t) = \pi \, A_F \, i \, w(t) + \pi \, GC \, x(t). \tag{2.20}$$

Therefore,

$$\begin{aligned}
x(t) \;&=\; i\pi \, S_F(t) \, i \, w_o + \int_0^t i\pi \, S_F(t-s) \, i\pi \, GC \, x(s) \, ds \;-\; z(t) \\
&=\; S_F(t) \, i \, w_o + \int_0^t S_F(t-s) \, GC \, x(s) \, ds \;-\; z(t) \\
&=\; S(t) \, i \, w_o + \int_0^t S(t-s) \left[B \, F \, i \, w(s) + GC \, x(s) \right] ds \\
&\quad - S(t) \, z_0 + \int_0^t S(t-s) \, B \, u(s) \, ds \\
&=\; S(t) \, x(0) + \int_0^t S(t-s) \, GC \, x(s) \, ds,
\end{aligned}$$

which implies that $x(t) = S_G(t) \, x(0)$. The stability of $x(t)$, $w(t)$ and $z(t)$ follows.

3 Optimal Projection Theory

Consider the steady-state fixed-order dynamic compensator problem, defined by the infinite-dimensional control system

$$\dot{z}(t) = Az(t) + Bu(t) + H_1\eta(t) \tag{3.1}$$

with measurements

$$y(t) = Cz(t) + H_2\eta(t). \tag{3.2}$$

The objective is to design a finite-dimensional fixed-order dynamic compensator

$$\begin{aligned}
\dot{z}_c(t) \;&=\; A_c z_c(t) + B_c y(t) \tag{3.3} \\
u(t) \;&=\; C_c z_c(t) \tag{3.4}
\end{aligned}$$

which minimizes the steady-state performance criterion

$$J(A_c, B_c, C_c) \overset{\text{def}}{=} \lim_{t \to \infty} \frac{1}{t} \int_0^t \mathrm{E}\left[\langle R_1 z(s),\, z(s)\rangle + u(s)^T R_2 u(s)\right] ds \quad (3.5)$$

where the operators A, B and C satisfy all the assumptions given in the previous section and $\mathrm{E}[\cdot]$ is the expectation. In addition, assume that the state and measurements are corrupted by a white noise $\eta(t)$ in the Hilbert space $\tilde{H}$, with zero-mean Gaussian, $H_1 \in \mathcal{L}(\tilde{H}, H)$, $H_2 \in \mathcal{L}(\tilde{H}, \Re^\ell)$, $R_1 \in \mathcal{L}(H)$ is self-adjoint and nonnegative definite, and that R_2 is an $m \times m$ symmetric positive-definite matrix. We assume that the disturbance and measurements are independent, i.e., $H_1 H_2^* = 0$, $V_1 = H_1 H_1^* \in \mathcal{L}(H)$ is nonnegative definite and of trace class, and that $V_2 = H_2 H_2^* \in \Re^{\ell \times \ell}$ is positive definite. Also, it is assumed that the initial state $z(0) = z_o$ is Gaussian and independent of $\eta(\cdot)$. The compensator will be assumed to be of fixed, finite order N_c (i.e., $z_c(t) \in \Re^{N_c}$) and the optimization is performed over $A_c \in \Re^{N_c \times N_c}$, $B_c \in \Re^{N_c \times \ell}$ and $C_c \in \Re^{m \times N_c}$.

If one introduces the augmented state space $\mathcal{H} = H \times \Re^{N_c}$, then the closed–loop system becomes a linear system on $\mathcal{H}$. Consequently, define the closed-loop operator $\mathcal{A} : \mathcal{D}(\mathcal{A}) \subseteq \mathcal{H} \to \mathcal{H}$ on the dense domain $\mathcal{D}(\mathcal{A}) = \mathcal{D}(A) \times \Re^{N_c}$ by

$$\mathcal{A} = \begin{bmatrix} A & BC_c \\ B_c C & A_c \end{bmatrix} = \begin{bmatrix} A & 0 \\ 0 & 0 \end{bmatrix} + \begin{bmatrix} 0 & BC_c \\ B_c C & A_c \end{bmatrix}.$$

Since the operator

$$\begin{bmatrix} A & 0 \\ 0 & 0 \end{bmatrix} : \mathcal{D}(\mathcal{A}) \to \mathcal{H},$$

generates an analytic semigroup

$$\begin{bmatrix} e^{At} & 0 \\ 0 & I_{n_c} \end{bmatrix} \qquad t \geq 0,$$

then conditions (2.7)-(2.8) imply that $\mathcal{A}$ is also closed and generates an analytic semigroup $e^{\mathcal{A}t}$ on $\mathcal{H}$ (see [10]). To guarantee that J is finite and independent of initial conditions we restrict our attention to the set of admissible compensators defined by

$$S = \{(A_c, B_c, C_c) : e^{\mathcal{A}t} \text{ is exponentially stable}\}. \quad (3.6)$$

If $(A_c, B_c, C_c) \in S$, then there exist $\alpha \geq 1$ and $\beta > 0$ such that

$$\|e^{\mathcal{A}t}\| \leq \alpha e^{-\beta t} \qquad t \geq 0. \quad (3.7)$$

Moreover, we know from Theorem 2.2 above that S is non-empty. We now state some results found in [1] and [2].

Lemma 3.1 *If $\hat{Q}$ and $\hat{P} \in \mathcal{L}(H)$ have finite rank and are nonnegative definite, then $\hat{Q}\hat{P}$ is nonnegative semi-simple. Furthermore, if rank $(\hat{Q}\hat{P}) = N_c$, then there exist G and $\Gamma \in \mathcal{L}(H, \Re^{N_c})$ and a positive semi-simple matrix $M \in \Re^{N_c \times N_c}$ such that*

$$\hat{Q}\hat{P} = G^* M \Gamma \tag{3.8}$$

$$\Gamma G^* = I_{N_c}. \tag{3.9}$$

Proof: Bernstein and Hyland give a complete proof of this result in [1]. Here we outline their proof in order to illustrate the form of the factorization of $\hat{Q}\hat{P}$ and to provide a description of the operators G and Γ. Since $\hat{Q}$ and $\hat{P}$ have finite rank, there exists a finite dimensional subspace $\mathcal{Z} \subset H$ such that $\hat{Q}\mathcal{Z} \subset \mathcal{Z}$, $\hat{Q}\mathcal{Z}^\perp = 0$, $\hat{P}\mathcal{Z} \subset \mathcal{Z}$ and $\hat{P}\mathcal{Z}^\perp = 0$. Hence there exists an orthonormal basis for H and in this basis $\hat{Q}$ and $\hat{P}$ have the infinite matrix representations

$$\hat{Q} = \left[\begin{array}{cc} \hat{Q}_1 & 0 \\ 0 & 0 \end{array} \right], \qquad \hat{P} = \left[\begin{array}{cc} \hat{P}_1 & 0 \\ 0 & 0 \end{array} \right],$$

where $\hat{Q}_1$, $\hat{P}_1 \in \Re^{r \times r}$ and $r = \dim \mathcal{Z}$. Consequently, there exists an invertible $\Phi \in \Re^{r \times r}$ such that $\hat{\Lambda} = \Phi^{-1}\hat{Q}_1\hat{P}_1\Phi$ is nonnegative and diagonal and $\hat{Q}\hat{P}$ is nonnegative and semi-simple. If rank $(\hat{Q}\hat{P}) = N_c$, then it is clear that Φ can be chosen so that

$$\hat{\Lambda} = \left[\begin{array}{cc} \Lambda & 0 \\ 0 & 0 \end{array} \right]$$

where $\Lambda \in \Re^{N_c \times N_c}$ is positive and diagonal. Hence,

$$\hat{Q}\hat{P} = \left[\begin{array}{cc} \Phi & 0 \\ 0 & I_\infty \end{array} \right] \left[\begin{array}{c} \left[\begin{array}{c} I_{N_c} \\ 0 \\ 0 \end{array} \right] \end{array} \right] \Lambda \left[\begin{array}{cc} \left[\begin{array}{cc} I_{N_c} & 0 \end{array} \right] & 0 \end{array} \right] \left[\begin{array}{cc} \Phi^{-1} & 0 \\ 0 & I_\infty \end{array} \right],$$

and if we define G, M and Γ by

$$G = \left[\begin{array}{cc} \left[\begin{array}{cc} S^\mathsf{T} & 0 \end{array} \right] & 0 \end{array} \right] \left[\begin{array}{cc} \Phi^\mathsf{T} & 0 \\ 0 & I_\infty \end{array} \right]$$

$$\Gamma = \left[\begin{array}{cc} \left[\begin{array}{cc} S^{-1} & 0 \end{array} \right] & 0 \end{array} \right] \left[\begin{array}{cc} \Phi^{-1} & 0 \\ 0 & I_\infty \end{array} \right]$$

$$M = S^{-1}\Lambda S,$$

for any invertible $S \in \Re^{N_c \times N_c}$, then G, Γ and M provide the desired factorization and this completes the proof.

Throughout the paper we will refer to G, Γ and M satisfying the above lemma as a $(G - M - \Gamma)$ – factorization of $\hat{Q}\hat{P}$. For convenience we define

$\Sigma = BR_2^{-1}B^*$ and $\overline{\Sigma} = C^*V_2^{-1}C$ and let I_{N_c} and I_H denote respectively the $N_c \times N_c$ identity matrix and the identity operator on H, respectively. We state Bernstein's and Hyland's main theorem which provides a set of necessary conditions that characterize the optimal steady-state fixed order dynamic compensator for bounded input and output operators (see [1]).

Theorem 3.2 *Let B and C be bounded operators and let N_c be given and suppose that there exists a controllable and observable dynamic compensator $(A_c, B_c, C_c) \in S$ of order N_c which minimizes J given by (3.5), then there exist nonnegative definite operators Q, P, $\hat{Q}$, and $\hat{P}$ on H such that A_c, B_c, and C_c are given by*

$$
\begin{aligned}
A_c &= \Gamma(A - Q\overline{\Sigma} - \Sigma P)G^* &\qquad (3.10)\\
B_c &= \Gamma Q C^* V_2^{-1} &\qquad (3.11)\\
C_c &= -R_2^{-1}B^* P\Gamma^* &\qquad (3.12)
\end{aligned}
$$

for some $(G - M - \Gamma)$ - factorization of $\hat{Q}\hat{P}$ and such that, with $\tau = G^*\Gamma \in \mathcal{L}(H)$, the following conditions are satisfied:

$$
\begin{array}{llll}
Q : & \mathcal{D}(A^*) \to \mathcal{D}(A) & P & : \mathcal{D}(A) \to \mathcal{D}(A^*)\\
\hat{Q} : & H \to \mathcal{D}(A) & \hat{P} & : H \to \mathcal{D}(A^*)\\
& rank\,(\hat{Q}) = & rank\,(\hat{P}) & = rank\,(\hat{Q}\hat{P})
\end{array}
$$

and

$$
\begin{aligned}
0 &= (A - \tau Q\overline{\Sigma})Q + Q(A - \tau Q\overline{\Sigma})^* + V_1 + \tau Q\overline{\Sigma}Q\tau^* &\qquad (3.13)\\
0 &= (A - \Sigma P\tau)^* P + P(A - \Sigma P\tau) + R_1 + \tau^* P\Sigma P\tau &\qquad (3.14)\\
0 &= \left[(A - \Sigma P)\hat{Q} + \hat{Q}(A - \Sigma P)^* + Q\overline{\Sigma}Q\right]\tau^* &\qquad (3.15)\\
0 &= \left[(A - Q\overline{\Sigma})^*\hat{P} + \hat{P}(A - Q\overline{\Sigma}) + P\Sigma P\right]\tau. &\qquad (3.16)
\end{aligned}
$$

Note that these necessary conditions consist of a system of four operator equations, including a pair of modified Riccati equations and a pair of modified Lyapunov equations which are coupled by the operator $\tau \in \mathcal{L}(H)$. The operator τ is idempotent, since $\tau^2 = \tau\tau = G^*\Gamma G^*\Gamma = G^*I_{n_c}\Gamma = G^*\Gamma = \tau$. In general τ is an oblique projection and may not be orthogonal since there is no requirement that τ be self-adjoint. Moreover, we note that in view of Lemma 3.1, Theorem 3.2 applies to $(SA_cS^{-1}, SB_c, C_cS^{-1})$ for any invertible $S \in \Re^{N_c \times N_c}$, since the $(G - M - \Gamma)$-factorization of $\hat{Q}\hat{P}$, used to determine A_c, B_c and C_c, is not unique. However, the operator τ remains invariant over the class of factorizations. An easy computation yields the following identities:

$$
\hat{Q} = \tau\hat{Q} \qquad \text{and} \quad \hat{P} = \hat{P}\tau. \qquad (3.17)
$$

It is helpful to have an alternative form of the optimal projection e-
quations to actually compute the optimal fixed-order compensator of the
approximating finite-dimensional plant. The following result for bounded
input bounded output operators may be found in [1].

Proposition 3.1 *If B and C are bounded, then the optimal projection e-
quations (3.13)–(3.16) are equivalent, respectively, to*

$$0 = AQ + QA^* + V_1 - Q\overline{\Sigma}Q + \tau_\perp Q\overline{\Sigma}Q\tau_\perp^* \tag{3.18}$$

$$0 = A^*P + PA + R_1 - P\Sigma P + \tau_\perp^* P\Sigma P\tau_\perp \tag{3.19}$$

$$0 = A_p\hat{Q} + \hat{Q}A_p + Q\overline{\Sigma}Q - \tau_\perp Q\overline{\Sigma}Q\tau_\perp^* \tag{3.20}$$

$$0 = A_q^*\hat{P} + \hat{P}A_q + P\Sigma P - \tau_\perp^* P\Sigma P\tau_\perp \tag{3.21}$$

where

$$\tau_\perp = I_H - \tau, \qquad A_p = A - \Sigma P \quad and \quad A_q = A - Q\overline{\Sigma}. \tag{3.22}$$

This form of the optimal projection equations shows that there is a
connection between Theorem 3.2 and the standard LQG result when dim
$H = N < \infty$. In this case, we note that the $(G - M - \Gamma)$-factorization of
$\hat{Q}\hat{P}$ when $N_c = N$ is given by $G = \Gamma = I_N$ and $M = \hat{Q}\hat{P}$. Since $\tau = I_N$
and $\tau_\perp = 0$, it follows that (3.18)–(3.19) reduce to the standard observer
and regulator Riccati equations.

To obtain a geometric interpretation of the optimal projection we introduce the "quasi-full-state" estimate

$$\hat{z}(t) = G^* z_c(t) \in H, \tag{3.23}$$

so that $\tau\hat{z}(t) = \hat{z}(t)$ and $z_c(t) = \Gamma\hat{z}(t)$. Hence, the closed-loop system can
be written as

$$\dot{z}(t) = Az(t) + B\hat{C}_c\tau\hat{z}(t) \tag{3.24}$$

$$\dot{\hat{z}}(t) = \tau(A + B\hat{C}_c - \hat{B}_cC)\tau\hat{z}(t) + \tau\hat{B}_cCz(t) \tag{3.25}$$

where

$$\hat{B}_c = QC^*V_2^{-1} \qquad and \qquad \hat{C}_c = -R_2^{-1}B^*P. \tag{3.26}$$

This shows that the geometric structure of the quasi-full-order compensator is dictated by the projection τ. Sensor inputs $\tau\hat{B}_cCz$ are annihilated
unless they are contained in $\mathcal{R}(\tau^*) = \mathcal{N}(\tau)^\perp$, while $\tau\hat{z}$ employed in the
control input is contained in $\mathcal{R}(\tau)$. Consequently, $\mathcal{R}(\tau)$ and $\mathcal{R}(\tau^*)$ are
the control and observation subspaces of the compensator, respectively. In
order to modify the previous results so that they will apply directly to unbounded B and C operators, care must be exercised to precisely define the
weak forms of (3.13)-(3.16) and (3.18)-(3.21). We shall not consider this
problem in this short note. However, we shall use these systems to guide
the approximations below.

4 Finite Dimensional Approximation

In general, the optimal projection equations (3.18)–(3.21) are infinite dimensional operator equations. To actually use these equations to compute the optimal fixed-finite-order compensator, a finite dimensional approximation is needed (see [2] for details).

Let H^N for $N = 1, 2, \cdots$, be a sequence of finite dimensional linear subspaces of H and let $\mathcal{P}^N : H \to H^N$ be the canonical orthogonal projections. Let $A^N \in \mathcal{L}(H^N)$, $B^N \in \mathcal{L}(\Re^m, H^N)$, $C^N \in \mathcal{L}(H^N, \Re^\ell)$, $R_1^N \in \mathcal{L}(H^N)$ and $V_1^N \in \mathcal{L}(H^N)$ be given and consider the approximating system

$$\dot{z}^N(t) = A^N z^N(t) + B^N u^N(t) + H_1^N \eta^N(t) \tag{4.1}$$

$$y^N(t) = C^N z^N(t) + H_2^N \eta^N(t). \tag{4.2}$$

The goal is to design a sequence of finite-dimensional dynamic compensators of fixed order N_c of the form

$$\dot{z}_c^N(t) = A_c^N z_c^N(t) + B_c^N y^N(t) \tag{4.3}$$

$$u^N(t) = C_c^N z_c^N(t), \tag{4.4}$$

which minimizes the performance criterion

$$J^N(A_c^N, B_c^N, C_c^N) \overset{\text{def}}{=} \lim_{t \to \infty} \frac{1}{t} \int_0^t \mathbf{E}\left[\langle R_1^N z^N(s), z^N(s) \rangle + u(s)^T R_2 u(s) \right] ds. \tag{4.5}$$

Now, for each $N = 1, 2, \cdots$, let $\left\{ \phi_j^N \right\}_{j=1}^{k^N}$ be a basis for H^N. Also, for any linear operator F^N with domain and range in H^N, unless otherwise noted, we use the same symbol F^N for its matrix representation with respect to the basis chosen. Let Ψ^N denote the k^N-square Gram matrix corresponding to the basis $\left\{ \phi_j^N \right\}_{j=1}^{k^N}$ (e.g., $\Psi^N = \left[\langle \phi_i^N, \phi_j^N \rangle_{H^N} \right]$). Note that

$$(A^N)^* = (\Psi^N)^{-1}(A^N)^\top \Psi^N \qquad (B^N)^* = (B^N)^\top \Psi^N \tag{4.6}$$

$$(C^N)^* = (\Psi^N)^{-1}(C^N)^\top \qquad (\Sigma^N) = B^N R_2^{-1}(B^N)^\top \Psi^N \tag{4.7}$$

$$(\tau_\perp^N)^* = (\Psi^N)^{-1}(\tau_\perp^N)^\top \Psi^N \qquad (\overline{\Sigma}^N) = (\Psi^N)^{-1}(C^N)^\top V_2^{-1} C^N \tag{4.8}$$

and if we define the $k^N \times k^N$ nonnegative definite matrices

$$Q_0^N \overset{\text{def}}{=} Q^N(\Psi^N)^{-1} \qquad P_0^N \overset{\text{def}}{=} \Psi^N P^N$$

$$\hat{Q}_0^N \overset{\text{def}}{=} \hat{Q}^N(\Psi^N)^{-1} \qquad \hat{P}_0^N \overset{\text{def}}{=} \Psi^N \hat{P}^N$$

$$V_0^N \overset{\text{def}}{=} V_1^N(\Psi^N)^{-1} \qquad R_0^N \overset{\text{def}}{=} \Psi^N R_1^N$$

$$\Sigma_0^N \overset{\text{def}}{=} B^N R_2^{-1}(B^N)^\top \qquad \overline{\Sigma}_0^N \overset{\text{def}}{=} (C^N)^\top V_2^{-1} C^N,$$

then the matrix equivalence of the operator equation (3.18)–(3.21) become

$$
\begin{aligned}
0 \; = \; & A^N Q_0^N + Q_0^N (A^N)^\top + V_0^N - Q_0^N \Sigma_0^N Q_0^N \\
& + \tau_\perp^N Q_0^N \overline{\Sigma}_0^N Q_0^N (\tau_\perp^N)^\top
\end{aligned}
\tag{4.9}
$$

$$
\begin{aligned}
0 \; = \; & (A^N)^\top P_0^N + P_0^N A^N + R_0^N - P_0^N \Sigma_0^N P_0^N \\
& + (\tau_\perp^N)^\top P_0^N \Sigma_0^N P_0^N \tau_\perp^N
\end{aligned}
\tag{4.10}
$$

$$
\begin{aligned}
0 \; = \; & A_{p_o}^N \hat{Q}_0^N + \hat{Q}_0^N (A_{p_o}^N)^\top + Q_0^N \overline{\Sigma}_0^N Q_0^N \\
& - \tau_\perp^N Q_0^N \overline{\Sigma}_0^N Q_0^N (\tau_\perp^N)^\top
\end{aligned}
\tag{4.11}
$$

$$
\begin{aligned}
0 \; = \; & (A_{q_o}^N)^\top \hat{P}_0^N + \hat{P}_0^N A_{q_o}^N + P_0^N \Sigma_0^N P_0^N \\
& - (\tau_\perp^N)^\top P_0^N \Sigma_0^N P_0^N \tau_\perp^N .
\end{aligned}
\tag{4.12}
$$

The approximating optimal dynamic compensator (A_c^N, B_c^N, C_c^N) of order N_c is then given by

$$
\begin{aligned}
A_c^N \; &= \; \Gamma_0^N (A^N - Q_0^N \overline{\Sigma}_0^N - \Sigma_0^N P_0^N)(G_0^N)^\top \tag{4.13}\\
B_c^N \; &= \; \Gamma_0^N Q_0^N (C^N)^\top V_2^{-1} \tag{4.14}\\
C_c^N \; &= \; -R_2^{-1}(B^N)^\top P_0^N (\Gamma_0^N)^\top \tag{4.15}
\end{aligned}
$$

where Γ_0^N, $G_0^N \in \Re^{N_c \times k^N}$ and $M_0^N \in \Re^{N_c \times N_c}$ provide a $(G_0^N - M_0^N - \Gamma_0^N)$-factorization of $\hat{Q}_0^N \hat{P}_0^N$.

We turn now to an example. Consider Burgers' equation, with Neumann boundary control given by

$$
\frac{\partial}{\partial t} z(t,x) \; = \; \epsilon \frac{\partial^2}{\partial x^2} z(t,x) - z(t,x) \frac{\partial}{\partial x} z(t,x), \quad 0 < x < 1, \quad t > 0
\tag{4.16}
$$

$$
z(0,x) \; = \; z_o(x)
\tag{4.17}
$$

$$
\frac{\partial}{\partial x} z(t,0) \; = \; -u_1(t), \qquad \frac{\partial}{\partial x} z(t,1) = u_2(t),
\tag{4.18}
$$

and observations

$$
\begin{aligned}
y_1(t) \; &= \; z(t,0) \tag{4.19}\\
y_2(t) \; &= \; z(t,1), \tag{4.20}
\end{aligned}
$$

where $\epsilon = \frac{1}{\mathrm{Re}} > 0$ and Re is the Reynolds number. Initially, we consider the linearized Neumann boundary control problem

$$
\frac{\partial}{\partial t} z(t,x) \; = \; \epsilon \frac{\partial^2}{\partial x^2} z(t,x), \qquad 0 < x < 1, \quad t > 0
\tag{4.21}
$$

$$
z(0,x) \; = \; z_0(x)
\tag{4.22}
$$

$$
\frac{\partial}{\partial x} z(t,0) \; = \; -u_1(t), \qquad \frac{\partial}{\partial x} z(t,1) = u_2(t).
\tag{4.23}
$$

We will apply the linearized feedback control laws constructed from this model to the nonlinear Burgers' equation. System (4.21)-(4.23) can be placed into the standard state space framework by defining the operator A_ϵ on $H = L_2(0,1)$ by

$$A_\epsilon \phi = \epsilon \phi'' \tag{4.24}$$

for all $\phi \in \mathcal{D}(A_\epsilon) = \{\phi \in H^2(0,1) : \phi'(0) = \phi'(1) = 0\}$. Define $W = V^* = H^{\frac{3}{2}}(0,1) = \mathcal{D}(\hat{A}_\epsilon^{\frac{3}{4}})$ and let $B : \Re^2 \to V$ be defined by $B = \hat{A}\mathcal{N}$ where $\hat{A} = -A_\epsilon + wI$ and we assume that w is not an eigenvalue of A_ϵ with homogeneous Neumann boundary conditions, so that $\hat{A}$ is boundedly invertible on $L_2(0,1)$. The Neumann map $\mathcal{N}$ is defined by the boundary system given in [11, pages 53–56]. Let $C : W \to \Re^2$ defined by

$$C\phi = \begin{bmatrix} \phi(0) \\ \phi(1) \end{bmatrix}, \tag{4.25}$$

The boundary control problem (4.21)-(4.23) can be represented by a differential equation

$$\frac{d}{dt} = A_\epsilon z(t) + Bu, \qquad z(0) = z_o \tag{4.26}$$

$$y(t) = Cz(t). \tag{4.27}$$

It is well known that A_ϵ generates an analytic semigroup $S(t)$ on H. Moreover, the spectrum $\sigma(A_\epsilon)$ of A_ϵ consists of all eigenvalues λ_n, $n = 0, 1, 2, \cdots$ given by $\lambda_n = -\epsilon n^2 \pi^2$ and for each eigenvalue λ_n the corresponding eigenfunction ϕ_n is given by

$$\phi_0(x) = 1 \qquad \phi_n(x) = \sqrt{2}\cos(n\pi x), \qquad 0 < x < 1. \tag{4.28}$$

One can easily verify conditions (2.7)-(2.8), by taking $\gamma = \frac{1}{4}$ (see [11]). It is straightforward to show that (H-1)-(H-5) are valid.

5 Numerical Results

Now, we formulate a specific approximation scheme for the boundary control problem (4.28). For each $N = 2, 3, \cdots$ let divide the unit interval $[0,1]$ into N equal subinterval $[x_i, x_{i+1}]$, $x_i = \frac{i-1}{N+1}$, $i = 1, 2, \cdots, N+1$. Let $H^N = \text{Span} \{h_i^N\}_{i=0}^{N+1}$ where $h_i^N(\cdot)$ are the standard hat functions defining continuous piecewise linear splines (see [3]). Note that $k^N = \dim H^N = N+2$ and let the approximate solution $z^N(t,x)$ of $z(t,x)$ for equation (4.26)-(4.27) be given by

$$z^N(t,x) = \sum_{i=0}^{N+1} z_i^N(t)h_i^N(x) \tag{5.1}$$

for some $z_i^N(t) \in \Re$, $i = 0, 1, \cdots, N + 1$. Standard finite element approximations yield the ODE system

$$\frac{d}{dt} z^N(t) = A_\epsilon^N z^N(t) + B^N u(t), \qquad z^N(0) = z_0^N \tag{5.2}$$

$$y^N(t) = C^N z^N(t) \tag{5.3}$$

where the matrices A_ϵ^N, B^N, C^N can be easely computed by using the Ritz-Galerkin approximation.

For our numerical example, we set $\alpha = \frac{1}{60}$, the initial condition $z_o(x) = \sin(\pi x)$, $r_1 = v_1 = 1$ and $r_2 = v_2 = 10^{-3}$. Also, $R_1 = r_1 I_H$, $R_2 = r_2 I_m$, $V_1 = v_1 I_H$, and $V_2 = v_2 I_\ell$. Therefore, it follows from Section 3 that $R_o^N = r_1 \Psi^N$ and $V_o^N = v_1 (\Psi^N)^{-1}$ where Ψ^N is the Gram matrix. In this numerical example we will compare the approximating optimal LQG (i.e., $N_c = N + 2$) with the dynamic compensators of various order N_c. The optimal projection equations (4.9)-(4.12) were solved using the homotopic continuation algorithm described in [16]. The approximating controllers defined by the linear fixed-order compensator (B_c^N and C_c^N) were applied to Burgers' equation (4.16)-(4.20).

We note that $N = 32$ produces converged optimal LQG designs. Hence, the reduced order compensators were tested on both the linear and nonlinear problem using the $N = 32$ order finite element model.

In the full order case $N = 32$ and $N_c = 34$, the converged feedback and observer functional gains are given in Figures 5.1 and 5.2, respectively. Since we are controlling the flux at each end point $x = 0$ and $x = 1$, we have two feedback functional gains, the one plotted with solid line is the flux control gain at the origin and the one plotted with dashed line is the flux control at the end point $x = 1$. Similarly, since we are sensing the flow at the origin and the end point, we have two observer gains (solid line for observer gain at the origin and dashed line for the observer gain at $x = 1$). Next, we applied the full order controller to Burgers' equation resulting in the nonlinear closed-loop trajectory given in Figure 5.3.

In the fixed-order case, we considered the accuracy of the impulse and step responses of the various reduced order compensator designs compared to the corresponding responses of the full order LQG design. Figure 5.4 illustrates the linear closed-loop impulse response for the full-order LQG and reduced order compensator (of order $N_c = 16$) designs. The impulse response of the linear closed-loop system for the 16th-order compensator is in perfect agreement with the LQG response. Note that in Figure 5.4 we see only one plot for both designs because both plots are essentially the same. Similar trends are seen (Figure 5.5) in the comparisons of the step responses (for the same design case) with the corresponding LQG responses.

For the nonlinear closed-loop response, the 16th-order compensator was applied to Burgers' equation and we see (in Figure 5.6) excellent agreement with the full order closed-loop trajectory response. Hence, replacing the 32nd-order optimal LQG controller by a 16th-order compensator produces a closed-loop system with minor performance degradation.

We also compared the performances of the closed-loop system of the 4th-order compensator with the full order LQG responses. Figures 5.7 and 5.8 are the impulse and step responses of the linear closed-loop system, respectively. If one compares these responses with the corresponding responses for the full order LQG controller shown in Figures 5.4 and 5.5, then it is clear that the 4th-order compensator performs almost as well as the full order LQG controller. Similar comments hold for the nonlinear closed-loop responses. For example, the 4th-order compensator response (Figure 5.9, solid line) compares well to the LQG response (Figure 5.9, Dashed lines), especially after time $T = 1.0$.

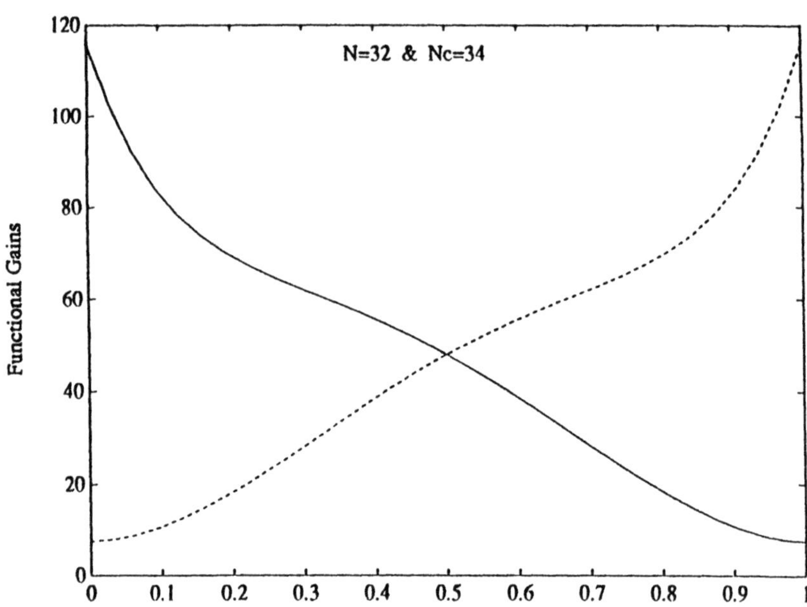

Figure 5.1: Functional Control Gains for the LQG Case, N=32

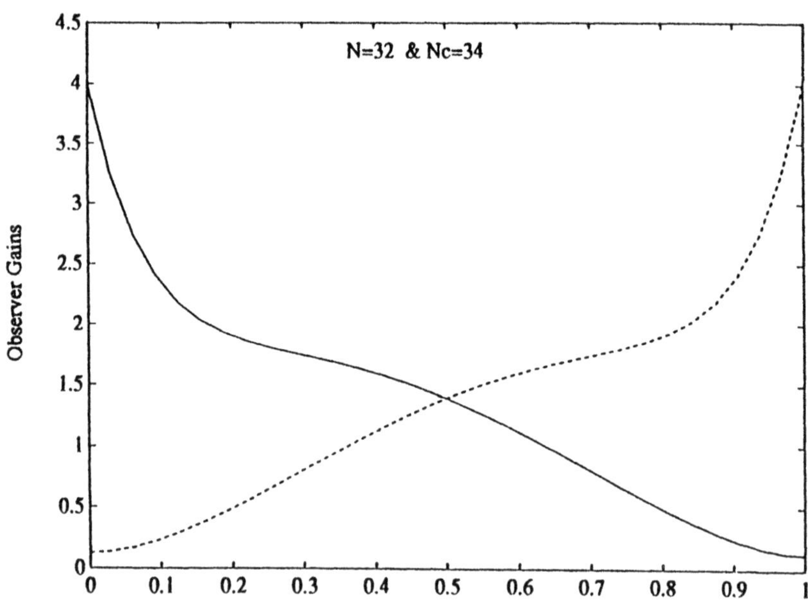

Figure 5.2: Functional Observer Gains for the LQG Case, N=32

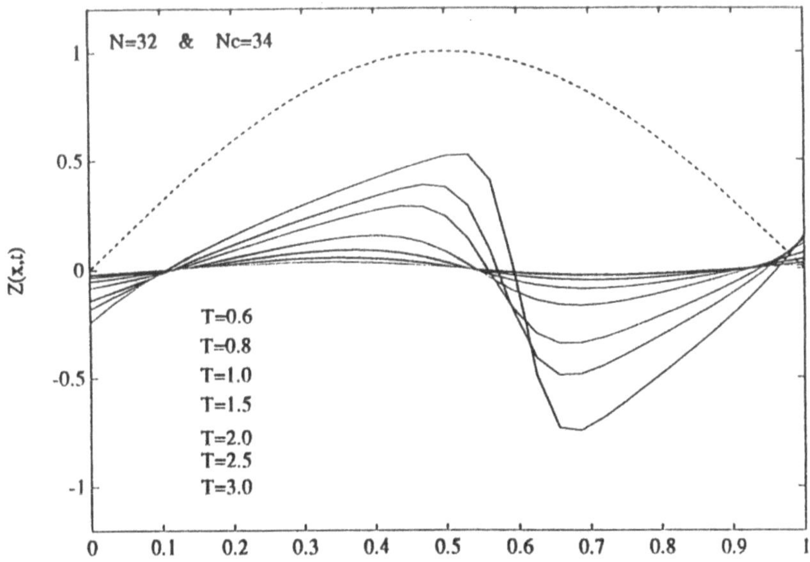

Figure 5.3: Closed-Loop Trajectory for the LQG Case, N=32

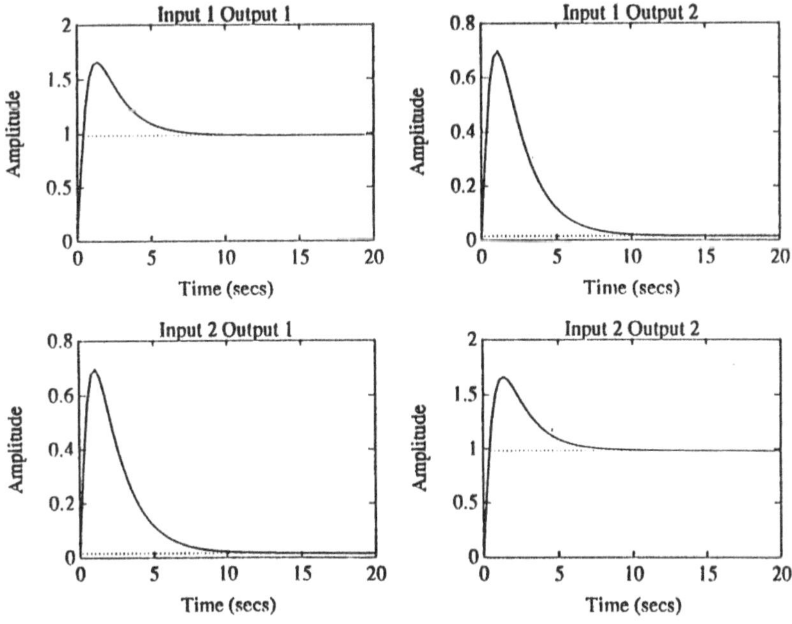

Figure 5.4: Impulse Response for LQG & Fixed-Order Cases, N=32, $N_c=16$

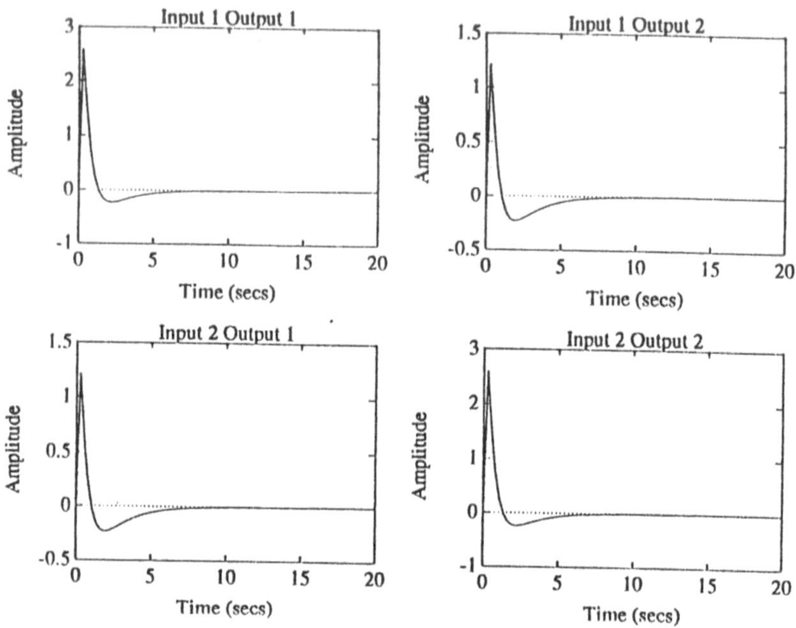

Figure 5.5: Step Response for the LQG & Fixed-Order Cases, N=32, N_c=16

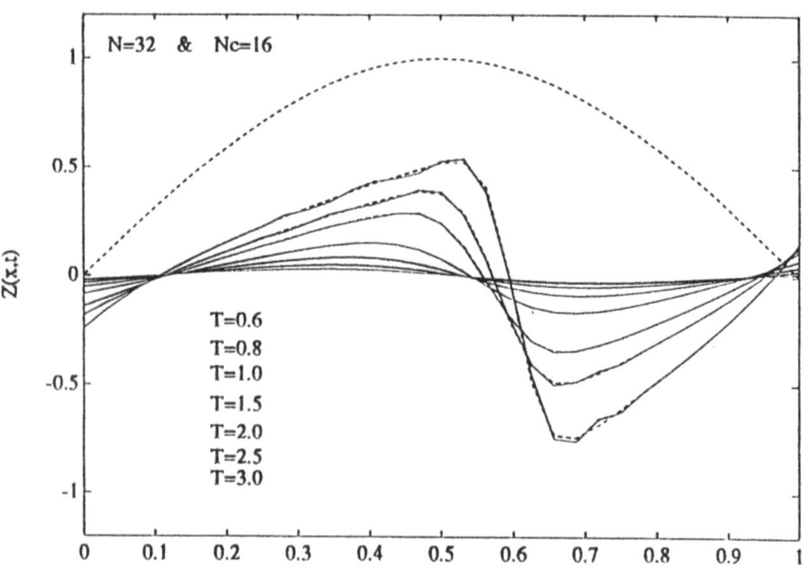

Figure 5.6: Closed-Loop for the LQG & Fixed-Order Case, N=32, N_c=16

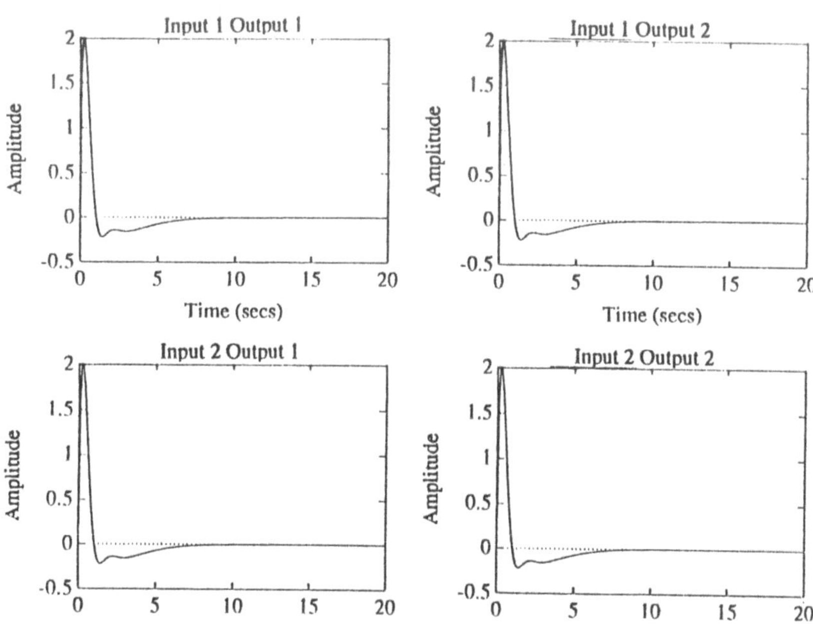

Figure 5.7: Impulse Response for the Fixed-Order Case, $N=32$, $N_c=4$

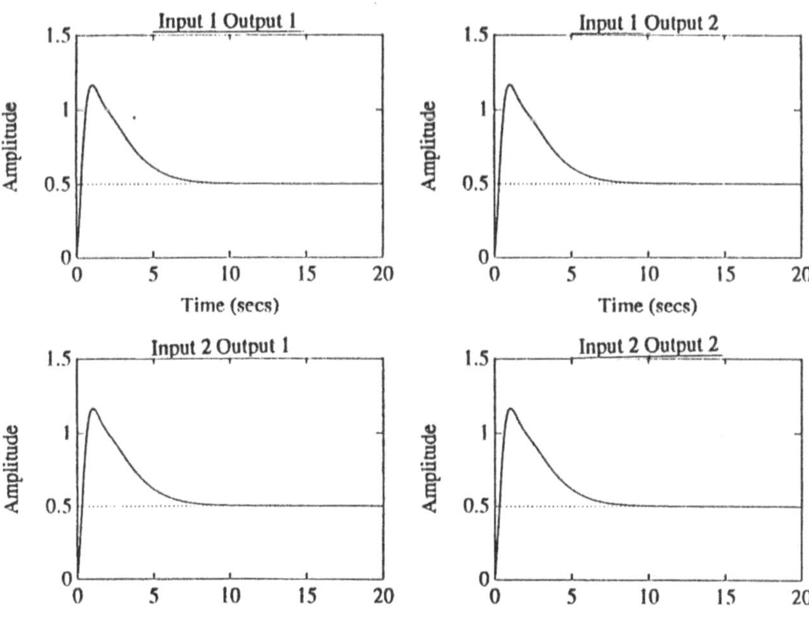

Figure 5.8: Step Response for the Fixed-Order Case, $N=32$, $N_c=4$

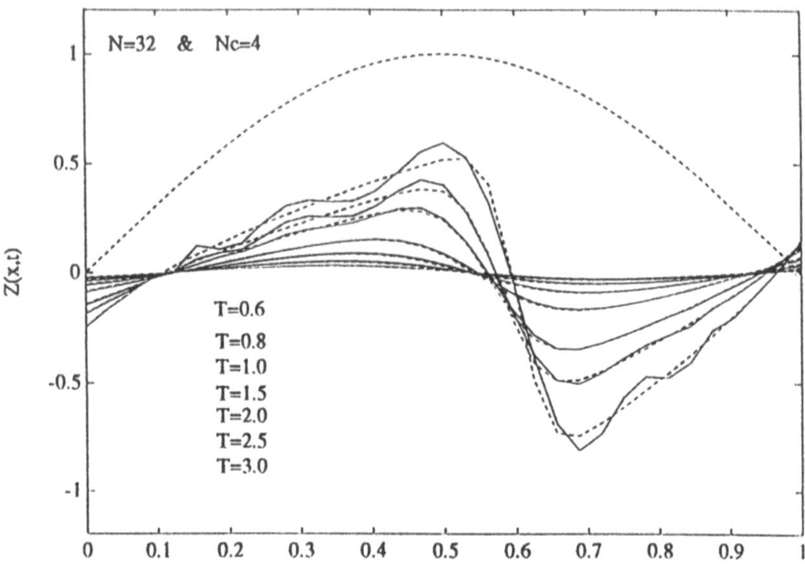

Figure 5.9: Closed-Loop for the LQG & Fixed-Order Cases, N=32, N_c=4

6 Conclusion

The purpose of this note was to show that finite dimensional dynamic compensators could be used to control a nonlinear partial differential equation without significant loss in performance. Although there is considerable theoretical and numerical work for bounded input and bounded output operators, numerical results for the unbounded control and observation operators are not as fully developed as the theory. For example, several authors have considered questions of existence of stabilizing dynamic compensators (even for nonlinear plants [6]) for boundary control problems. However, approaches, such as the optimal projection method, that result in a "computable" fixed order compensator have not been applied to more general boundary control problems. Although the numerical results presented here show that the optimal projection method can produce excellent designs for problems with boundary control and observation, there are a number of theoretical and numerical issues that need to be resolved in order to extend this approach to practical problems of this type.

References

[1] BERNSTEIN, D. S., and HYLAND, D. C., "The Optimal Projection Equations for Finite Dimensional Fixed Order Dynamic Compensator of Infinite Dimensional Systems," *SIAM J. Control and Optimization*, Vol. 24, 1986, pp. 122–151.

[2] BERNSTEIN, D. S., and ROSEN, I. G., "Finite Dimensional Approximation for Optimal Fixed Order Compensation of Distributed Parameter Systems," *J. Optimal Cont. Appl. and Meth*, June 1988, pp. 1–20.

[3] BURNS, A. J., and KANG, S., "A Stabilization Problem for Burgers' Equation with Unbounded Control and Observation," *International Series of Numerical Mathematics*, Vol. 100, Birkhauser Verlag Basel, 1991.

[4] CURTAIN, R. F., "Finite Dimensional Compensator Design for Parabolic Distributed Systems with Point Sensors and Boundary Input," *IEEE Trans. Automat. Control*, AC-26 (1982), pp. 98–104.

[5] CURTAIN, R. F., "Finite Dimensional Compensator for Parabolic Distributed Systems with Unbounded Control and Observation," *SIAM J. Control and Optimization*, Vol. 22, no. 2, 1984, pp. 255–275.

[6] CURTAIN, R. F., "Stabilizability of Semilinear Evolution Equations in Hilbert Space," *J. Math. Pures et Appl.*, Vol. 63, 1984, pg. 121–128.

[7] CURTAIN, R. F., and SALAMON, D., "Finite Dimensional Compensators for Infinite Dimensional Systems with Unbounded Input Operators," *SIAM J. Control and Optimization*, Vol. 24, no. 4, 1986, pp. 255–275.

[8] FUJII, N., "Feedback Stabilization of Distributed Parameter Systems by a Functional Observer," *SIAM J. Control and Optimization*, Vol. 18 (1980), pp. 108–120.

[9] JOHNSON, T. L., "Optimization of Low Order Compensators for Infinite Dimensional Systems," *Proc. 9-th IFIP Symposium on Optimization Techniques*, Warsau, Poland, Sept 1979, pp. 394–401.

[10] KATO, T., *Perturbation Theory for Linear Operators*, Springer-Verlag, NewYork, 1966.

[11] LASIECKA, I., and TRIGGIANI, R., "Differential and Algebraic Riccati Equations with Application to Boundary/Point Control Problems: Continuous Theory and Approximation Theory," *Lecture Notes in Control and Information Sciences*, Vol. 164, Springer-Verlag, New York, 1991.

[12] NAMBU, T., "Feedback Stabilization of Diffusion Equations by a Functional Observer," *J. Diff. Equation*, 43 (1982), pp. 257–280.

[13] PEARSON, R. K., "Optimal Fixed Form Compensators for Large Space Structures," in *ACOSS SIX, RADC-TR-81-289*, Rome Air Development Center, Griffiss Air Force Base, New York, 1981.

[14] PRITCHARD, A. J., and SALAMON, D., "The Linear Quadratic Control Problem for Infinite Dimensional System with Unbounded Input and Output Operators," *SIAM J. Control and Optimization*, Vol. 25, 1987, pp. 121–144.

[15] SCHUMACHER, J. M., "A Direct Approach to Compensator Design for Distributed Parameter Systems," *SIAM J. Control and Optimization*, Vol. 21 (1983), pp. 823–836.

[16] DRAGAN ZIGIC, WATSON, L.T., EMANUEL, G., COLLINS, Jr., and BERNSTEIN, D. S., "Homotopy Approaches to the H_2 Reduced Order Model Problem," To Appear.

BOUNDARY CONTROL AND STABILIZATION FOR A VISCOUS BURGERS' EQUATION

Christopher I. Byrnes [*]
Department of Systems Science and Mathematics
Washington University
St. Louis, Missouri 63130

David S. Gilliam [†]
Department of Mathematics
Texas Tech University
Lubbock, Texas 79409

1   Introduction

This paper is concerned with the use of a boundary control to stabilize
a viscous Burgers' equation on a finite interval. The main result of the
paper parallels a result found in [1]. The main differences are that our
original uncontrolled system is not asymptotically stable. Indeed, using
a center manifold argument, it can be shown that the trajectories of the
uncontrolled problem approach constant equilibria as t goes to infinity. By
tuning a boundary control parameter (gain) we show that the resulting
controlled system is exponentially stable for initial data small enough in
the so-called "zero dynamics" subspace. Following [7], we define the zero
dynamics to be the system corresponding to a positive infinite value of the
gain parameter. Furthermore the margin of stability is monotone increasing
with increasing k and converges to that obtained for the zero dynamics (i.e.,
infinite gain). Thus, we obtain that the controlled Burgers' system is locally
exponentially stable uniformly in $k \to \infty$.

2   The Burgers' System and Zero Dynamics

Consider the uncontrolled viscous Burgers' system

$$
\begin{aligned}
\frac{\partial w}{\partial t}(x,t) &= A_\epsilon w(x,t) + F(w(x,t)), \quad x \in (0,1),\, t > 0 \\
0 &= -\frac{\partial w}{\partial x}(0,t), \\
0 &= \frac{\partial w}{\partial x}(1,t), \\
w(x,0) &= w_0(x)
\end{aligned}
\tag{2.1}
$$

[*] Supported in part by Grants from AFOSR and NSF
[†] Supported in part by AFOSR and Texas ARP

where A_ϵ is the unbounded selfadjoint operator

$$A_\epsilon = \epsilon \frac{d^2}{dx^2} \quad \epsilon > 0 \tag{2.2}$$

with dense domain in $L^2(0,1)$

$$D(A_\epsilon) = \{f \in H^2 \mid f'(0) = f'(1) = 0\} \tag{2.3}$$

($' = d/dx$) and

$$F(w) = -\frac{dw}{dx} w. \tag{2.4}$$

It is easy to see that the system (2.1) is not asymptotically stable. In particular, the initial condition $w_0(x) = 1$ is in the domain of the spatial operator A_ϵ and for this initial data the unique classical solution for all time is given by $w(x,t) = 1$.

Nevertheless, by an infinite dimensional version of the center manifold theorem (cf. [9, 11]), solutions do approach constant equilibria as $t \to \infty$. Namely, since $\mathcal{D}(-A_\epsilon)^{1/2}) = H^1(0,1)$ and

$$F : H^1(0,1) \to Z, \quad F(w) = -ww'$$

is C^1 with $F(0) = 0$, $F'(0) = 0$, And since $0 \in \sigma_p(A_\epsilon)$ is a simple eigenvalue, we see that $Z = L^2(0,1)$ has the natural decomposition $Z = Z_1 \oplus Z_2$ where

$$Z_1 = \text{Null}\,(A_\epsilon) = C, \quad Z_2 = Z_1^\perp.$$

Now Z_1 is invariant for both A_ϵ and F and we can apply the results from [9, 11] to obtain that there exists β, $\rho > 0$ so that for initial data $w_0 \in H^1(0,1)$ with $\|w_0\|_{H^1} \leq \rho$

$$w = z_1 + z_2, \quad z_1 \in Z_1, \quad z_2 \in Z_2$$

and there is a $z_\infty \in C$ so that

$$|z_1(t) - z_\infty| + \|z_2(t)\|_{1/2} = O\left(e^{-\beta t}\right).$$

The Controlled Burgers' system is defined by

$$
\begin{aligned}
\frac{\partial w}{\partial t}(x,t) &= A_{\epsilon,k} w(x,t) + F(w(x,t)), \quad x \in (0,1),\ t > 0 \\
-\frac{\partial w}{\partial x}(0,t) &+ kw(0,t) = 0 \\
\frac{\partial w}{\partial x}(1,t) &= 0 \\
w(x,0) &= w_0(x)
\end{aligned} \tag{2.5}
$$

where $A_{\epsilon,k}$ is given by the unbounded operator

$$A \equiv A_{\epsilon,k} = \epsilon\,\frac{d^2}{dx^2} \tag{2.6}$$

with dense domain in $L_2(0,1)$

$$\mathcal{D}(A_{\epsilon,k}) = \{f \in H^2(0,1): \ f'(0) - kf(0) = 0, \ f'(1) = 0\}. \tag{2.7}$$

Here we note that for $k = 0$, the uncontrolled system (2.1) is obtained and, in this paper, we consider the parameter $k > 0$ as a boundary control parameter or gain. Our goal is to establish stability of the controlled system on the zero dynamics subspace (cf. Definition 3.2).

We define the zero dynamics for the controlled Burgers' system to be the system obtained for infinite value of the gain parameter k. This corresponds to the system

$$\begin{aligned}
\frac{\partial w}{\partial t}(x,t) &= A_{\epsilon,\infty}w(x,t) + F(w(x,t)), \quad x \in (0,1), \ t > 0\\
\frac{\partial w}{\partial x}(1,t) &= 0\\
y(t) &= w(0,t) = 0\\
w(x,0) &= w_0(x)
\end{aligned} \tag{2.8}$$

where $A_{\epsilon,\infty}$ is given by the unbounded operator

$$A \equiv A_{\epsilon,\infty} = \epsilon\,\frac{d^2}{dx^2} \tag{2.9}$$

with dense domain in $L_2(0,1)$

$$\mathcal{D}(A_{\epsilon,\infty}) = \{f \in H^2(0,1): \ f(0) = 0, \ f'(1) = 0\}. \tag{2.10}$$

3 The Linearization about Zero

The linearization about zero of (2.5) is the controlled heat equation

$$\begin{aligned}
\frac{\partial w}{\partial t}(x,t) &= A_{\epsilon,k}w(x,t) \quad x \in (0,1), \ t > 0\\
-\frac{\partial w}{\partial x}(0,t) + kw(0,t) &= 0\\
\frac{\partial w}{\partial x}(1,t) &= 0\\
w(x,0) &= w_0(x)
\end{aligned} \tag{3.1}$$

For $k = 0$, neither (2.5) nor the linearization (3.1) of (2.5) is asymptotically stable, but for $k > 0$ the linearization (3.1) is asymptotically stable. In fact, $A_{\epsilon,k}$ is a strictly negative selfadjoint operator for $k > 0$, so that the Lumer-Phillips Theorem asserts that $A_{\epsilon,k}$ generates a contraction semigroup $S_{\epsilon,k}(t)$ satisfying

$$\|S_{\epsilon,k}(t)\| \leq e^{\lambda_1(k)t}$$

where $\lambda_1(k)$ is the first eigenvalue of $A_{\epsilon,k}$. In this case, it is easy to show that

$$0 > \lambda_1(k) \to -\epsilon \left(\frac{\pi}{2}\right)^2, \quad k \to \infty.$$

Thus the system (3.1) has solution $w(t)$ satisfying

$$\|w(t)\| = \|S_{\epsilon,k}(t)w_0\| \leq e^{\lambda_1(k)t}\|w_0\|.$$

The spectrum of $A_{\epsilon,k}$ is given by the numbers $\lambda = -\epsilon\mu^2$ as the roots of the characteristic equation

$$k\cos(\mu) = \mu\sin(\mu) \tag{3.2}$$

This equation has infinitly many zeros $\{\mu_j\}_{j=1}^{\infty}$ satisfying

$$(j-1)\pi < \mu_j < (j-1/2)\pi, \quad j = 1, 2, \cdots.$$

Corresponding to the eigenvalues $\lambda_j = -\epsilon\mu_j^2$ we have the complete orthonormal system of eigenfunctions

$$\phi_j(x) = \kappa_j\cos(\mu_j(1-x)), \quad j = 1, 2, \cdots, \tag{3.3}$$

$$\kappa_j = \sqrt{\frac{2k}{k + \sin^2(\mu_j)}} \tag{3.4}$$

$$= \sqrt{\frac{2(k^2 + \mu_j^2)}{k^2 + \mu_j^2 + k}}$$

in $L^2(0,1)$. The last equality in (3.4) follows from the characteristic equation, namely we have

$$\sin^2(\mu_j) = \frac{k^2}{k^2 + \mu_j^2}, \tag{3.5}$$

$$\cos^2(\mu_j) = \frac{\mu_j^2}{k^2 + \mu_j^2}$$

The eigenvalues $\lambda_j = -\epsilon\mu_j^2$, are simple and for all $f \in H = L^2(0,1)$

$$f = \sum_{j=1}^{\infty} f_j\phi_j, \quad f_j = <f,\phi_j>.$$

Furthermore,

$$D(A_{\epsilon,k}) = \{f \in L^2(0,1) \mid \sum_{j=1}^{\infty} \lambda_j^2 |f_j|^2 < \infty\}$$

and for $f \in D(A_{\epsilon,k})$ the spectral theorem asserts that

$$A_{\epsilon,k}f = \sum_{j=1}^{\infty} \lambda_j f_j \phi_j.$$

In addition, for $f \in H$ we have the representation for the resolvent

$$(\lambda I - A_{\epsilon,k})^{-1}f = \sum_{j=1}^{\infty} \frac{1}{(\lambda - \lambda_j)} f_j \phi_j$$

and again by the spectral theorem the selfadjoint operator $A_{\epsilon,k}$ is the infinitesimal generator of an analytic semigroup $S_{\epsilon,k}(t)$ given by

$$S_{\epsilon,k}(t)f = \sum_{j=1}^{\infty} e^{\lambda_j t} f_j \phi_j$$

from which it is easy to show that

$$\|S_{\epsilon,k}(t)\| \le e^{\lambda_1 t}.$$

The square root of minus the spatial operator, $(-A_{\epsilon,k})^{1/2}$, satisfies the following properties (cf. [11, 13]).

Lemma 3.1   1. *The operator $(-A_{\epsilon,k})^{1/2}$ with domain*

$$D((-A_{\epsilon,k})^{1/2}) = \{f \in H \mid \sum_{j=1}^{\infty} \epsilon \mu_j^2 |f_j|^2 < \infty\} = H^1(0,1)$$

is a densely defined selfadjoint operator in H with bounded inverse (since $0 \notin \sigma((-A_{\epsilon,k})^{1/2}.)$

2. *For $f \in D((-A_{\epsilon,k})^{1/2})$*

$$(-A_{\epsilon,k})^{1/2}f = \sum_{j=1}^{\infty} \sqrt{\epsilon}\mu_j f_j \phi_j$$

3. $\|(-A_{\epsilon,k})^{1/2}f\|^2 = \sum_{j=1}^{\infty} \epsilon \mu_j^2 |f_j|^2$

4. $S_{\epsilon,k}(t): H \to D((-A_{\epsilon,k})^{1/2})$ *for every* $t > 0$ *and*

5. *For every* $f \in D((-A_{\epsilon,k})^{1/2})$,

$$S_{\epsilon,k}(t)(-A_{\epsilon,k})^{1/2}f = (-A_{\epsilon,k})^{1/2}S_{\epsilon,k}(t)f.$$

Definition 3.2 *We define the zero dynamics subspace of* $L^2(0,1)$ *to be the natural domain for the square root* $(-A_{\epsilon,\infty})^{1/2}$ *of the spatial operator* $(-A_{\epsilon,\infty})$. *In particular, we have*

$$D((-A_{\epsilon,\infty})^{1/2}) = \tilde{H}_0^1(0,1) \equiv H^1(0,1) \cap \{f \mid f(0) = 0\}.$$

For the remainder of the paper we will use the notation

$$\|f\|_1 = \|(-A_{\epsilon,k})^{1/2}f\| \tag{3.6}$$

for $f \in D((-A_{\epsilon,k})^{1/2}) = H^1(0,1)$.

Note that since $(-A_{\epsilon,k})^{1/2}$ is closed, the graph is closed and hence the subspace $D((-A_{\epsilon,k})^{1/2})$ with the graph norm is a Banach space. Further, since $(-A_{\epsilon,k})^{1/2}$ is invertible, we have

$$\|f\|_1 \le \|f\| + \|f\|_1 \le c\|f\|_1$$

so the norm $\|\cdot\|_1$ is complete on $D((-A_{\epsilon,k})^{1/2})$.

One approach to establishing the local existence and uniqueness is to verify the hypothesis in Theorem 3.3.3 in [11], which we now state.

Theorem 3.3 *If the mapping* $F(z) = -zz'$ *from an open subset*

$$U \subset \mathcal{R} \times D((-A_{\epsilon,k})^{1/2})$$

to H *is locally Hölder continuous in* t, *locally Lipschitzian in* z, *then for any* $(t_0, z_0) \in U$ *there is a* $t_1 > t_0$ *such that (2.5) has a unique classical solution* z *on* $[t_0, t_1)$ *and such that* $z(t_0) = z_0$.

4 Some Basic Estimates

In order to estimate the L^2 norm of the nonlinear term F in the controlled Burgers' system (2.5), it suffices to establish norm estimates relating $\|f\|$, $\|f'\|$ and $\|(-A_{\epsilon,k})^{1/2}f\|$.

An immediate consequence of the fact that 0 is in the resolvent set of $(-A_{\epsilon,k})^{1/2}$ is that we obtain an analog of the Poincare inequality for H^1; viz, there exists a positive constant c such that for $f \in H^1$

$$\|f\| \le c\|(-A_{\epsilon,k})^{1/2}f\|.$$

However, this estimate can be made more precise using spectral theory, which yields a more explicit constant c.

BOUNDARY CONTROL FOR BURGERS' EQUATION

Theorem 4.1 *For fixed ϵ, $k > 0$ and $f \in D((-A_{\epsilon,k})^{1/2}) = H^1(0,1)$, we have*

$$\|f\| \le c_0 \|(-A_{\epsilon,k})^{1/2} f\|.$$

with

$$c_0 = \frac{1}{\sqrt{\epsilon \mu_1(k)}}$$

$$0 < \mu_1(k) \to \frac{\pi}{2}, \quad k \to \infty.$$

Remark 4.2 The following are standard estimates which can be found in, for example, [10] page 153, [12], page 192.

1. On $H^1(0,1)$ the operator $D = d/dx$ is closed and the norms

$$\|f\|_{H^1} = \|f\| + \|Df\|,$$

$$\|f\|_a = \|f\|_{H^1} + \sup_{x \in [0,1]} |f(x)|$$

make H^1 a Banach space with

$$\|f\|_{H^1} \le \|f\|_a$$

for all $f \in H^1$. By the open mapping theorem there are constants c_1, c_2 such that

$$c_1 \|f\|_a \le \|f\|_{H^1} \le c_2 \|f\|_a.$$

and hence the linear functional given by point evaluation is a continuous linear functional on H^1, i.e.,

$$\ell_x(f) = f(x)$$

defines a continuous linear functional on H^1 since for every $x \in [0,1]$ we have

$$|f(x)| = |\ell_x(f)| \le \|f\|_a \le \frac{1}{c_1} \|f\|_{H^1}.$$

2. Using a different approach, this result is obtained in Kato [12] (Example 1.8 page 192) where the following inequalities are given for any $x \in [0,1]$ and any positive integer n

$$|f(x)| \le \frac{1}{\sqrt{2n+3}} \|f'\| + \frac{n+1}{\sqrt{2n+1}} \|f\|$$

showing that the coefficient of the derivative term can be made arbitrarily small.

C.I. BYRNES AND D.S. GILLIAM

Simple estimates relating $\|f'\|$, and $\|(-A_{\epsilon,k})^{1/2}f\|$ are obtained using symmetric forms $h_k(\cdot)$ generated by the square root of the operator $(-A_{\epsilon,k})$. Namely, we define

$$h_k(f) = \|(-A_{\epsilon,k})^{1/2}f\|^2, \quad \mathcal{D}(h_k) = H^1 = \mathcal{D}((-A_{\epsilon,k})^{1/2}), \ 0 < k < \infty,$$

and

$$h_\infty(f) = \|(-A_{\epsilon,\infty})^{1/2}f\|^2, \quad \mathcal{D}(h_\infty) = \tilde{H}_0^1 = \mathcal{D}((-A_{\epsilon,\infty})^{1/2}).$$

Recall that form inequality $h_1 \leq h_2$ (for forms bounded from below) means

$$\mathcal{D}(h_2) \subset \mathcal{D}(h_1), \quad h_1(f) \leq h_2(f), \ f \in \mathcal{D}(h_2)$$

(cf. Kato [12], page 330).

Theorem 4.3 *For $\epsilon > 0$, $k > 0$, we have for $f \in H^1$, with the norm*

$$\|f\|_1 = \|(-A_{\epsilon,k})^{1/2}f\|$$

that

$$\|f'\| \leq \frac{1}{\sqrt{\epsilon}}\|f\|_1 \tag{4.1}$$

and for $f \in \tilde{H}_0^1$,

$$\|f'\| = \frac{1}{\sqrt{\epsilon}}\|f\|_1 \tag{4.2}$$

Proof: For $f \in \mathcal{D}((-A_{\epsilon,k}))$, by integration by parts,

$$h_k(f) = \|(-A_{\epsilon,k})^{1/2}f\|^2 = \epsilon\left(k|f(0)|^2 + \|f'\|^2\right) \geq \epsilon\|f'\|^2$$

Now since $\mathcal{D}((-A_{\epsilon,k}))$ is a core in $\mathcal{D}((-A_{\epsilon,k})^{1/2})$ we have for each $f \in \mathcal{D}((-A_{\epsilon,k})^{1/2})$ (and hence $f' \in L_2$) there is a sequence $\{f_n\} \subset \mathcal{D}((-A_{\epsilon,k}))$ such that

$$\|f_n - f\| \to 0, \quad \|(-A_{\epsilon,k})^{1/2}(f_n - f)\| \to 0, \ n \to \infty$$

Thus we have

$$\sqrt{\epsilon}\|f_n' - f_m'\| \leq \|(-A_{\epsilon,k})^{1/2}(f_n - f_m)\| \to 0$$

which implies that $\{f_n'\}$ is Cauchy in L_2. Hence there is a g such that

$$f_n \to f, \quad f_n' \to g$$

BOUNDARY CONTROL FOR BURGERS' EQUATION

Now $D = d/dx$ is a closed operator on $H^1 = \mathcal{D}((-A_{\epsilon,k})^{1/2})$ so that $g \in H^1$ and $g = f'$. With this we have

$$\|(-A_{\epsilon,k})^{1/2} f_n\| \geq \sqrt{\epsilon} \|f_n'\|$$

and hence

$$\|(-A_{\epsilon,k})^{1/2} f\| \geq \sqrt{\epsilon} \|f'\|$$

which is the first part of Theorem 4.3.

The other half of the theorem is obtained from a more careful look at the first inequality of the proof, namely, for $f \in \mathcal{D}((-A_{\epsilon,k}))$

$$\|(-A_{\epsilon,k})^{1/2} f\|^2 = \sqrt{\epsilon}\left(k|f(0)|^2 + \|f'\|^2\right)$$

and by the previous computation we see that for the sequence $\{f_n\} \subset \mathcal{D}((-A_{\epsilon,k}))$ (given above)

$$\|(-A_{\epsilon,k})^{1/2}(f_n - f_m)\|^2 = \sqrt{\epsilon}\left(k|f_n(0) - f_m(0)|^2 + \|f_n' - f_m'\|^2\right).$$

Thus, in addition to the conclusion that $\{f_n'\}$ is Cauchy, we see that $\{f_n(0)\}$ converges to a constant c which must be $f(0)$ since the linear functional $\ell_0(f) = f(0)$ (of point evaluation) is a continuous linear functional on H^1. Hence for $f \in H^1$

$$\|(-A_{\epsilon,k})^{1/2} f\|^2 = \sqrt{\epsilon}\left(k|f(0)|^2 + \|f'\|^2\right)$$

So, if in addition, $f \in \widetilde{H}_0^1$, (i.e., $f(0) = 0$) then we obtain the second conclusion of the theorem:

$$\|(-A_{\epsilon,k})^{1/2} f\|^2 = \sqrt{\epsilon}\left(\|f'\|^2\right).$$

$\square$

Recall that the usual norm on H^1 is

$$\|f\|_{H^1} = \|f\| + \|f'\|$$

but on $\widetilde{H}_0^1$ we have

$$\|f\| \leq \|f'\| \tag{4.3}$$

so that for $f \in \widetilde{H}_0^1$

$$\|f'\| \leq \|f\|_{H^1} \leq 2\|f'\|.$$

With these estimates we can prove that F is locally Lipschitz, i.e., for $\|z_1\|_1, \|z_2\|_1 < M$, there exists a constant, C depending only on M, ϵ and k but otherwise independent of z_1, z_2 such that

$$\|F(z_1) - F(z_2)\| \leq C\|z_1 - z_2\|_1.$$

This allows us to apply the basic local existence result found in [11] and [13]. For Burgers' equation we have for z_1, z_2 in $\mathcal{D}((-A_{\epsilon,k})^{1/2}) = H^1$ with $\|z_1\|_1$, $\|z_2\|_1 < M$

$$
\begin{aligned}
\|F(z_1) - F(z_2)\| &= \|z_1 z_1' - z_2 z_2'\| \\
&\leq \|z_1 z_1' - z_1 z_2' + z_1 z_2' - z_2 z_2'\| \\
&\leq \|z_1 z_1' - z_1 z_2'\| + \|z_1 z_2' - z_2 z_2'\| \\
&\leq \|z_1\|_\infty \|z_1' - z_2'\| + \|z_2'\|\|z_1 - z_2\|_\infty \\
&\leq \frac{c_0}{\sqrt{\epsilon}} \left(\|z_1\|_1 \|z_1 - z_2\|_1 \right. \\
&\qquad \left. + \|z_2\|_1 \|z_1 - z_2\|_1 \right) \\
&= \frac{c_0}{\sqrt{\epsilon}} \left((\|z_1\|_1 + \|z_2\|_1) \|z_1 - z_2\|_1 \right. \\
&\leq C \|z_1 - z_2\|_1
\end{aligned}
$$

where

$$
C = \frac{2 c_0 M}{\sqrt{\epsilon}}.
$$

Global existence of classical solutions is essentially demonstrated in Theorem 6.3.3, page 199 of [11] and Lemma 5.6.7, page 159 of [13], for initial data in $D((-A_{\epsilon,k})^{1/2})$.

5   The Main Stability Result

Theorem 5.1 *For every β satisfying $0 < \beta < \epsilon \mu_1(k)^2 = \gamma < \epsilon \pi^2/4$, there exists $\rho = \rho_{\epsilon,\beta} > 0$ such that for $z_0 \in \bar{H}_0^1(0,1)$ with*

$$
\|z_0\|_{H^1} \leq \rho/2
$$

there is a unique solution $z(t) \in H^1$ of

$$
\begin{aligned}
\frac{\partial z}{\partial t}(t) &= A_{\epsilon,k} z(t) + F(z(t)) \\
z(0) &= z_0
\end{aligned}
$$

for all $t > 0$ and, moreover, the solution satisfies

$$
\|z(t)\|_{H^1} \leq 2 e^{-\beta t} \|z_0\|_{H^1}
$$

In fact ρ can be taken to be any number such that

$$
\rho \left(\frac{\sqrt{2\pi}}{\sqrt{\epsilon(\gamma - \beta)}} + \frac{7}{\gamma - \beta} \right) < \frac{1}{2}
\tag{5.1}
$$

Lemma 5.2 *For $z \in H^1$ and $F(z) = -zz'$ we have*

$$\|F(z)\| \le (\|z\|_{H^1})^2 \qquad (5.2)$$

Proof:

$$
\begin{aligned}
\|F(z)\| &= \|zz'\| \\
&\le \|z\|_\infty \|z'\| \\
&\le \|z\|_{H^1} \|z'\| \\
&\le (\|z\| + \|z'\|)^2 \\
&= (\|z\|_{H^1})^2
\end{aligned}
$$

Lemma 5.3 *The following estimates hold*

1. For $f \in \tilde{H}_0^1$ and all $t > 0$, we have

$$\|S_{\epsilon,k}(t)f\|_{H^1} \le e^{-\gamma t}\|f\|_{H^1} \qquad (5.3)$$

where $\gamma = \epsilon \mu_1^2$.

2. For $f \in H^1$ and all $t > 0$

$$\|S_{\epsilon,k}(t)f\|_{H^1} \le \left(\frac{\sqrt{2}}{\sqrt{\epsilon t}} + 7\right) e^{-\gamma t}\|f\| \qquad (5.4)$$

Proof:
For part one we have for $f \in \tilde{H}_0^1$,

$$
\begin{aligned}
\|S_{\epsilon,k}(t)f\|_{H^1} &= \|S_{\epsilon,k}(t)f\| + \|(S_{\epsilon,k}(t)f)'\| \\
&\le \|S_{\epsilon,k}(t)f\| + \frac{1}{\sqrt{\epsilon}}\|(-A_{\epsilon,k})^{1/2}S_{\epsilon,k}(t)f\| \\
&\le \|S_{\epsilon,k}(t)\|\left(\|f\| + \frac{1}{\sqrt{\epsilon}}\|(-A_{\epsilon,k})^{1/2}f\|\right) \\
&= \|S_{\epsilon,k}(t)\|(\|f\| + \|f'\|) \\
&\le e^{-\gamma t}\|f\|_{H^1}
\end{aligned}
$$

where the last equality follows from (4.2). Note that no such estimate is possible on all of $H^1(0,1)$.

As for part two, with $f \in H^1$ one computes

$$\|S_{\epsilon,k}(t)f\|_{H^1} = \|S_{\epsilon,k}(t)f\| + \|(S_{\epsilon,k}(t)f)'\|$$

$$\leq \quad \|S_{\epsilon,k}(t)\|\|f\| + \frac{1}{\sqrt{\epsilon}}\|(-A_{\epsilon,k})^{1/2}S_{\epsilon,k}(t)f\|$$

$$\leq \quad e^{-\gamma t}\|f\| + \frac{1}{\sqrt{\epsilon}}\|(-A_{\epsilon,k})^{1/2}S_{\epsilon,k}(t)f\|$$

$$= \quad e^{-\gamma t}\|f\| + \frac{1}{\sqrt{\epsilon}}\|\sum_{j=1}^{\infty}\sqrt{\epsilon}\mu_j e^{-\epsilon\mu_j^2 t}<f,\phi_j>\phi_j\|$$

$$= \quad e^{-\gamma t}\|f\| + e^{-\gamma t}\left(\sum_{j=1}^{\infty}\mu_j^2 e^{-2\epsilon(\mu_j^2-\mu_1^2)t}|<f,\phi_j>|^2\right)^{1/2}$$

$$\leq \quad e^{-\gamma t}\|f\| + \sup_{j\geq 1,t>0}\left(\mu_j e^{-\epsilon(\mu_j^2-\mu_1^2)t}\right)e^{-\gamma t}\left(\sum_{j=1}^{\infty}|<f,\phi_j>|^2\right)^{1/2}$$

$$= \quad e^{-\gamma t}\|f\| + \sup_{j\geq 1,t>0}\left(\mu_j e^{-\epsilon(\mu_j^2-\mu_1^2)t}\right)e^{-\gamma t}\|f\|$$

$$= \quad \left[1 + \sup_{j\geq 1,t>0}\left(\mu_j e^{-\epsilon(\mu_j^2-\mu_1^2)t}\right)\right]e^{-\gamma t}\|f\|$$

Recall that

$$(j-1)\pi \leq \mu_j \leq (j-1/2)\pi,$$

and note that the first term in the sup is just μ_1 for all time, so we consider

$$
\begin{aligned}
\sup_{j\geq 2,t>0}\left(\mu_j e^{-\epsilon(\mu_j^2-\mu_1^2)t}\right) &\leq \sup_{j\geq 2,t>0}\left((j-1/2)\pi e^{\epsilon(1/4-(j-1)^2)\pi^2 t}\right) \\
&\leq \sup_{j\geq 1,t>0}\left((j+1/2)\pi e^{(1/4-j^2)\epsilon\pi^2 t}\right) \\
&\leq \sup_{j\geq 1,t>0}\left((3j/2)\pi e^{(1/4-j^2)\epsilon\pi^2 t}\right)
\end{aligned}
$$

Accordingly we define

$$f(\xi) = \xi e^{(1/4-\xi^2)\epsilon\pi^2 t}$$

for which a straightforward computation shows that the maximum of f occurs at

$$\xi_0 = \frac{1}{\pi\sqrt{2\epsilon t}}.$$

Furthermore f is increasing from 0 to ξ_0 and decreasing thereafter. Since the domain of interest is $\xi \geq 1$, we must consider two cases: 1) $\xi_0 < 1$ and 2) $\xi_0 > 1$.

A simple calculation shows that

1. for $\xi_0 < 1$, the max of f occurs at $\xi = 1$. Also $\xi_0 < 1$ implies

$$t > \frac{1}{2\epsilon\pi^2}$$

and in this case

$$\sup_{j\geq 1, t>0} \left((3j/2)\pi e^{(1/4-j^2)\epsilon\pi^2 t} \right) \leq \frac{3\pi}{2}.$$

2. for $\xi_0 > 1$, the max of f occurs at ξ_0 and $\xi_0 > 1$ implies

$$t < \frac{1}{2\epsilon\pi^2}$$

and in this case we have

$$\sup_{j\geq 1, t>0} \left((3j/2)\pi e^{(1/4-j^2)\epsilon\pi^2 t} \right) \leq \frac{3}{2\sqrt{2t\epsilon}}.$$

From an analysis of these cases we have

$$\sup_{j\geq 1, t>0} \left((3j/2)\pi e^{(1/4-j^2)\epsilon\pi^2 t} \right) \leq \left(\frac{3}{2\sqrt{2t\epsilon}} + \frac{3\pi}{2} \right)$$

so that finally we have

$$\|S_{t,k}(t)f\|_{H^1} \leq \left[1 + \frac{3\pi}{2} + \frac{3}{2\sqrt{2t\epsilon}} \right] \|f\|$$

$$\leq \left[7 + \frac{\sqrt{2}}{\sqrt{t\epsilon}} \right] \|f\|$$

$\square$

Proof of Theorem 5.1:

Using the estimates of the previous section and Theorem 3.3 we obtain local existence of classical solutions on some small time interval. Thus under the assumptions,

$$\|z_0\| \leq \rho/2, \quad 0 < \beta < \gamma = \epsilon\mu_1^2$$

we know there is a unique local solution. As in [11], it is easy to see that the solution is actually a global solution provided ρ is sufficiently small. In order to verify this and obtain the stability result we first employ the

C.I. BYRNES AND D.S. GILLIAM

variation of parameters formula to express (2.5) in the form of an integral equation,

$$z(t) = S_{\epsilon,k}(t)z_0 + \int_0^t S_{\epsilon,k}(t-\tau)F(z(\tau))\,d\tau, \tag{5.5}$$

$$\|z(\cdot,t)\|_{H^1} \leq \|S_{\epsilon,k}(t)z_0\|_{H^1} + \int_0^t \|S_{\epsilon,k}(t-\tau)F(z(\tau))\|_{H^1}\,d\tau, \tag{5.6}$$

$$\leq e^{-\gamma t}\|z_0\|_{H^1} + \int_0^t \left(\frac{\sqrt{2}}{\sqrt{\epsilon(t-\tau)}}+7\right)e^{-\gamma(t-\tau)}\|F(z(\tau))\|\,d\tau$$

$$\leq e^{-\gamma t}\|z_0\|_{H^1} + \int_0^t \left(\frac{\sqrt{2}}{\sqrt{\epsilon(t-\tau)}}+7\right)e^{-\gamma(t-\tau)}\|z(\tau)\|_{H^1}^2\,d\tau$$

Let t_1 be the largest possible time for which $\|z(t)\|_{H^1} < \rho$. Note that on $[0,t_1)$ we have

$$\|F(z(t))\| \leq (\|z(t)\|_{H^1})^2 \leq \rho^2$$

Thus we have either $\|z(t_1)\|_{H^1} = \rho$ or $t_1 = \infty$. From Lemma 5.3, we have

$$\|z(\cdot,t)\|_{H^1} \leq e^{-\gamma t}\|z_0\|_{H^1} + \rho^2 \int_0^t \left(\frac{\sqrt{2}}{\sqrt{\epsilon(t-\tau)}}+7\right)e^{-\gamma(t-\tau)}\,d\tau$$

$$\leq \frac{\rho}{2}+\rho^2\left\{\int_0^t \left(\frac{\sqrt{2}e^{-\gamma(t-\tau)}}{\sqrt{\epsilon(t-\tau)}}\right)d\tau + 7\int_0^t e^{-\gamma(t-\tau)}\,d\tau\right\}$$

$$\leq \frac{\rho}{2}+\rho^2\left\{\int_0^t \left(\frac{\sqrt{2}e^{-\gamma s}}{\sqrt{\epsilon s}}\right)ds + 7\int_0^t e^{-\gamma s}\,ds\right\}$$

$$\leq \frac{\rho}{2}+\rho^2\left\{\int_0^\infty \left(\frac{\sqrt{2}e^{-\gamma s}}{\sqrt{\epsilon s}}\right)ds + 7\int_0^\infty e^{-\gamma s}\,ds\right\}$$

$$= \frac{\rho}{2}+\rho^2\left\{\left(\frac{\sqrt{2\pi}}{\sqrt{\epsilon\gamma}}\right)+\frac{7}{\gamma}\right\}$$

$$< \frac{\rho}{2}+\frac{\rho}{2}=\rho$$

provided

$$2\rho < \left\{\left(\frac{\sqrt{2\pi}}{\sqrt{\epsilon\gamma}}\right)+\frac{7}{\gamma}\right\}^{-1}$$

which follows from (5.1). Thus $t_1 = \infty$ and we have a global solution satisfying $\|z(t)\|_{H^1} < \rho$ for $t \in [0,\infty)$ provided $\|z_0\|_{H^1} \leq \rho/2$ and $z_0 \in \tilde{H}_0^1$.

Now let

$$w(t) = \sup_{0 \leq s \leq t} \|z(s)\|_{H^1}e^{\beta s}$$

and consider

$$\|z(\cdot,t)\|_{H^1}e^{\beta t} \le e^{\beta t}\left(\|S_{\epsilon,k}(t)z_0\|_{H^1} + \int_0^t \|S_{\epsilon,k}(t-\tau)F(z(\tau))\|_{H^1}\,d\tau\right),$$

$$\le e^{\beta t}\left(e^{-\gamma t}\|z_0\|_{H^1} + \int_0^t \left(\frac{\sqrt{2}}{\sqrt{\epsilon(t-\tau)}}+7\right)e^{-\gamma(t-\tau)}\|F(z(\tau)\|\,d\tau\right)$$

$$\le e^{-(\gamma-\beta)t}\|z_0\|_{H^1} + \rho\int_0^t \left(\frac{\sqrt{2}}{\sqrt{\epsilon(t-\tau)}}+7\right)e^{-(\gamma-\beta)(t-\tau)}e^{\beta\tau}\|z(\tau)\|_{H^1}\,d\tau$$

$$\le \|z_0\|_{H^1} + \rho\left\{\int_0^t\left(\frac{\sqrt{2}}{\sqrt{\epsilon s}}+7\right)e^{-(\gamma-\beta)s}\,ds\right\}w(t)$$

$$\le \|z_0\|_{H^1} + \rho\left\{\int_0^\infty\left(\frac{\sqrt{2}}{\sqrt{\epsilon s}}+7\right)e^{-(\gamma-\beta)s}\,ds\right\}w(t)$$

$$\le \|z_0\|_{H^1} + \rho\left\{\frac{\sqrt{2\pi}}{\sqrt{\epsilon(\gamma-\beta)}}+\frac{7}{(\gamma-\beta)}\right\}w(t)$$

$$\le \|z_0\|_{H^1} + \frac{1}{2}w(t)$$

by our choice of ρ in (5.1). Hence we have

$$w(t) \le 2\|z_0\|_{H^1}$$

and therefore

$$\|z(t)\|_{H^1} \le 2e^{-\beta t}\|z_0\|_{H^1}.$$

$\square$

References

[1] J.A. BURNS AND S. KANG, "A control problem for Burgers' equation with bounded input/output," *Nonlinear Dynamics*, (1991), Vol. 2, 235-262.

[2] J.A. BURNS AND S. KANG, "A Stabilization problem for Burgers' equation with unbounded control and observation," *International Series of Numerical Mathematics*, (1991), Vol. 100, 51-72.

[3] C.I. BYRNES, D.S. GILLIAM, "Asymptotic properties of root locus for distributed parameter systems," *Proc. of 27th IEEE Conf. on Dec. and Control*, Austin, 1988.

[4] C.I. BYRNES, D.S. GILLIAM, "Boundary feedback stabilization of distributed parameter systems", *Proceedings of 9th MTNS* (1989).

[5] C.I. BYRNES, D.S. GILLIAM, J. HE, "A root locus methodology for parabolic distributed parameter systems," *Computation and Control, Progress in Systems and Control Theory*, Birkhäuser, 135-150, (1991).

[6] C.I. BYRNES, D.S. GILLIAM, J. HE, "Root locus and boundary feedback design for distributed paramter systems," Submitted to *SIAM J. Cont. Opt.*

[7] C.I. BYRNES, D.S. GILLIAM, "Stability of certain distributed parameter systems by low dimensional controllers: a root locus approach," *Proceedings 29th IEEE International Conference on Decision and Control.*

[8] C.I. BYRNES, D.S. GILLIAM, "Root locus and boundary feedback design for nonlinear distributed parameter systems," *Proceedings 30th IEEE International Conference on Decision and Control.*

[9] J. CARR, "Applications of the center manifold theorem," Springer-Verlag, 1980.

[10] S. GOLDBERG, "Unbounded Linear Operators: Theory and Applications," Dover Publications, Inc., New York, 1985.

[11] D. HENRY, "Geometric Theory of Semilinear Parabolic Equations," Springer-Verlag, Lec. Notes in Math., Vol 840, 1981.

[12] T. KATO, "Perturbation theory for linear operators," Springer-Verlag, New York, (1966).

[13] A. PAZY, "Semigroups of Linear Operators and Applications to Partial Differential Equations," Springer-Verlag, 1983.

A SINC-GALERKIN METHOD FOR CONVECTION DOMINATED TRANSPORT

Timothy S. Carlson, John Lund, and Kenneth L. Bowers

Department of Mathematical Sciences
Montana State University
Bozeman, Montana 59717

1   Introduction

The study of boundary layer equations in fluid flow leads one to consider parabolic partial differential equations with large Reynolds numbers. That is to say the ratio of the convective term to the diffusive term is large. Analytically this can be modeled by the differential equation

$$
\begin{aligned}
\mathcal{L}u(x,t) &= f(x,t) \ , \quad 0 < x < 1, \quad t > 0 \\
\mathcal{L}u(x,t) &\equiv u_t(x,t) - u_{xx}(x,t) + p(x,t)u_x(x,t) \\
u(0,t) &= u(1,t) = 0 \ , \quad t \geq 0 \\
u(x,0) &= 0 \ , \quad 0 \leq x \leq 1
\end{aligned}
\tag{1.1}
$$

where $|p(x,t)u_x(x,t)| \gg |u_{xx}(x,t)|$. Besides the numerical difficulties associated with the semi–infinite temporal domain, the boundary layer calculations in the spatial operator present an equally difficult challenge. For the Sinc–Galerkin method, the temporal difficulties have been addressed in [9] and [10]. The goal in this work is to describe the Sinc–Galerkin method for the highly convective spatial problem and show that this spatial calculation can be accurately carried forward in time via the approach in [9].

In Section 2 the Sinc-Galerkin method first developed by Stenger [15] is reviewed with the intention of considering problems of the form

$$
\mathcal{S}u(x) \equiv -u''(x) + p(x)u'(x) = f(x) \ , \quad 0 < x < 1
\tag{1.2}
$$

$$
u(0) = u(1) = 0 \ .
$$

Numerous different approaches to this problem have been proposed in [1], [5], [6], and [12]. A review of the error terms involved in applying Stenger's method to (1.2) is discussed. These error terms, when balanced with regard to exponential order, define an optimal mesh spacing in the case that no boundary layer is present. In the present work, a parameter δ is introduced which, via the error terms, manages the error due to the boundary layer. This parameter is incorporated into the mesh spacing and thereby defines a mesh reallocation which is shown to more adequately resolve the boundary layer. It is also pointed out that as $\delta \to 0$ (i.e. as the boundary layer becomes negligible), this new method reduces to Stenger's method. Two numerical examples are given illustrating the use of this new method.

In Section 3 the attention is turned back to (1.1). The obligation of the temporal discretization is to maintain the accuracy obtained in the spatial discretization. A numerical example is given demonstrating a fully Sinc-Galerkin approach to a problem of the form (1.1). This approach has its roots in [11] and has been recently used in [17] and [18].

2 The ordinary differential equation

A Galerkin method for the approximation of (1.2) may be briefly summarized as follows. Select a set of basis functions

$$\{S_k(x)\}$$

and define the approximate solution by

$$u_m(x) = \sum_{k=-M}^{N} u_k S_k(x) \quad , \quad m = M + N + 1 \quad . \tag{2.1}$$

The coefficients $\{u_i\}$ in (2.1) are determined by orthogonalizing the residual with respect to the basis functions. This yields the discrete Galerkin system

$$(\mathcal{S}u_m - f, S_j) = 0 \quad , \quad -M \leq j \leq N \tag{2.2}$$

with the inner product on the interval $(0, 1)$ given by

$$(u, v) = \int_0^1 u(x)v(x)w(x)dx \tag{2.3}$$

where $w(x)$ is, for the moment, an unspecified weight function.

Introducing the Sinc-Galerkin method requires a few preliminary definitions.

Definition 2.1 *In the Sinc-Galerkin method, the basis functions are derived from the Whittaker cardinal (sinc) function*

$$\operatorname{sinc}(x) = \frac{\sin(\pi x)}{\pi x} \quad , \quad -\infty < x < \infty$$

and its translates

$$S(j, h)(x) = \operatorname{sinc}\left(\frac{x - jh}{h}\right) \quad , \quad h > 0 \quad .$$

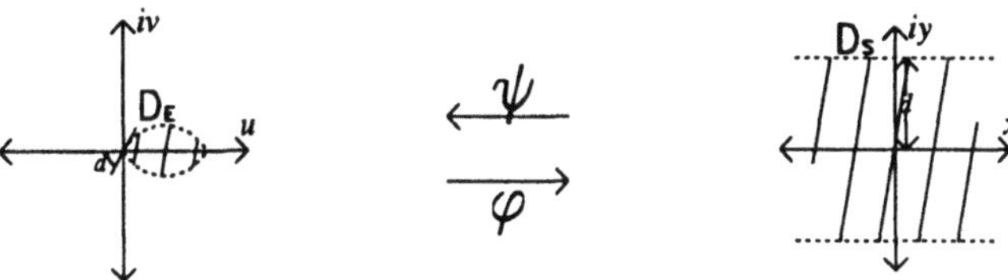

Figure 2.1: The domains D_E and D_S

In order to have the sinc translates in Definition 2.1 defined on the interval $(0, 1)$, consider the conformal map

$$\phi(z) = \ell n \left(\frac{z}{1-z} \right) \ . \tag{2.4}$$

This map carries the eye-shaped region

$$D_E = \left\{ z = u + iv : \left| \arg\left(\frac{z}{1-z} \right) \right| < d \leq \frac{\pi}{2} \right\}$$

onto the infinite strip

$$D_S = \left\{ w = x + iy : |y| < d \leq \frac{\pi}{2} \right\} \ .$$

These regions are depicted in Figure 2.1.

The basis functions used in (2.1) are then defined by

$$S_k(x) = S(k, h) \circ \phi(x). \tag{2.5}$$

Definition 2.2 *Let $B(D)$ denote the class of functions analytic in a simply connected domain D which satisfy for some constant a with $0 \leq a < 1$,*

$$\int_{\psi(x+L)} |F(w)dw| = \mathcal{O}(|x|^a), \quad x \to \pm\infty \tag{2.6}$$

where $L = \{iy : |y| < d\}$ and for γ a simple closed contour in D

$$N(F, D) \equiv \lim_{\gamma \to \partial D} \int_\gamma |F(w)dw| < \infty. \tag{2.7}$$

Substituting (2.1) into (2.2) leads, after integrating by parts the terms involving derivatives of the dependent variable, to the discrete linear system

$$\left\{ \frac{-1}{h^2} I^{(2)} D(\phi' w) - \frac{1}{h} I^{(1)} D\left(\frac{\phi'' w}{\phi'} + 2w' + pw \right) \right.$$

$$\left. - D\left(\frac{w'' + (pw)'}{\phi'} \right) \right\} \vec{u} = D\left(\frac{w}{\phi'} \right) \vec{f}. \tag{2.8}$$

Here $D(\gamma)$ is an $m \times m$ diagonal matrix with $\gamma(x)$ evaluated at the sinc nodes and the $m \times m$ matrices

$$I^{(\ell)} \equiv \left[\delta_{jk}^{(\ell)} \right], \quad \ell = 0, 1, 2$$

where

$$\delta_{jk}^{(0)} \equiv [S(j, h) \circ \phi(x)]\Big|_{x=x_k} = \left\{ \begin{array}{ll} 1, & j = k \\ 0, & j \neq k, \end{array} \right.$$

$$\delta_{jk}^{(1)} \equiv h \frac{d}{d\phi} [S(j, h) \circ \phi(x)]\Big|_{x=x_k} = \left\{ \begin{array}{ll} 0, & j = k \\ \frac{(-1)^{k-j}}{k-j}, & j \neq k, \end{array} \right.$$

and

$$\delta_{jk}^{(2)} \equiv h^2 \frac{d^2}{d\phi^2} [S(j, h) \circ \phi(x)]\Big|_{x=x_k} = \left\{ \begin{array}{ll} \frac{-\pi^2}{3}, & j = k \\ \frac{-2(-1)^{k-j}}{(k-j)^2}, & j \neq k, \end{array} \right.$$

with $x_k = \phi^{-1}(kh)$.

Implicit in the development of (2.8) is the assumption that the boundary terms arising from the integration by parts vanish. The selection

$$w(x) = \frac{1}{\phi'(x)} \tag{2.9}$$

guarantees that the boundary terms vanish and has the added feature of allowing one to handle problems of the form (1.2) when the differential equation has regular singular points. A complete discussion on the choice of weight functions is given in [9]. Solutions obtained from (2.8) then have the exponential convergence rate guaranteed by the following theorem.

Theorem 2.3 *Assume that the functions p and f in*

$$Su(x) \equiv -u''(x) + p(x)u'(x) = f(x), \quad 0 < x < 1$$
$$u(0) = u(1) = 0$$

and the unique solution u are analytic in the simply connected domain D_E. Let ϕ be the conformal one-to-one map of D_E onto D_S given in (2.4). Assume also that $f/\phi' \in B(D_E)$, and $uF \in B(D_E)$ where

$$F = \phi', \quad (p/\phi')', \quad p. \tag{2.10}$$

Suppose there are positive constants C_1, C_2, α, and β so that

$$|u(x)| \leq \begin{cases} C_1 x^\alpha, & x \in (0, 1/2) \\ C_2 (1-x)^\beta, & x \in [1/2, 1). \end{cases} \tag{2.11}$$

If the $\{u_k\}_{k=-M}^{N}$ in

$$u_m(x) = \sum_{k=-M}^{N} u_k S(k, h) \circ \phi(x)$$

are determined by solving (2.8) then

$$\begin{aligned}
\|u - u_m\|_\infty &\leq & C_\alpha M \exp(-\alpha M h) \\
&& + \; C_\beta N \exp(-\beta N h) \\
&& + \; N^2 C_I \exp(-\pi d/h)
\end{aligned} \tag{2.12}$$

where C_α, C_β, and C_I are independent of M, N, and h. Making the choices

$$M = \left[\left| \frac{\beta}{\alpha} \, N \right| \right], \quad h = \left(\frac{\pi d}{\beta N} \right)^{1/2} \tag{2.13}$$

yields

$$\|u - u_m\|_\infty \leq K N^2 \exp(-(\pi d \beta N)^{1/2}) \tag{2.14}$$

where K is independent of N and h.

∎

Notice that the selections in (2.13) balance the error from the three error terms on the right–hand side of (2.12) with regards to their exponential contribution. If all $C_\gamma (\gamma = \alpha, \beta, I)$ are of the same order of magnitude, the right–hand side of (2.14) is an accurate measure of the error between u and u_m. Briefly, the $C_\gamma (\gamma = \alpha, \beta)$ bound the behavior of the solution u in the neighborhood of zero and one, respectively. The C_I does not lend itself to as simple a verbal description, however the numerical evidence in the following two examples indicates that for these convection dominated problems, C_I is on the same order as C_α.

Example 2.1 For positive κ, consider the model problem

$$-u''(x) + \kappa u'(x) = \kappa, \quad 0 < x < 1$$

$$u(0) = u(1) = 0.$$

(2.15)

This problem exhibits a steep front near $x = 1$ for $\kappa \gg 1$. The true solution to this problem is given by

$$u(x) = x - \frac{\exp(\kappa x) - 1}{\exp(\kappa) - 1}.$$

(2.16)

A finite element approach for this problem is discussed in [4]. Figure 2.2 displays (2.16) for increasing values of κ. For $\kappa = 1000$ the solution graphically appears to be discontinuous and not much different from $\kappa = 100$ and is therefore not plotted.

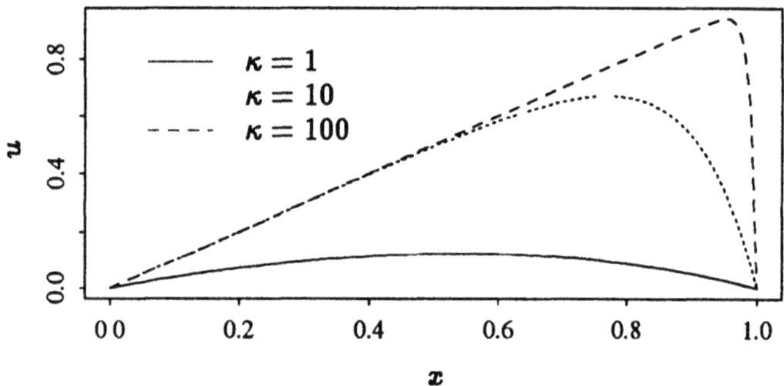

Figure 2.2: Solution of (2.15) for $\kappa = 1, 10, 100$

For this example, with the choice of $w(x) = 1/\phi'(x) = x(1 - x)$, the discrete linear system (2.8) reduces to

$$\left\{ \frac{-1}{h^2} I^{(2)} + \frac{1}{h} I^{(1)} D\left(2x - 1 - \kappa x(1 - x)\right) \right.$$

$$\left. + D\left(2x(1 - x) + \kappa x(1 - x)(2x - 1)\right) \right\} \vec{u}$$

$$= D\left(\kappa(x(1 - x))^2\right) \vec{1}.$$

(2.17)

An inspection of (2.16), or appealing to a local analysis of (2.15), shows that for x near 1

$$u(x) \approx \kappa(1 - x) \tag{2.18}$$

and for x near zero

$$u(x) \approx x. \tag{2.19}$$

This shows that $\alpha = \beta = 1$ are appropriate choices for exponents in (2.11). Designating h by h_s and balancing the exponential contributions to the error (i.e. equating the exponents $\beta N h_s$, $\alpha M h_s$, and $\pi d/h_s$ on the right–hand side of (2.12)) yields (as given in (2.13))

$$h_s = \left(\frac{\pi d}{\beta N} \right)^{1/2} = \frac{\pi}{\sqrt{2N}}, \tag{2.20}$$

when using $d = \pi/2$. Also from (2.13), choosing $N = M$ yields an exponential convergence rate of $\exp(-N h_s) = \exp(-\pi\sqrt{N/2})$. This exponential convergence rate is shown in Table 2.1. These choices of M and N are independent of κ, but conventional wisdom dictates that this could be improved. From (2.19) $C_\alpha \approx 1$ but from (2.18) $C_\beta \approx \kappa$ so that a more accurate error representation ensues if κ is factored from the C_β. Incorporating κ in the balancing of the error terms by writing $\kappa = \exp(\delta \ell n(10))$ and equating $-\alpha M h_s$ with $-\beta N h_s + \delta \ell n(10)$ leads to

$$M_1 = \left[\left| \frac{\beta}{\alpha} N - \frac{\delta \ell n(10)}{\alpha h_s} \right| \right] \leq M, \quad \delta \geq 0. \tag{2.21}$$

Using (2.21) instead of M as defined in (2.13) yields an exponential error contribution $\exp(-M_1 h_s)$ which is tabulated in Table 2.2. A comparison

N	$\exp(-\pi d/h_s) = \exp(-N h_s)$
16	$1.3834e - 04$
32	$3.4873e - 06$
64	$1.9139e - 08$

Table 2.1: Expected exponential errors using h_s as given in (2.20)

of Table 2.1 and Table 2.2 shows that $\exp(-M_1 h_s)$ is a more reliable error representation than $\exp(-N h_s)$.

The errors reported in Table 2.2 are those at the sinc grid points. The error is measured by

$$\|\vec{u}_t - \vec{u}_m\|_\infty \equiv \max_j |u(x_j) - u_j|,$$

and the quantity x_c is the critical point at which the solution of (2.15) attains its maximum value. The error reported in the third column corresponds to picking $M = N$, while the error in the fourth column corresponds to the M_1 defined by (2.21).

$\kappa = 10 \quad x_c = 0.769737$					
$M = N$	h_s	$\|\vec{u}_t - \vec{u}_m\|_\infty$	$\|\vec{u}_t - \vec{u}_m\|_\infty$	$\exp(-M_1 h_s)$	M_1
16	0.5554	$7.7190e-04$	$7.8656e-04$	$1.2756e-03$	12
32	0.3927	$2.2354e-05$	$2.2352e-05$	$2.4844e-05$	27
64	0.2777	$1.3551e-07$	$1.3876e-07$	$1.7647e-07$	56
$\kappa = 100 \quad x_c = 0.953948$					
$M = N$	h_s	$\|\vec{u}_t - \vec{u}_m\|_\infty$	$\|\vec{u}_t - \vec{u}_m\|_\infty$	$\exp(-M_1 h_s)$	M_1
16	0.5554	$8.4024e-03$	$8.4418e-03$	$1.1762e-02$	8
32	0.3927	$2.4592e-04$	$2.4592e-04$	$2.6213e-04$	21
64	0.2777	$1.4911e-06$	$1.4912e-06$	$1.6272e-06$	48
$\kappa = 1000 \quad x_c = 0.993092$					
$M = N$	h_s	$\|\vec{u}_t - \vec{u}_m\|_\infty$	$\|\vec{u}_t - \vec{u}_m\|_\infty$	$\exp(-M_1 h_s)$	M_1
16	0.5554	$7.7792e-02$	$8.1400e-02$	$1.0845e-01$	4
32	0.3927	$2.4751e-03$	$2.4769e-03$	$2.7656e-03$	15
64	0.2777	$1.5042e-05$	$1.5042e-05$	$1.5004e-05$	40

Table 2.2: Errors in the solution of (2.15) while using a smaller system size (corresponding to M_1).

Since $\exp(-Nh_s) = \exp(-\pi d/h_s)$ is smaller than $\exp(-M_1 h_s)$ one is led to investigate a rebalancing of all the error contributions. The error contributions arise from $\exp(-\alpha M h)$, $\exp(-\beta N h + \delta \ell n(10))$ and $\exp(-\pi d/h)$. Equating the exponents in the last two terms, one finds a different h, dependent upon δ and denoted by h_δ, where

$$h_\delta = \frac{\delta \ell n(10) + \sqrt{(\delta \ell n(10))^2 + 4\pi d \beta N}}{2\beta N} \geq h_s, \quad \delta \geq 0. \tag{2.22}$$

By the same procedure leading to (2.21) M_δ is defined by

$$M_\delta = \left[\left|\frac{\beta}{\alpha}N - \frac{\delta \ell n(10)}{\alpha h_\delta}\right|\right], \quad \delta \geq 0. \tag{2.23}$$

The third column of Table 2.3 represents the error using the "standard" M and h_s while the fourth column is the error using (2.23) coupled with (2.22). Note that the selection h_δ has placed more sinc nodes $\frac{\exp(jh_\delta)}{\exp(jh_\delta)+1}$ near the boundary layer at $x = 1$. From Figure 2.2 this is geometrically the correct

$\kappa = 10$ $x_c = 0.769737$					
$M = N$	h_s	$\|\vec{u}_t - \vec{u}_m\|_\infty$	$\|\vec{u}_t - \vec{u}_m\|_\infty$	h_δ	M_δ
16	0.5554	$7.7190e - 04$	$2.1495e - 04$	0.6320	13
32	0.3927	$2.2354e - 05$	$6.5156e - 06$	0.4303	27
64	0.2777	$1.3551e - 07$	$4.0620e - 08$	0.2963	57
$\kappa = 100$ $x_c = 0.953948$					
$M = N$	h_s	$\|\vec{u}_t - \vec{u}_m\|_\infty$	$\|\vec{u}_t - \vec{u}_m\|_\infty$	h_δ	M_δ
16	0.5554	$8.4024e - 03$	$1.0616e - 03$	0.7176	10
32	0.3927	$2.4592e - 04$	$1.9415e - 05$	0.4712	23
64	0.2777	$1.4911e - 06$	$1.2532e - 07$	0.3160	50
$\kappa = 1000$ $x_c = 0.993092$					
$M = N$	h_s	$\|\vec{u}_t - \vec{u}_m\|_\infty$	$\|\vec{u}_t - \vec{u}_m\|_\infty$	h_δ	M_δ
16	0.5554	$7.7792e - 02$	$6.5190e - 02$	0.8117	8
32	0.3927	$2.4751e - 03$	$1.1853e - 04$	0.5152	19
64	0.2777	$1.5042e - 05$	$3.7434e - 07$	0.3368	44

Table 2.3: Improved results for (2.15) using M_δ in (2.23) and h_δ in (2.22)

thing to do. The nodes for h_δ are nearer both 0 and 1 because $h_\delta \geq h_s$. Thus the solution at the nodes more accurately reflects the true solution as shown in Figure 2.3. A by–product of this analysis, as evidenced from the errors in Table 2.3 is the increased accuracy obtained in the calculation of the approximate solution. This is especially appealing since the only change in the system (2.17) is the location of the points at which the diagonal matrices are evaluated – the form of the system is not changed!

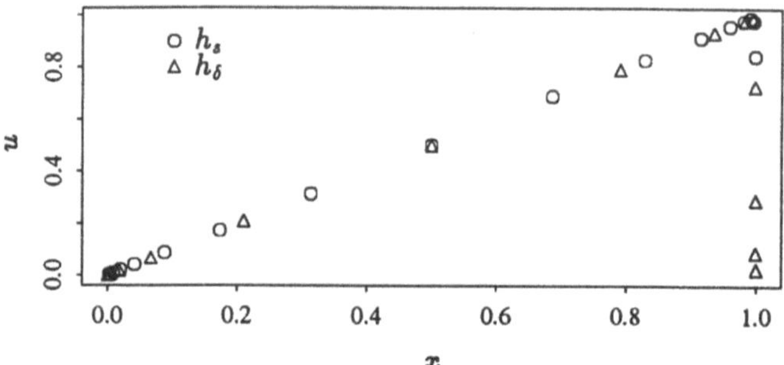

Figure 2.3: Effect of new node placement for $N = 8$ and $\kappa = 1000$. The nodes are plotted along the true solution to more graphically illustrate the repositioning.

The method outlined for the previous example performs equally well in the case of non–constant $p(x)$. This is illustrated via the following example.

Example 2.2 Consider the differential equation (for $\kappa > 0$) defined by

$$-u''(x) + \frac{\kappa}{x}u'(x) = \kappa(\kappa + 1)x^{\kappa - 1} \quad 0 < x < 1 \tag{2.24}$$

$$u(0) = u(1) = 0.$$

This problem has the dual difficulty represented by a regular singular point at $x = 0$ and a boundary layer $x = 1$ ($\kappa \gg 0$). The true solution of (2.24) is given by

$$u(x) = -\kappa x^{\kappa + 1}\ln(x) \tag{2.25}$$

so that the intensity of the singularity at $x = 0$ is amplified for small κ and the boundary layer at $x = 1$ is difficult to resolve for large κ. The case of small κ is treated in [9]. The solution (2.25) is illustrated in Figure 2.4 for increasing values of κ. As in the last example the linear system (2.17), which for the present example takes the form,

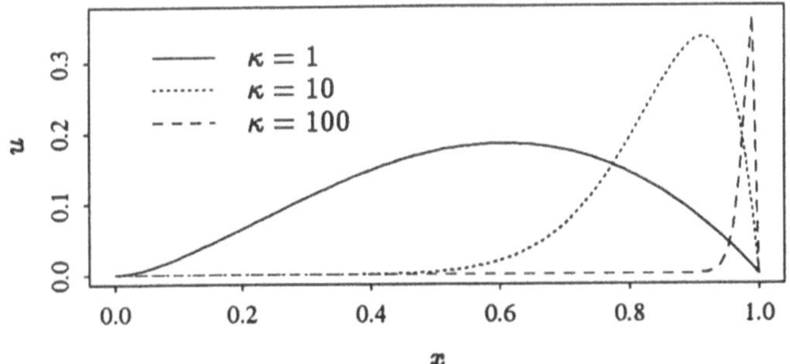

Figure 2.4: Solution of (2.24) for $\kappa = 1, 10, 100$

$$\left\{\frac{-1}{h^2}I^{(2)} \;+\; \frac{1}{h}I^{(1)}D\left(2x - 1 - \kappa(1-x)\right)\right.$$

$$\left. +\quad D\big(x(1-x)(2+\kappa)\big)\right\}\vec{u} \qquad (2.26)$$

$$=\quad D\left(\kappa(\kappa+1)x^{\kappa-1}(x(1-x))^2\right)\vec{1},$$

is as easily constructed for h_s as it is for h_δ. Indeed, $h_s = h_\delta$ for $\kappa = 1(\delta = 0)$, and these results are displayed in Table 2.4.

	$\kappa = 1$	
$M = N$	h_s	$\|\vec{u}_t - \vec{u}_m\|_\infty$
16	0.5554	$8.5831e - 05$
32	0.3927	$2.4838e - 06$
64	0.2777	$1.5055e - 08$

Table 2.4: The case $\kappa = 1$ whereby $M = M_\delta$, $h_s = h_\delta$

The solution exhibits the asymptotic behavior described by (2.18), and therefore, using the values of M_δ and h_δ as dictated by (2.23) and (2.22), respectively, is appropriate. Table 2.5 is in the same format as Table 2.2 and the results are quite similar to those in Example 2.1. Whether using

$\kappa = 10$					
$M = N$	h_s	$\|\vec{u}_t - \vec{u}_m\|_\infty$	$\|\vec{u}_t - \vec{u}_m\|_\infty$	h_δ	M_δ
16	0.5554	$9.0663e-04$	$4.4623e-04$	0.6320	13
32	0.3927	$2.4801e-05$	$7.5488e-06$	0.4303	27
64	0.2777	$1.5045e-07$	$4.4898e-08$	0.2963	57
$\kappa = 100$					
$M = N$	h_s	$\|\vec{u}_t - \vec{u}_m\|_\infty$	$\|\vec{u}_t - \vec{u}_m\|_\infty$	h_δ	M_δ
16	0.5554	$8.7344e-03$	$1.3946e-03$	0.7176	10
32	0.3927	$2.5415e-04$	$4.1999e-05$	0.4712	23
64	0.2777	$1.5029e-06$	$1.8509e-07$	0.3160	50
$\kappa = 1000$					
$M = N$	h_s	$\|\vec{u}_t - \vec{u}_m\|_\infty$	$\|\vec{u}_t - \vec{u}_m\|_\infty$	h_δ	M_δ
16	0.5554	$7.4852e-02$	$2.6979e-03$	0.8117	8
32	0.3927	$2.4845e-03$	$1.1747e-04$	0.5152	19
64	0.2777	$1.5016e-05$	$1.2987e-06$	0.3368	44

Table 2.5: Results for solutions of (2.24) via (2.26)

h_s in (2.20) or h_δ defined by (2.22), the selection of M_1 given by (2.21) or M_δ give by (2.23) reduces the system size. As in Example 2.1, h_δ has the effect of distributing the nodes near the boundary layer.

Based on these examples, the authors recommend the selection h_δ in (2.22) for all $\delta \geq 0$. In the case $\delta = 0$, (2.22) reduces to Stenger's original choice (2.13) which works very well for problems with no boundary layer.

3   The partial differential equation

Having satisfactorily resolved the boundary layer in the ordinary differential equation, the goal now is to carry the calculations forward in time. The numerical solution of the transient convection–diffusion equation has been reviewed in [7] and specific techniques presented in, for example, [2], [3], [13], [14], [16], [17], and [18]. A brief review of fully Sinc-Galerkin methods for partial differential equations is given here since one can find a thorough discussion in [9].

This method for problems of the form

$$
\begin{aligned}
\mathcal{L}u(x,t) &= f(x,t) \ , \quad 0 < x < 1, \quad t > 0 \\
\mathcal{L}u(x,t) &\equiv u_t(x,t) - u_{xx}(x,t) + p(x,t)u_x(x,t) \\
u(0,t) &= 0, \quad t \geq 0 \\
u(1,t) &= 0, \quad t \geq 0 \\
u(x,0) &= 0, \quad 0 \leq x \leq 1
\end{aligned}
\tag{3.1}
$$

is outlined. A fully Sinc-Galerkin approach to (3.1) can be summarized as follows. Choose an approximation of the form

$$
\begin{aligned}
u_{mn}(x,t) &= \sum_{q=-M_t}^{N_t} \sum_{p=-M}^{N} u_{pq} S_p(x) S_q^*(t) \\
m &= M + N + 1, n = M_t + N_t + 1
\end{aligned}
$$

where the sinc basis function in the temporal domain is

$$
S_q^*(t) \equiv S(q,h) \circ \Upsilon(t) = \operatorname{sinc}\left(\frac{\Upsilon(t) - qh}{h}\right)
$$

and $(0, \infty)$ is mapped to the real line by the conformal map

$$
\Upsilon(z) = \ell n(z)
$$

which carries the wedge shaped region

$$
D_W = \{z = re^{i\theta}: \ |\theta| < d \leq \pi/2\}
$$

onto the infinite strip D_S. This conformal mapping is depicted in Figure 3.1. The basis functions $S_p(x)$ are those given in (2.5). Orthogonalize the residual to determine the coefficients u_{pq},

$$
\begin{aligned}
(\mathcal{L}u_{mn} - f, S_i S_j^*) &= 0 \ , \\
i &= -M, -M+1, \ldots, N \ , \\
j &= -M_t, -M_t+1, \ldots, N_t.
\end{aligned}
\tag{3.2}
$$

The inner product is

$$
(f,g) = \int_0^\infty \int_0^1 f(x,t)g(x,t)\frac{1}{\phi'(x)}\sqrt{\Upsilon'(t)}\, dx\, dt.
$$

Again $1/\phi'(x)$ is used as a weight in the spatial domain and now $\sqrt{\Upsilon'(t)}$ is the weight in the temporal domain. A discussion of the reason for this

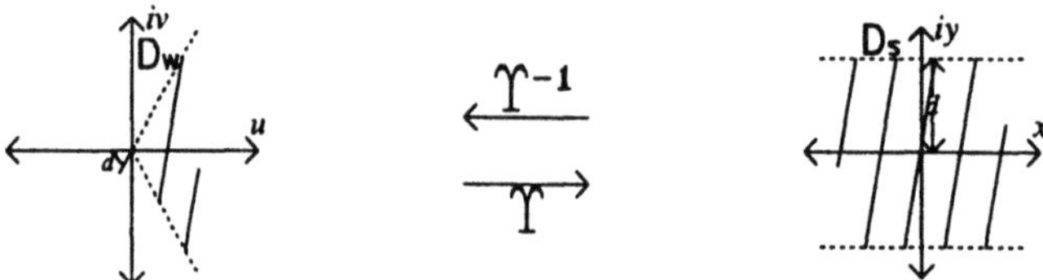

Figure 3.1: The domains D_W and D_S

choice of weight function in the temporal domain can be found in [8] and [9]. Evaluating (3.2) yields, via a rather lengthy but straightforward calculation, the following matrix system:

$$AUD\left(\frac{1}{\sqrt{\Upsilon'}}\right) + D\left(\frac{1}{(\phi')^2}\right)UB^T = \mathcal{F}. \tag{3.3}$$

Here the $m \times m$ matrix A is given by

$$A = \left\{\frac{-1}{h^2}\,I^{(2)} + \frac{1}{h}\,I^{(1)}D\left(\frac{\phi''}{(\phi')^2} - \frac{p}{\phi'}\right) - D\left(\frac{(1/\phi')'' + (p/\phi')'}{\phi'}\right)\right\} \tag{3.4}$$

while the $n \times n$ matrix B is

$$B = \left[\frac{-1}{h}I^{(1)} + D\left(\frac{1}{2}\right)\right]D(\sqrt{\Upsilon'}) \tag{3.5}$$

and the $m \times n$ matrix $\mathcal{F}$ is

$$\mathcal{F} = D\left(\frac{1}{(\phi')^2}\right)FD\left(\frac{1}{\sqrt{\Upsilon'}}\right). \tag{3.6}$$

The jk-th entry of the $m \times n$ matrix F, for $j = -M,\ldots,N$ and $k = -M_t,\ldots,N_t$ contains the point evaluations of the function $f(x,t)$ and U contains the coefficients u_{jk}.

Example 3.1 Consider

$$
\begin{aligned}
u_t(x,t) &= u_{xx}(x,t) - \kappa u_x(x,t) + f(x,t), \quad 0 < x < 1, \quad t > 0 \\
u(0,t) &= 0, \quad t > 0 \\
u(1,t) &= 0, \quad t > 0 \\
u(x,0) &= 0, \quad 0 < x < 1
\end{aligned}
\tag{3.7}
$$

where $f(x,t)$ is such that the true solution is

$$
u(x,t) = t\exp(-t+1)\left(x - \frac{\exp(\kappa x) - 1}{\exp(\kappa) - 1} \right).
$$

This solution is the product of the temporal function $(t\exp(-t+1))$ with the solution to the problem given in Example 2.1. The true solution of (3.7) for $\kappa = 100$ is plotted out to the time level $t = 5$ in Figure 3.2 and Figure 3.3. The two different perspectives illustrate the boundary layer at $x = 1$.

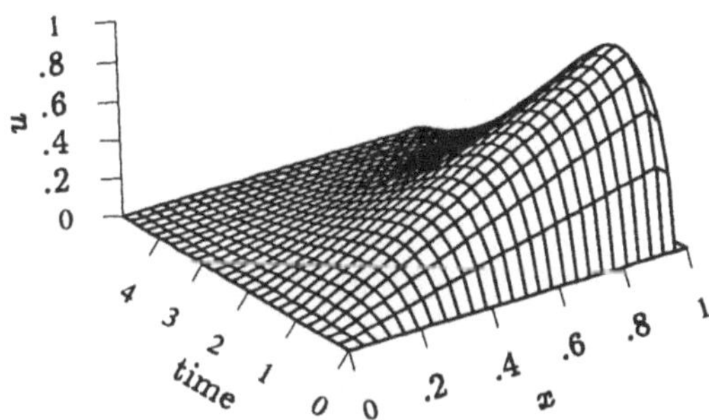

Figure 3.2: True solution of (3.7) for $\kappa = 100$

The purpose of this example is to show that the spatial accuracy achieved in Example 2.1 can be carried over to the temporal domain. To do this, one must choose a sufficient number of basis functions in the temporal domain to maintain the order of accuracy obtained in Example 2.1. Since the solution decays exponentially in time, the method in [9] shows that with $M_t = 24$ and $N_t = 6$ the error in the temporal domain is approximately 10^{-5}.

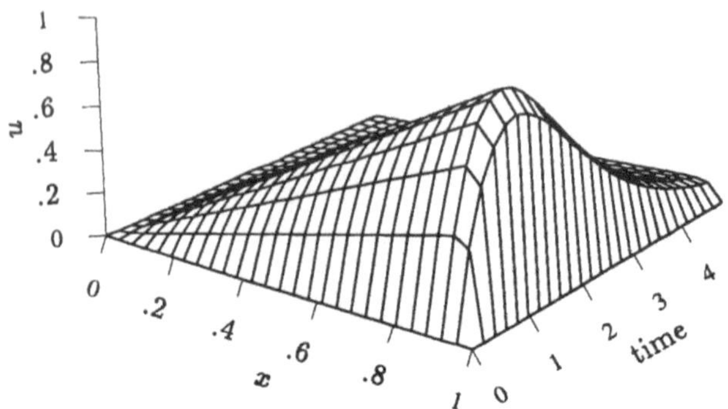

Figure 3.3: True solution of (3.7) for $\kappa = 100$ from a different perspective illustrating the boundary layer

Results for (3.7) are displayed in Table 3.1. The first three columns represent the "standard" choice of M and h as defined by (2.13) with the error being reported in the third column. The remainder of the table represents the choices (2.23) and (2.22) with the error reported in the fourth column. The quantity SS is the size of the matrix U in (3.3). The error is measured by

$$\|u_t - u_m\| \equiv \max_{i,j} |u(x_i, t_j) - u_{ij}| \quad .$$

Upon comparing the results in Table 3.1 with those in Table 2.3, one sees that the accuracy, whether using h_s or h_δ, obtained in the spatial domain is preserved in the temporal calculation.

$\kappa = 10$					
$M = N$	SS	$\| u_t - u_m \|$	$\| u_t - u_m \|$	SS	M_δ
16	31×33	$7.7191e - 04$	$2.1495e - 04$	31×30	13
32	31×65	$2.4743e - 05$	$2.4706e - 05$	31×60	27
$\kappa = 100$					
$M = N$	SS	$\| u_t - u_m \|$	$\| u_t - u_m \|$	SS	M_δ
16	31×33	$8.4024e - 03$	$1.0615e - 03$	31×27	10
32	31×65	$2.4592e - 04$	$3.4972e - 05$	31×56	23
$\kappa = 1000$					
$M = N$	SS	$\| u_t - u_m \|$	$\| u_t - u_m \|$	SS	M_δ
16	31×33	$7.7791e - 02$	$6.5189e - 02$	31×25	8
32	31×65	$2.4750e - 03$	$1.1853e - 04$	31×52	19

Table 3.1: Errors in the solution of (3.7).

References

[1] G. F. Carey and A. Pardhanani. Multigrid solution and grid redistribution for convection-diffusion. *Internat. J. Numer. Methods Engrg.*, 27:655–664, 1989.

[2] T. F. Chan. Stability analysis of finite difference schemes for the advection-diffusion equation. *SIAM J. Numer. Anal.*, 21:272–284, 1984.

[3] P. A. B. DeSampaio. A Petrov-Galerkin/modified operator formulation for convection-diffusion problems. *Internat. J. Numer. Methods Engrg.*, 30:331–347, 1990.

[4] C. A. J. Fletcher. *Computational Galerkin Methods*. Springer-Verlag, New York, 1984.

[5] J. Freund and E.-M. Salonen. A logic for simple Petrov-Galerkin weighting functions. *Internat. J. Numer. Methods Engrg.*, 34:805–822, 1992.

[6] C. I. Günther. Conservative versions of the locally exact consistent upwind scheme of second order (LECUSSO-scheme). *Internat. J. Numer. Methods Engrg.*, 34:793–804, 1992.

[7] J. C. Heinrich and O. C. Zienkiewicz. The finite element method and "upwinding" techniques in the numerical solution of convection dominated flow problems. In T. J. R. Hughes, editor, *Finite Elements for Convection Dominated Flow*, volume 34, pages 105–136. ASME Publ. AMD, 1979.

[8] D. L. Lewis, J. Lund, and K. L. Bowers. The space-time Sinc-Galerkin method for parabolic problems. *Internat. J. Numer. Methods Engrg.*, 24(9):1629–1644, 1987.

[9] J. Lund and K. L. Bowers. *Sinc Methods for Quadrature and Differential Equations*. SIAM, Philadelphia, 1992.

[10] J. Lund, K. L. Bowers, and T. S. Carlson. Fully Sinc-Galerkin computation for boundary feedback stabilization. *J. Math. Systems, Estimation, and Control*, 1(2):165–182, 1991.

[11] J. T. Oden. A general theory of finite elements, II. Applications. *Internat. J. Numer. Methods Engrg.*, 1:247–259, 1969.

[12] E. Pardo and D. C. Weckman. A fixed grid finite element technique for modelling phase change in steady-state conduction-advection problems. *Internat. J. Numer. Methods Engrg.*, 29:969–984, 1990.

[13] A. Rigal. Numerical analysis of two-level finite difference schemes for unsteady diffusion-convection problems. *Internat. J. Numer. Methods Engrg.*, 28:1001–1021, 1989.

[14] A. Rigal. Numerical analysis of three-time-level finite difference schemes for unsteady diffusion-convection problems. *Internat. J. Numer. Methods Engrg.*, 30:307–330, 1990.

[15] F. Stenger. A Sinc-Galerkin method of solution of boundary value problems. *Math. Comp.*, 33:85–109, 1979.

[16] J. J. Westerink and D. Shea. Consistent higher degree Petrov-Galerkin methods for the solution of the transient convection-diffusion equation. *Internat. J. Numer. Methods Engrg.*, 28:1077–1101, 1989.

[17] C.-C. Yu and J. C. Heinrich. Petrov-Galerkin methods for the time-dependent convective transport equation. *Internat. J. Numer. Methods Engrg.*, 23:883–901, 1986.

[18] J. R. Yu and T. R. Hsu. On the solution of diffusion-convection equations by the space-time finite element method. *Internat. J. Numer. Methods Engrg.*, 23:737–750, 1986.

DISCRETE OBSERVABILITY OF THE WAVE EQUATION ON BOUNDED DOMAINS IN EUCLIDEAN SPACE

Alisa DeStefano

Department of Mathematics *
College of the Holy Cross
Worcester, Massachusetts 01610

1 Introduction

The study of observability is the study of deducing information about the state of a dynamical system from incomplete measurements. Discrete observability asks if we can recover the initial data with only a discrete set of measurements. Discrete observability of the heat equation has been studied for bounded domains in Euclidean space and compact homogeneous spaces by Gilliam, Li and Martin [4], and Wallace and Wolf [6]. This paper examines discrete observability of the wave equation on bounded domains in Euclidean space. The solutions to the wave equation involve an additional initial condition, and the terms of the series solution no longer decay with time, so a new approach is needed.

This paper is organized as follows. In the second section, we give the background and notation for the wave equation. We then discuss our sampling scheme, i.e. our discrete observations. Here we consider Type II sampling. This consists of sampling the displacement at an infinite but discrete set of spatial points for two times. See [3] for another type of sampling. Finally, we give a precise definition of discrete observability for this setting. In the third section, we show that the general problem of discrete observability (determining under what conditions these samples uniquely determine the solution of the wave equation) depends only on the spectrum of the elliptic operator in the wave equation. The main result is that the wave equation is discretely observable if a certain condition involving the times and eigenvalues is satisfied. We also show that there exists a sample set for which the system is discretely observable. The fourth and final section demonstrates the results with an example of a system which is discretely observable for a large class of sample points and times.

2 The Wave Equation: Background and Notation

In order to study the wave equation on a bounded domain in Euclidean space and the observability properties of its solutions, we must first establish

---

*This paper is based on a chapter in the author's Ph.D. thesis, which was completed at Dartmouth College.

some notation Let Ω be a bounded domain in $\mathbb{R}^n$ satisfying fairly minimal smoothness conditions on $\partial\Omega$, the boundary of Ω (see [4]) Consider a linear hyperbolic partial differential equation

$$\frac{\partial^2}{\partial t^2} f(x\ \ t) = \Delta f(x\ \ t) - q(x) f(x\ \ t) \tag{2 1}$$

on $\Omega \times \mathbb{R}$, where $\Delta = \sum_{i=1}^{n} \frac{\partial^2}{\partial x_i^2}$ is the Laplacian, $t \geq 0$, and we assume that $q(x) \geq 0$ is Holder continuous on the compact domain $\overline{\Omega} = \Omega \bigcup \partial\Omega$ The function $f(x\ \ t)$ represents the displacement at time t from Ω for the given initial conditions We are interested in the solutions that for each $t > 0$ have finite energy in the sense that

$$\int_\Omega |f(x\ \ t)|^2 \, dx < \infty$$

That is, $f(x\ \ t) \in L^2(\Omega)$, the Hilbert space with inner product given by

$$\langle f, g \rangle = \int_\Omega f(x) g(x) \, dx$$

We will consider boundary conditions for which $A(x, \Delta) = \Delta - q(x)$ is a self-adjoint operator In particular, we consider classical boundary conditions

$$\alpha(x) f(x\ \ t) + (1 - \alpha(x)) f_\nu(x\ \ t) = 0 \tag{2 2}$$

for $x \in \partial\Omega$, $\alpha(x) \in C^2$ such that $0 \leq \alpha(x) \leq 1$, and f_ν is the normal derivative to Ω The initial conditions are the initial displacement,

$$\lim_{t \to 0} f(x\ \ t) = a(x) \in L^2(\Omega) \tag{2 3}$$

and the initial velocity,

$$\lim_{t \to 0} \frac{\partial f}{\partial t}(x\ \ t) = b(x) \in L^2(\Omega) \tag{2 4}$$

We will consider classical solutions to equation 2 1 with the given boundary and initial conditions The existence of such solutions can be proven if we make sufficient assumptions on the bounded domain and the initial conditions (see [2]) With the above boundary conditions, the operator $A(x, \Delta)$ is an unbounded, self-adjoint operator with dense domain $\mathcal{D} \equiv \mathcal{D}(A(x, \Delta)) \subset L^2(\Omega)$ The spectrum of $A(x, \Delta)$ consists of a sequence of distinct eigenvalues $\{\nu_j\}$ satisfying

$$0 < \nu_1 < \nu_2 <$$

and $\nu_j \to \infty$ as $j \to \infty$. We will consider the case where each eigenvalue ν_j has finite multiplicity m_j. Associated with each eigenvalue ν_j there is an orthonormal set of eigenfunctions $\{\phi_{jk}\}_{k=1}^{m_j}$ in $L^2(\Omega)$, such that

$$A(x, \Delta)\phi = -\nu_j \phi.$$

Now consider the sequence of eigenvalues counted with multiplicity. That is,

$$\lambda_1 \leq \lambda_2 \leq \cdots$$

and

$$\lim_{n \to \infty} \lambda_n = \infty$$

where each eigenvalue has finite multiplicity and in the listing is repeated the same number of times as its multiplicity. We introduce the sum $n_r = m_1 + m_2 + \cdots + m_r$. Hence the first n_r eigenvalues correspond to the first r distinct eigenvalues repeated with multiplicity. We also reindex the eigenfunctions to correspond to this ordering of eigenvalues.

From the general theory, any function in $L^2(\Omega)$ can be expanded in a Fourier series using the orthonormal eigenfunctions $\{\phi_j\}$ described above. Hence there is a filtration of $L^2(\Omega)$ given by

$$L^2(\Omega) = \bigcup_{r>0} E_r(\Omega)$$

where $E_r(\Omega)$ is the union of the first r eigenspaces and $\dim E_r = n_r$. That is, for a function $g(x) \in E_r(\Omega)$, we have

$$g(x) = \sum_{j=1}^{n_r} g_j \phi_j(x).$$

With this notation, the solution to the wave equation with initial displacement $a(x) = \sum_{j=1}^{\infty} a_j \phi_j(x)$ and initial velocity $b(x) = \sum_{j=1}^{\infty} b_j \phi_j(x)$ becomes

$$f(x : t) = \sum_{j=1}^{\infty} \left(a_j \cos(\mu_j t) + \frac{b_j}{\mu_j} \sin(\mu_j t) \right) \phi_j(x) \qquad (2.5)$$

where $\mu_j = \sqrt{\lambda_j}$.

Observability is the study of which samples of $f(x : t)$ uniquely determine the function $f(x : t)$. More generally, one can consider samples of f taken via a scalar function. We consider the case of discrete observation, in which one asks the question: Does sampling preserve observability? That is, if we have some discrete set of measurements of a system, do these uniquely determine the solution of the system? In order to answer this question, we must describe our sampling scheme.

Type II Sampling: We consider a discrete infinite set of spatial points $\{x_k\}_{k=1}^\infty$ and sample the displacement $f(x : t)$ at two times, t_0 and t_1, obtaining the measurements

$$d_k(t_0) = f(x_k : t_0) \tag{2.6}$$
$$d_k(t_1) = f(x_k : t_1) \tag{2.7}$$

for all k.

This sampling scheme is based on that of Wallace and Wolf [6] for the heat equation on a compact homogeneous space. They sample the solution (temperature) $u(x : t)$ at a discrete infinite set of spatial points $\{x_k\}_{k=1}^\infty$ and at any time $t \geq 0$. In our situation, we need additional samples due to the second initial condition.

To determine the conditions for which the wave equation is discretely observable, we first need a precise definition of discrete observability for Type II sampling.

Definition 2.1 *The wave equation*

$$\frac{\partial^2}{\partial t^2} f(x : t) = \Delta f(x : t) - q(x) f(x : t)$$

with initial conditions

$$\lim_{t \to 0} f(x : t) = a(x)$$
$$\lim_{t \to 0} f_t(x : 0) = b(x)$$

is discretely observable for times t_0, t_1 and spatial points $\{x_k\}_{k=1}^\infty$ *if the following holds: if f is the solution to the wave equation with the given initial data, and if for a fixed t, $f_r(\cdot : t)$ denotes the orthogonal projection of $f(\cdot : t)$ to $E_r(\Omega)$, then the $f_r(x_k : t)$, $1 \leq k \leq n_r$, $t = t_0, t_1$, determine f_r.*

In other words, there is a sequence of locations $\{x_k\}_{k=1}^\infty$ such that for every r, the solution matching the observed values

$$\{f_r(x_1 : t_0), f_r(x_2 : t_0), \ldots, f_r(x_{n_r} : t_0)\}$$

and

$$\{f_r(x_1 : t_1), f_r(x_2 : t_1), \ldots, f_r(x_{n_r} : t_1)\}$$

is unique.

Remark: The solution to the wave equation is uniquely defined by the coefficients $\{a_j\}, \{b_j\}$; to decide if the samples uniquely determine the function $f(x : t)$, we need only show that the samples uniquely determine the coefficients $\{a_j\}, \{b_j\}$. Henceforth, we will refer to the coefficients $\{a_j\}, \{b_j\}$ as the **solution coefficients**.

3 Discrete Observability

Now that we have a precise definition of discrete observability, we would like to see under what conditions the wave equation is discretely observable for Type II sampling. First, we need the following lemma.

Lemma 3.1 *Given the eigenfunctions $\{\phi_j\}_{j=1}^{\infty}$ which are an orthonormal basis of $L^2(\Omega)$, then there exists a sequence of spatial points $\{x_k\}_{k=1}^{\infty}$ such that for every integer n the following matrix is invertible:*

$$Q_n = \begin{pmatrix} \phi_1(x_1) & \phi_2(x_1) & \cdots & \phi_n(x_1) \\ \phi_1(x_2) & \phi_2(x_2) & \cdots & \phi_n(x_2) \\ \vdots & \vdots & \ddots & \vdots \\ \phi_1(x_n) & \phi_2(x_n) & \cdots & \phi_n(x_n) \end{pmatrix}$$

Proof: We will prove the lemma by induction. First, since ϕ_1 is an eigenfunction of the operator $A(x, \Delta)$, it is not identically zero. Hence, there exists a point $x_1 \in \Omega$ such that $\phi_1(x_1) \neq 0$. Now we assume that there exist points $\{x_k\}_{k=1}^{n-1}$ such that Q_{n-1} is invertible and show that there exists a point $x_n \in \Omega$ such that for the sequence of spatial points $\{x_k\}_{k=1}^{n}$, Q_n is invertible.

For notational purposes, we divide the Q_n matrix as follows:

$$Q_n = \left(\begin{array}{cccc|c} \phi_1(x_1) & \phi_2(x_1) & \cdots & \phi_{n-1}(x_1) & \phi_n(x_1) \\ \phi_1(x_2) & \phi_2(x_2) & \cdots & \phi_{n-1}(x_2) & \phi_n(x_2) \\ \vdots & \vdots & \ddots & \vdots & \vdots \\ \phi_1(x_{n-1}) & \phi_2(x_{n-1}) & \cdots & \phi_{n-1}(x_{n-1}) & \phi_n(x_{n-1}) \\ \hline \phi_1(x_n) & \phi_2(x_n) & \cdots & \phi_{n-1}(x_n) & \phi_n(x_n) \end{array} \right)$$

$$= \left(\begin{array}{c|c} Q_{n-1} & R \\ \hline V(x_n) & \phi_n(x_n) \end{array} \right)$$

Now by the induction hypothesis there exist $\{x_k\}_{k=1}^{n-1}$ for which Q_{n-1} is invertible. We need to show that there exists a point x_n such that the columns of Q_n are linearly independent. Since the columns of Q_{n-1} are linearly independent, we know that the first $n-1$ columns of Q_n are also linearly independent. We need now only show that the last column is independent of the first $n-1$.

The induction hypothesis implies that there is only one linear combination of the columns of Q_{n-1} that equals R. Now if we can find a point x_n such that the same linear combination of elements of $V(x_n)$ does not equal $\phi_n(x_n)$, then no linear combination of the first $n-1$ columns of Q_n equals the nth column, and Q_n is invertible.

$$A. \text{ DESTEFANO}$$

We have that there exists a unique vector $c \in \mathbb{R}^{n-1}$ where

$$c = \begin{pmatrix} c_1 \\ c_2 \\ \vdots \\ c_{n-1} \end{pmatrix}$$

such that $R = Q_{n-1} \cdot c$. Since $\{\phi_j\}_{j=1}^{\infty}$ are a basis in $L^2(\Omega)$, they are linearly independent as functions. Hence no linear combination of the $\{\phi_i(x)\}_{i=1}^{n-1}$ equals $\phi_n(x)$. In particular,

$$\phi_n(x) \neq \sum_{i=1}^{n-1} c_i \phi_i(x) = V(x) \cdot c$$

where $V(x) = (\phi_1(x), \phi_2(x), \cdots, \phi_{n-2}(x), \phi_{n-1}(x))$. Thus, there exists a point $x_n \in \Omega$ such that

$$\phi_n(x_n) \neq V(x_n) \cdot c$$

and so

$$\begin{pmatrix} Q_{n-1} \\ V(x_n) \end{pmatrix} \cdot c \neq \begin{pmatrix} R \\ \phi_n(x_n) \end{pmatrix}$$

This says that the columns of Q_n are linearly independent. Therefore, Q_n is invertible for the sequence $\{x_k\}_{k=1}^{n}$. $\qquad\square$

Now we will show that the wave equation is observable for Type II sampling when a certain condition involving the sample times and eigenvalues holds.

Theorem 3.2 *The solution $f(x : t)$ of the wave equation is discretely observable at times $t_0, t_1 \geq 0$ if the following condition holds:*

$$\text{For every } r, \ \sin(\mu_j(t_1 - t_0)) \neq 0 \text{ for every } j = 1, \ldots, r. \qquad (3.8)$$

If this is the case, then there exist spatial points $\{x_k\}_{k=1}^{\infty}$ such that each partial sum

$$f_r(x : t) = \sum_{1 \leq j \leq n_r} a_j \cos(\mu_j t)\phi_j(x) + \frac{b_j}{\mu_j} \sin(\mu_j t)\phi_j(x)$$

is determined by the "observations" $f_r(x_k : t)$, $1 \leq k \leq n_r$ and $t = t_0, t_1$.

Proof: Given any set of spatial points $\{x_k\}_{k=1}^{\infty}$, we sample at the first n_r points to get the following two matrix equations corresponding to the two sample times t_0, t_1:

$$D_r(t_0) = Q_r \cdot V_r(t_0)$$
$$D_r(t_1) = Q_r \cdot V_r(t_1)$$

where

$$D_r(t) = \begin{pmatrix} f_r(x_1 : t) \\ f_r(x_2 : t) \\ \vdots \\ f_r(x_{n_r} : t) \end{pmatrix}$$

$$Q_r = \begin{pmatrix} \phi_1(x_1) & \phi_2(x_1) & \cdots & \phi_{n_r}(x_1) \\ \phi_1(x_2) & \phi_2(x_2) & \cdots & \phi_{n_r}(x_2) \\ \vdots & \vdots & \ddots & \vdots \\ \phi_1(x_{n_r}) & \phi_2(x_{n_r}) & \cdots & \phi_{n_r}(x_{n_r}) \end{pmatrix}$$

and

$$V_r(t) = \begin{pmatrix} \alpha_1(t) \\ \alpha_2(t) \\ \vdots \\ \alpha_{n_r}(t) \end{pmatrix}$$

for

$$\alpha_j(t) = a_j \cos(\mu_j t) + \frac{b_j}{\mu_j} \sin(\mu_j t).$$

There are two steps in determining under what conditions the system is observable. First, we need that there exist spatial points for which Q_r is invertible (for every r), so that if $D_r(t_0)$ and $D_r(t_1)$ are known for every r, then $V_r(t_0)$ and $V_r(t_1)$ are uniquely determined. This is true by Lemma 3.1. Next, we need to impose restrictions on t_0, t_1 so that the Fourier coefficients of the initial conditions (the solution coefficients) are uniquely determined by $V_r(t_0)$ and $V_r(t_1)$. For this, we combine the vectors $V_r(t_0)$, $V_r(t_1)$ into one vector

$$W_r = \begin{pmatrix} \alpha_1(t_0) \\ \alpha_1(t_1) \\ \alpha_2(t_0) \\ \alpha_2(t_1) \\ \vdots \\ \alpha_{n_r}(t_0) \\ \alpha_{n_r}(t_1) \end{pmatrix}$$

for

$$\alpha_j(t) = a_j \cos(\mu_j t) + \frac{b_j}{\mu_j} \sin(\mu_j t).$$

Now let

$$\overline{R}_r = \begin{pmatrix} R_1 & & & \\ & R_2 & & \\ & & \ddots & \\ & & & R_{n_r} \end{pmatrix}$$

where

$$R_j = \begin{pmatrix} \cos(\mu_j t_0) & \frac{1}{\mu_j}\sin(\mu_j t_0) \\ \cos(\mu_j t_1) & \frac{1}{\mu_j}\sin(\mu_j t_1) \end{pmatrix}$$

and

$$C_r = \begin{pmatrix} a_1 \\ b_1 \\ a_2 \\ b_2 \\ \vdots \\ a_{n_r} \\ b_{n_r} \end{pmatrix}.$$

Then

$$W_r = \overline{R}_r \cdot C_r.$$

Now for observability to hold, $\overline{R}_r$ must be invertible (i.e. we can completely determine the solution coefficients C_r). Since

$$\det \overline{R}_r = \det R_1 \cdot \det R_2 \cdots \det R_{n_r}$$

and

$$\begin{aligned} \det R_j &= \frac{1}{\mu_j}(\cos(\mu_j t_0)\sin(\mu_j t_1) - \sin(\mu_j t_0)\cos(\mu_j t_1)) \\ &= \frac{1}{\mu_j}\sin(\mu_j(t_1 - t_0)), \end{aligned}$$

we find that $\det \overline{R}_r \neq 0$ if $\sin(\mu_j(t_1 - t_0)) \neq 0$ for every $j = 1, \ldots, r$.

If this is true for every r (condition 3.8 holds), then the wave equation is discretely observable. $\qquad\square$

4   Example: The Wave Equation on a String

We would like to give an explicit example of a system which is discretely observable for a large class of sample points and sample times. First we need the following lemma which is interesting in its own right. The proof is based on one given by William Heller.

Consider the following matrix:

$$Q_n = \begin{pmatrix} \sin x_1 & \sin 2x_1 & \cdots & \sin nx_1 \\ \sin x_2 & \sin 2x_2 & \cdots & \sin nx_2 \\ \vdots & \vdots & \ddots & \vdots \\ \sin x_n & \sin 2x_n & \cdots & \sin nx_n \end{pmatrix}$$

Lemma 4.1 *For any distinct points $\{x_k\}_{k=1}^n$ which are in the open interval $(0, \pi)$, Q_n is invertible.*

Proof: By contradiction. Assume we have distinct points $\{x_k\}_{k=1}^n \in (0, \pi)$ such that Q_n is not invertible. Hence, $\det Q_n = 0$. Therefore, there exists an n vector

$$
C = \begin{pmatrix} c_1 \\ c_2 \\ \vdots \\ c_n \end{pmatrix}
$$

such that

$$
Q_n \cdot C = 0
$$

Now we can write $Q_n \cdot C = F$, where

$$
F = \begin{pmatrix} f(x_1) \\ f(x_2) \\ \vdots \\ f(x_n) \end{pmatrix}
$$

for

$$
f(x) = c_1 \sin x + c_2 \sin 2x + \cdots + c_n \sin nx
$$

By our assumption, $f(x_k) = 0$ for $\{x_k\}_{k=1}^n$. So in the interval $[0, \pi]$, $f(x)$ has $n + 2$ zeros (the chosen points $\{x_k\}$ and the points 0 and π).

Now we will use the following trigonometric identity to show that $f(x)$ can have at most $n + 1$ zeros in the interval $[0, \pi]$.

$$
\sin nx = \sin x (a_1 \cos^{n-1} x - a_3 \cos^{n-3} x + a_5 \cos^{n-5} x - a_7 \cos^{n-7} x + \ldots)
$$

where a_j are constants. In other words, $\sin nx$ is just the product of $\sin x$ and a polynomial of degree $\leq n - 1$ in $\cos x$. Letting $P_j(y)$ denote a polynomial of degree at most j in the variable y, we have

$$
\begin{aligned}
f(x) &= c_1 \sin x + c_2 \sin x P_1(\cos x) + \cdots + c_n \sin x P_{n-1}(\cos x) \\
&= \sin x \tilde{P}_{n-1}(\cos x)
\end{aligned}
$$

where $\tilde{P}_{n-1}(y)$ is just another polynomial of degree $n - 1$. Now $\tilde{P}_{n-1}(y)$ has at most $n - 1$ distinct real roots $r_1, r_2, \ldots, r_{n-1}$, and on the interval $[0, \pi]$, $\cos x = r_i$ at most once. Therefore, $\tilde{P}_{n-1}(\cos x)$ has at most $n - 1$ zeros, and since $\sin x = 0$ only at 0 and π, $f(x)$ has at most $n + 1$ zeros on the interval $[0, \pi]$, contradicting our assumptions, which implied that $f(x)$ has $n + 2$ zeros. $\qquad\square$

Theorem 3.2 and Lemma 4.1 allow us to give an explicit example of a system which is discretely observable for a large class of sample points and sample times. Consider the wave equation on a string of length π:

$$\frac{\partial^2}{\partial t^2} f(x : t) = \Delta f(x : t) \tag{4.9}$$

with boundary conditions:

$$f(0 : t) = f(\pi : t) = 0$$

and initial conditions:

$$f(x : 0) \quad = \quad a(x) = \sum_{j=1}^{\infty} a_j \sin jx$$

$$f_t(x : 0) \quad = \quad b(x) = \sum_{j=1}^{\infty} b_j \sin jx$$

For this bounded domain, the eigenvalues are $\lambda_j = j^2$ and the eigenfunctions are $\sin jx$. Solving the system by separation of variables we have

$$f(x : t) = \sum_{j=1}^{\infty} (a_j \cos jt + \frac{b_j}{j} \sin jt) \sin jx \tag{4.10}$$

Now recall that the proof of Theorem 3.2 relied on the fact that there were points $\{x_k\}_{k=1}^{\infty}$ such that the matrix Q_n was invertible for every finite n. In our case,

$$Q_n = \begin{pmatrix} \sin x_1 & \sin 2x_1 & \cdots & \sin nx_1 \\ \sin x_2 & \sin 2x_2 & \cdots & \sin nx_2 \\ \vdots & \vdots & \ddots & \vdots \\ \sin x_n & \sin 2x_n & \cdots & \sin nx_n \end{pmatrix}$$

Although Lemma 3.1 gave us existence of such points, Lemma 4.1 shows us that we can, in fact, choose any sequence of points in the interval $(0, \pi)$ as our sample points.

For the system to be observable for the chosen spatial points, the condition from Theorem 3.2 requires that

$$\text{For every } r, \ \sin(\mu_j(t_1 - t_0)) \neq 0 \text{ for every } j = 1, \ldots, r. \tag{4.11}$$

In our case this becomes

$$\text{For every } r, \ \sin(j(t_1 - t_0)) \neq 0 \text{ for every } j = 1, \ldots, r. \tag{4.12}$$

and this is true for any times t_0, t_1 such that

$$t_1 - t_0 \neq m\pi \text{ for } m \in \mathbb{Z} \tag{4.13}$$

Thus, for this one-dimensional example, there are many choices of samples which make the given system discretely observable.

Acknowledgements

I would like to thank John Lund and Ken Bowers for their dedication and efforts in putting together such a wonderful conference. Also, thanks to Joe DeStefano for his helpful suggestions regarding the proof of Lemma 3.1. Finally, I would like to thank my Ph.D. thesis advisor, Dorothy Wallace, for her guidance and support.

References

[1] S. AGMOM, "Lectures on Elliptic Boundary Value Problems," D. Van Nostrand Company, Princeton, 1965.

[2] R. DENNEMEYER, "Introduction to Partial Differential Equations and Boundary Value Problems," McGraw-Hill, New York, 1968.

[3] A. DESTEFANO, Discrete Observability of the Wave Equation, Ph.D. Thesis, Dartmouth College, April, 1992.

[4] D.S. GILLIAM, Z. LI and C. MARTIN, "Discrete Observability of the Heat Equation on Bounded Domains," *International Journal of Control*, 1988, pp. 48:755–780.

[5] D.S. GILLIAM and C. MARTIN, "Discrete Observability and Dirichlet Series," *Systems and Control Letters*, 1987, pp. 9:345–348.

[6] D.I. WALLACE and J.A. WOLF, "Observability of Evolution Equations for Invariant Differential Operators,", *J. Math. Systems, Estimation, and Control*, 1991, pp. 1:29–44.

A NEW ALGORITHM FOR NONLINEAR FILTERING

Giovanni B. DiMasi,
Dipartimento di Matematica Pura ed Applicata
Università di Padova and LADSEB-CNR
Padova 35100, Italy

Diego Bricio Hernández
Centro de Investigación en Matemáticas, A.C.
36000 Guanajuato, Gto.
Apartado Postal 402
México

Thomas J. Taylor *
Department of Mathematics
Arizona State University
Tempe, AZ 85287-1804

1 Introduction and Results

The systems and control literature has displayed a long-standing interest in the recursive filtering or estimation problem going back to the solution of the linear filtering problem with the introduction of the Kalman-Bucy Filter in the 1960's. The nonlinear filtering problem has unfortunately not yielded so readily to analysis, to the extent that in most cases there does not exist any widely accepted method of approximating solutions which is both accurate and sufficiently computationally tractable for real time computation. It is the goal of this note to suggest a type of transformation of the discrete time nonlinear filtering equations providing a reformulation of the problem that, in certain circumstances, may be more easily treated.

Specifically our formulation replaces the discrete time unnormalized conditional density equation (here below sometimes called UCD equation), which is often regarded as an infinite dimensional bilinear model driven by the physical observations y_t, with a nonlinear finite dimensional stochastic difference equation driven in part by "internal" random forces and in part by the observations y_t. In our new setup, as also when using the unnormalized conditional density equation, conditional statistics are computed as a ratio of integrals against a density.

It is a significant aspect of our ideas that the finite dimensional processes of the transformed problem take values in a nilpotent Lie group, the

---

*Supported in part by AFOSR grant no. 88-0254 and a CNR-NATO Senior Research Fellowship.

G. B. DiMASI, D. B. HERNÁNDEZ AND T. J. TAYLOR

Heisenberg group, in a way which is natural to the mathematics of the filtering problem. Thus, the terminology "nilpotent approximation," takes a different meaning here than it has in other contexts (see, e.g., [2], and references within). The reason for our use of the Heisenberg group is that the unnormalized conditional density equation involves an interplay of multiplication and integral operators; the representation theory of the Heisenberg group has the peculiar property that convolutions along a certain pair of directions in the Heisenberg group, when restricted to an irreducible representation, become a multiplication and a convolution, respectively. Thus we may "lift" suitable multiplication operators to operators given by convolutions with probability measures: the interplay of multiplication operators with integral operators becomes the interplay of transition operators which are equivalent to the interplay of non-commutative random forces.

In the next section we introduce the necessary elementary non commutative harmonic analysis. In the third section we apply this machinery to the study of the nonlinear filtering of Markov processes taking values in the real line with linear observation functions corrupted by additive Gaussian white noise. We restrict ourselves below, for simplicity, to the consideration of one-dimensional problems with linear observations, the extension to $\mathbf{R}^n$ valued problems with linear observations is easy or to nonlinear observations as well, as we describe below. In the fourth section we describe our recursive Monte Carlo algorithm, and discuss its properties vís-à-vís finite difference approximations. In the last section we discuss the extension of our framework to the consideration of continuous time filtering problems; certain pathologies arise in this context, although these seem to be resolvable.

2 Harmonic Analysis on Heisenberg Groups

In this section we discuss some basic aspects of non-commutative harmonic analysis on the Heisenberg group. Our principal reference for this section is Auslander [1].

Let $N = \mathbf{R} \times \mathbf{R} \times \mathbf{R}$. We define a multiplication $N \times N \to N$ by $(t, \tau, a)(s, \sigma, b) = (t + s, \tau + \sigma, a + b + t\sigma)$ and an inverse operation $(t, \tau, a)^{-1} = (-t, -\tau, -a+t\tau)$. One may verify that these operations define a group structure on N with identity element $(0, 0, 0)$. This group structure is obviously analytic, so that N is a Lie group, which is called the Heisenberg-Weyl group. For each $(t, \tau, a) \in N$ the right multiplication map $R(t, \tau, a) :$ $N \to N$ is defined by $R(t, \tau, a)(x, \xi, c) = (x, \xi, c)(t, \tau, a)^{-1} = (x-t, \xi-t, c- a + t\tau - x\tau)$, and we may note that $R(t, \tau, a)R(s, \sigma, b) = R((t, \tau, a)(s, \sigma, b))$. We also note that the Jacobian determinant of these transformations is equal to 1 so that the Lebesgue measure λ on N is preserved by all the transformations $R(t, \tau, a)$. We define the operators $U(t, \tau, a)$ acting on Lebesgue

A NEW ALGORITHM FOR NONLINEAR FILTERING

measurable functions on N by $[U(t, \tau, a)f](x, \xi, c) = f(R(t, \tau, a)(x, \xi, c)) = f(x - t, \xi - \tau, c - a + t\tau - x\tau)$, and the invariance of λ implies that the operators $U(t, \tau, a)$ are unitary on $L^2(N, \lambda)$ and that $U(t, \tau, a)U(s, \sigma, b) = U((t, \tau, a)(s, \sigma, b))$. Thus the correspondence $(t, \tau, a) \to U(t, \tau, a)$ is a group homomorphism (in fact isomorphism) from N into the group of unitary operators on $L^2(N, \lambda)$, i.e., a *unitary representation of* N. Recall that a unitary group representation V on a Hilbert space H is said to be *irreducible* if there does not exist a non-trivial closed subspace $H' \subset H$ which is invariant under V. The irreducible unitary representations of N are characterized by the famous theorem of von Neumann (see [1], for example): each faithful irreducible representation is unitarily equivalent to some representation of the following form. We consider for each real number $z \neq 0$ the mapping $\tilde{U}_z$ from N into the unitary operators on $L^2(\mathbf{R})$ given by $[\tilde{U}_z(t, \tau, a)f](x) = e^{-iaz}e^{-i(x-t)z\tau}f(x - t)$. It follows from the fact that $(t, \tau, a) = (0, 0, a)(0, \tau, 0)(t, 0, 0)$ that $\tilde{U}_z$ is a representation of N, and from von Neumann [1] that this representation is irreducible.

Now, our representation U of N on $L^2(N, \lambda)$ is not irreducible. However, using Fourier analysis we may decompose $L^2(N, \lambda)$ into irreducible representations in a simple and explicit way. Let $\tilde{f}(x, \xi, c)$ be a function in $L^2(N, \lambda)$. We may take the *partial Fourier transform* of $\tilde{f}$ with respect to the (ξ, c) variables:

$$f(x, y, z) = \frac{1}{2\pi} \int \int e^{(iy\xi + zc)}\tilde{f}(x, \xi, c)\, d\xi\, dc.$$

The Plancheral theorem gives us that $f(x, y, z)$ is again in $L^2(\mathbf{R}^3)$ and that

$$\tilde{f}(x, \xi, c) = \frac{1}{2\pi} \int \int e^{-i(yx + zc)}f(x, y, z)\, dy\, dz.$$

Now, we note that

$$U(t, \tau, a)[f(x)e^{i(y\xi + zc)}] = e^{i(y\tau + za)}e^{-iz\tau(x-t)}[f(x - t)e^{i(y\xi + zc)}].$$

That is, the set of all functions $H_{y,z}$ on N of the form $f(x)e^{i(y\xi + zc)}$ with $f \in L^2(\mathbf{R})$ is invariant under the representation U. Moreover, $U \mid H_{y,z}$ is isomorphic to $\tilde{U}_z$, via the Lebesgue measure preserving group isomorphism of N given by the map $\varphi(t, \tau, a) = (t, \tau, a + \frac{y}{z}\tau)$. Note that in a suitable generalized sense the spaces $H_{y,z}, H_{y',z'}$ are orthogonal for $y, z \neq y', z'$. Thus we see that the representation U is an "orthogonal direct integral" of representations $\tilde{U}_z$; here "direct integral" is the continuous generalization of the notion of direct sum.

Now, the following considerations illustrate the value of these considerations in filtering theory. Namely, let $\hat{u}(\tau)$ be a function in $L^1(\mathbf{R})$. Then the integral $U(\hat{u}) = \int \hat{u}(\tau)U(0,\tau,0)\,d\tau$ defines a bounded operator in $L^2(N,\lambda)$ of convolution type, in particular, $U(\hat{u})$ is defined by: for $f \in L^2(N,\lambda)$

$$
\begin{aligned}
[U(\hat{u})f](x,\xi,c) &= \left[\int \hat{u}(t)U(0,t,0)\,dt\,f\right](x,\xi,c) \\
&= \int \hat{u}(\tau)f(x,\xi-\tau,c-x\tau)\,d\tau.
\end{aligned}
$$

However, if we look at the component of f in $H_{y,z}$, which is $f(x)e^{i(y\xi+zc)}$, we see that

$$
\begin{aligned}
U(\hat{u})f(x)e^{i(y\xi+zc)} &= \int \hat{u}(\tau)f(x)e^{i(y(\xi-\tau)+z(c-x\tau))}\,d\tau \\
&= u(zx+y)f(x)e^{i(y\xi+zc)},
\end{aligned}
$$

where u is the inverse Fourier transform of $\hat{u}$. In other words, the convolution operator $U(\hat{u})$ *looks like a multiplication operator*, when restricted to each irreducible component $H_{y,z}$.

3 Applications to the Discrete Time Nonlinear Filtering Problem

We consider the following general filtering problem. Let $\{x_t\}_{t\in\mathbb{N}}$ be a discrete time (Feller) Markov process taking values in the real line with transition density $P(x_t|x_{t-1})$ and initial density $P_0(x)$. We consider corrupted linear observations of x_t of the form

$$
y_t = h_t x_t + v_t, \tag{3.1}
$$

where the process v_t is $N(0,\sigma^2)$ discrete time Gaussian white noise independent of x_t, and h_t is a time-dependent real number. The filtering problem is to compute the probability distribution $P(x_t|\mathcal{Y}^t)$ of x_t conditioned on the history of observations $\mathcal{Y}^t = \{y_s\}_{s\leq t}$. In principle, the solution of this nonlinear filtering problem may be obtained recursively by solving the discrete time unnormalized conditional density equation [3]. In its most general form, this equation may be written as a difference equation

$$
\rho(x_t|\mathcal{Y}^t) = P(y_t|x_t)\int P(x_t|x_{t-1})\rho(x_{t-1}|\mathcal{Y}^{t-1})\,dx_{t-1} \tag{3.2}
$$

where $\rho(x_t|\mathcal{Y}^t)$ is an unnormalized conditional density, $P(y_t|x_t)$ is the conditional probability density of y_t given x_t and in terms of ρ, $P(x_t|\mathcal{Y}^t) = \rho(x_t|\mathcal{Y}^t)/\int \rho(x|\mathcal{Y}^t)\,dx$ is the conditional probability density of x_t, given $\mathcal{Y}^t$. Under our assumptions $P(y_t|x_t)$ is the Gaussian density $N(h_t x_t,\sigma^2)$ (as a function of y_t). In the above equation $P(y_t|x_t)$ is to be regarded as the operator of multiplication by $P(y_t|x_t)$ as a function of x; this is an un-normalized Gaussian centered at $\frac{y_t}{h_t}$ and with "variance" $\frac{\sigma^2}{h_t^2}$. In the special case that x_t is a Gaussian Markov process $P(x_t|\mathcal{Y}^t)$ is Gaussian if $P(x_0|\mathcal{Y}^0)$ is Gaussian and $P(x_t|\mathcal{Y}^t)$ is determined by its mean and variance; in this case (3.2) may be reduced to the discrete time Kalman filter.

A useful property of (3.2) is that it is recursive, i.e., to update ρ requires only ρ and the most recent value of y_t. On the other hand, in the non-linear and non-Gaussian regime, the equations (3.2) are often infinite dimensional in the sense that (3.2) is an equation on the infinite dimensional space of probability densities and that the number of parameters required to specify $P(x_t|\mathcal{Y}^t)$ increases unboundedly as $t \to \infty$. It is this infinite dimension-ality and increasing complexity which makes recursive non-linear filtering algorithms difficult to implement. Other non-linear filtering algorithms are based on direct implementation of the Kallianpur-Striebel formula via Monte Carlo simulation of Wiener measure; these are non-recursive meth-ods and require storage and manipulation of the entire past history $\{y_s\}_{s \le t}$ at each update.

We present here a reformulation of (3.2) which may facilitate the sim-ulation of solutions of the filtering problem. Namely, recall from section 2 that the convolution operator $U(\hat{u})$ corresponds to multiplication by the appropriate translation and dilation of u on $H_{y,k}$. First we reformulate (3.2) as

$$\rho(x_t|\mathcal{Y}^t) = T_{y_t/h_t} N(0,\sigma^2)(h_t x_t) T_{-y_t/h_t} P_x \rho(x_{t-1}|\mathcal{Y}^{t-1}) \qquad (3.3)$$

where P_x denotes the transition operator of our process x_t, and T_{y_t/h_t} de-notes the operator of translation by the quantity y_t/h_t. Next, note that the Fourier transform of $N(0,\sigma^2)(h_t x_t)$ is again Gaussian—but in general no longer a probability density. However, note that (3.3) is an *unnormalized* conditional density equation, so that there is no information lost in consid-ering instead of (3.3) an equation obtained by multiplying the right side of (3.3) with any time dependent constant α_t. In particular, we may choose α_t in order to obtain that the Fourier transform of $\alpha_t N(0,\sigma^2/h_t^2)$ is again a probability density (equal to $N(0,\frac{h_t^2}{\sigma^2})$).

Define operators $(\tilde{P}_x f)(x,\xi,c)$ and $\tilde{P}_\xi^t$ on $L^2(N,\lambda)$ by $(\tilde{P}_x f)(x,\xi,c) = \int P(x|x_{t-1})f(x_{t-1},\xi,c)\,dx_{t-1}$ and $\tilde{P}_\xi^t = \int N(0,h_t^2/\sigma^2)(\tau)U(0,\tau,0)\,d\tau$. We consider the following equation, which we shall see may be considered the "lift" of equation (3.3) to the Heisenberg group:

$$\Psi(x_t,\xi_t,c_t|\mathcal{Y}^t) \qquad (3.4)$$

$$= U\left(\frac{y_t}{h_t},0,0\right)\tilde{P}_\xi^t U\left(-\frac{y_t}{h_t},0,0\right)\tilde{P}_x \Psi(x_{t-1},\xi_{t-1},c_{t-1}\mid\mathcal{Y}^{t-1}).$$

$$\text{G. B. DiMASI, D. B. HERNÁNDEZ AND T. J. TAYLOR}$$

Note first of all that, for each $t \geq 0$, each of the three operators $\tilde{P}_x$, $\tilde{P}_\xi^t$, $U(\frac{y_t}{h_t}, 0, 0)$ has the property of being a Markov operator, i.e., preserves positivity, fixes the constants, and is a contraction on the space $C_0(N)$ (= bounded continuous functions on N vanishing at ∞). The product of Markov operators is a Markov operator, and such operators may always be represented as the transition function of a Markov process. It follows that $Q_t = U(\frac{y_t}{h_t}, 0, 0)\tilde{P}_\xi^t U(-\frac{y_t}{h_t}, 0, 0)\tilde{P}_x$ is the transition operator of a (non-homogeneous) Markov process on N. Evidently, from the form of Q_t, this Markov process may be viewed as the response to internal random forces along the x and ξ directions, and as driven by the observations y_t (note that here x and ξ have the "directions" that are non-commutative).

For functions Ψ in the space $H_{-y, -z}$, $\Psi(x, \xi, c) = \rho(x)e^{-i(y\xi + zc)}$, the operator Q_t takes the following form:

$$Q_t \rho(x)e^{-i(y\xi + zc)}$$
$$= \alpha_t N\left(0, \frac{\sigma^2}{h_t^2}\right)\left(z\left(x - \frac{y_t}{h_t}\right) + y\right)[P_x \rho](x)e^{-i(y\xi + zc)}.$$

We see, therefore, that in the space $H_{0, -1}$ equation (3.4) reproduces equation (3.3) precisely. We may call (3.4) the *lift* of (3.3).

Some remarks on the procedures discussed in this section are in order:

(1) There are Heisenberg groups of all odd dimensions larger than 1, and there is a corresponding lifting for filtering problems with linear observations on n-dimensional space to the $2n + 1$ dimensional Heisenberg group.

(2) There is no loss of generality in considering only linear observations. When the observations are nonlinear in the standard form $y_t = h(x_t, t) + v_t$, the introduction of t and $h(x_t, t)$ as additional spatial variables provides a filtering problem with linear observations, although of higher dimension.

4 A Recursive Monte Carlo Algorithm

The considerations in the last section may yield a new algorithm for the nonlinear filtering problem. This is based on the probabilistic interpretation of equation (3.4). It is rather easy to see that the operator Q_t is the transition operator of a system of stochastic difference equations on the Heisenberg group. These are the following simple equations:

$$
\begin{aligned}
x_t &\rightarrow\ x_{t+1}, \quad \text{according to the probability } P(x_{t+1}|x_t), \\
\xi_{t+1} &= \xi_t + w_t, \\
c_{t+1} &= c_t + \left(x_t - \frac{y_t}{h_t}\right)w_t,
\end{aligned}
\tag{4.1}
$$

where w_t is a discrete time Gaussian white noise $N(0, \frac{h^2}{\sigma^2})$, independent of (x_t, ξ_t, c_t).

In terms of the solutions of the UCD equation, the optimal estimate of a statistic $\zeta(x)$ is given by the conditional statistic

$$\hat{\zeta}(\mathcal{Y}^t) = \int \zeta(x)\rho(x|\mathcal{Y}^t)\,dx \Big/ \int \rho(x|\mathcal{Y}^t)\,dx. \qquad (4.2)$$

This conditional statistic may also be described in terms of a Markov process satisfying the difference equation (4.1) and having any initial probability density $\Psi_0(x, \xi, c)$ which is related to the initial probability $P_0(x)$ by

$$P_0(x) = \frac{\int e^{ic}\Psi_0(x, \xi, c)\,d\xi\,dc}{\int e^{ic}\Psi_0(x, \xi, c)\,dx\,d\xi\,dc},$$

for example, the density $P_0(x)e^{-(\xi^2+c^2)}$ suitably normalized. With respect to such a process, we may write for the optimal estimate $\hat{\zeta}$

$$\begin{aligned}
\hat{\zeta}(\mathcal{Y}^t) &= \frac{E(\zeta(x_t)e^{ic_t})}{E(e^{ic_t})} \\
&= \frac{\int \zeta(x)e^{ic}\Psi(x, \xi, c \mid \mathcal{Y}^t)\,dx\,d\xi\,dc}{\int e^{ic}\Psi(x, \xi, c \mid \mathcal{Y}^t)\,dx\,d\xi\,dc}.
\end{aligned} \qquad (4.3)$$

This suggests the following recursive Monte Carlo algorithm for solving nonlinear filtering problems with linear observation.

1) Generate a family of initial conditions $\{(x_0^k, \xi_0^k, c_0^k)\}_{k=1}^M$ with the property that for sufficiently smooth functions f the following approximate inequality holds:

$$\frac{1}{M}\sum_k f(x_0^k, \xi_0^k, c_0^k) \approx \int f(x, \xi, c)\Psi_0(x, \xi, c)\,dx\,d\xi\,dc \qquad (4.4)$$

2) Propagate this density on N one step by generating one Gaussian random number for w_0 and generating one random number x_1 according to the distribution $P(x_1|x_0)$, independently for each initial condition (x_0^k, ξ_0^k, c_0^k). Plug these numbers and y_0 into equation (4.1) to obtain a collection $\{(x_1^k, \xi_1^k, c_1^k)\}_{k=1}^M$ having an empirical distribution approximating the density at time 1, $\Psi_1(x, \xi, c)$, as in (4.4).

3) Compute the conditional statistic $\hat{\zeta}(\mathcal{Y}^1)$ according to the

$$\hat{\zeta}(\mathcal{Y}^1) = \left(\sum_k \zeta(x_1^k)e^{ic_1^k}\right)\left(\sum_k e^{ic_1^k}\right)^{-1} \qquad (4.5)$$

G. B. DiMASI, D. B. HERNÁNDEZ AND T. J. TAYLOR

4) Using the new collection of points $\{(x_t^k, \xi_t^k, c_t^k)\}_{k=1}^M$ as initial conditions, and the new observations y_t, iterate steps 2) and 3).

We give several remarks on the potential advantages of this method and its higher dimensional generalizations over other approaches to solving the filtering problem. First of all, the difficulty of solving or approximating (3.4) directly as an equation in an infinite dimensional space is as great as that of solving (3.2) directly; and the equations are isomorphic in the algebraic sense that the number of parameters needed to describe the solution grows at the same rate, i.e., exponentially in general. In particular, in spite of the fact that (4.1) are the equations of a finite dimensional Markov process, we will not generally have a filtering problem of the type that has been called a finite dimensional filtering problem.

One type of approximate nonlinear filtering that one often encounters in the literature is based on finite difference approximation of (3.2) (e.g., [4, 5, 6]). For such approximations, one may compute that the number of operations versus accuracy grows linearly in the number of discretization points M (assumed uniformly spaced). Indeed, the accuracy grows linearly in M while for finite difference methods the update is achieved essentially by matrix multiplication so that the cost grows as M^2 (at least in the absence of sparseness). On the other hand, for Monte Carlo methods the accuracy grows as $\sqrt{M}$, while the cost of the algorithm we describe grows as M. Thus, the cost-benefit of our method grows as $\sqrt{M}$. Applications typically require the real-time implementation of filtering algorithms, consequently obtaining accurate non-linear filtering has posed considerable difficulties. The Monte Carlo method we describe above may thus be computed more efficiently at the same level of accuracy and more perfectly suited to implementation on SIMD parallel machines in the sense that the only necessary communication between processors is in the computation of the average (4.5). Another advantage of our algorithm is that for finite difference methods the number of mesh points required to achieve a certain accuracy grows exponentially with the dimension, whereas Monte Carlo methods are independent of dimension. For example, the average (4.5) has a cost which is independent of the dimension of the state space of our original Markov process x_t, while the integrals in (4.2) have a cost which grows exponentially with the dimension (assuming a uniform mesh).

While the assessment of the relative value of our algorithm is still in process, it seems to provide a promising new idea for reducing the complexity of some non-linear filtering problems.

5   Some Ideas About Continuous Time Filtering

In this section we will discuss lifting the continuous time filtering problem to the Heisenberg group. The continuous time situation provides greater difficulties than the discrete time situation, so our discussion will be almost entirely formal, without resolving in any serious way the mathematical pathologies that arise. We will let x_t denote a continuous time Markov process, which satisfies the standard conditions; see [8] for details. We suppose that we observe

$$dy_t = hx\, dt + dv_t,$$

where v_t is a standard $N(0,1)$ Wiener process independent of x.

In this situation the continuous time UCD equation, more commonly called the Duncan-Morteson-Zakai equation, takes the form (using Stratonovich calculus)

$$d\rho_t = \left(D - \frac{1}{2}h^2 x^2\right)\rho_t\, dt + hx\rho_t\, dy_t. \tag{5.1}$$

Proceeding formally as in the discrete time case, we may lift this equation to the Heisenberg group. One may begin by asserting the following correspondence principle

$$\frac{d}{dx} \quad \leftrightarrow \quad \frac{\partial}{\partial x},$$

$$ix \quad \leftrightarrow \quad \frac{\partial}{\partial \xi} + x\frac{\partial}{\partial c},$$

$$i \quad \leftrightarrow \quad \frac{\partial}{\partial c},$$

where the operators on the left side of the correspondence are the values taken by those on the right when restricted to $H_{0,-1}$. Next, we recognize that (5.1) is an unnormalized conditional density equation, so that we have the freedom to add any differential which involves ρ linearly and has a process independent of ρ as a multiplicative factor. We proceed by abuse of this freedom to the extent of subtracting the "infinite" (i.e., undefined) process $\frac{1}{2}(\frac{dy_t}{dt})^2\rho\, dt$, obtaining the formal equation

$$d\rho = \left(D - \left[hx - \frac{dy_t}{dt}\right]^2\right)\rho\, dt.$$

Under the correspondence principle, this becomes the formal partial differential equation on N

$$d\Psi = \left(D_x + \left[h\left\{\frac{\partial}{\partial \xi} + x\frac{\partial}{\partial c}\right\} - \frac{dy_t}{dt}\frac{\partial}{\partial c}\right]^2\right)\psi\, dt. \tag{5.2}$$

Since $dy_t = hx\, dt + dv_t$, where v_t is standard Wiener process, we have that $\frac{dy_t}{dt} = hx_t + \frac{dv_t}{dt}$ is the sum of a well-defined process with a Gaussian white noise—a process which has meaning only as a process with distributional sample paths. This means that there are problems making rigorous sense of (5.2). Nevertheless, motivated by the concept of *renormalization* as described in the book by Nelson [7], we regard (5.2) as the forward equation of the following formal system of stochastic differential equations.

$$
\begin{aligned}
d\tilde{x}_t & \quad \text{according to } D, \\
d\xi_t & = h\,dw_t, \\
dc & = \left(h\tilde{x}_t - \frac{dy_t}{dt} \right) dw_t,
\end{aligned}
\tag{5.3}
$$

where $\tilde{x}_t$ is an independent copy of the Markov process x_t, and w_t is an independent Wiener process.

The only difficulty in interpreting (5.3) lies in the last equation, where we are forced to consider stochastic integrals such as

$$
\int h(\tilde{x}_t - x_t)\,dw_t + \int \frac{dv_t}{dt}\,dw_t.
\tag{5.4}
$$

Only the last integral presents difficulties of interpretation. We suggest that this be regarded as the limit of the Riemann sum

$$
\sum \frac{\Delta v_t}{\Delta t} \Delta w_t = \frac{1}{\Delta t} \sum \Delta v_t \Delta w_t.
\tag{5.5}
$$

The sum on the right converges as $\Delta t \to 0$ to the mutual variation of the independent Wiener processes v_t, w_t, which is zero. Even more, one may show that $\sum \Delta v_t \Delta w_t$ is $o(\Delta t)$ as $\Delta t \to 0$. Thus only the first integral of (5.4) makes a contribution to the stochastic differential equation. These ideas provide a way to define the equations (5.3) rigorously. Moreover, this interpretation of equations (5.3) is what is obtained ¿from equation (4.1) in the limit by considering the continuous time filtering problem as a limit of the discrete time problem. Details on these issues will appear elsewhere.

References

[1] L. AUSLANDER, "Lecture Notes in Nil-Theta Functions," *CBMS Lecture Notes*, v. 34, AMS, 1977.

[2] C. ROCKLAND, "Intrinsic Nilpotent Approximation," *Acta Applicandae Mathematica*, v. 8, 1987, pp. 731–740.

[3] A. JAZWINSKI, "Stochastic Processes and Filtering Theory," *Academic Press*, New York, 1970.

[4] H. J. KUSHNER, "Probability Methods for Approximations in Stochastic Control and for Elliptic Equations," *Academic Press*, New York, 1977.

[5] Y. YAVIN, "Numerical Studies in Nonlinear Filtering," *Lecture Notes in Control*, v. 65, Springer Verlag, 1985.

[6] D. TALAY, "Efficient Schemes for the Approximation of Expectations of Functionals of the Solution of an S.D.E. and Applications," in *Filtering and Control of Random Processes*, Korezlioglu, Mazziotto and Szperglas eds., Lecture Notes in Control and Information Science, v. 61, Springer Verlag, 1984.

[7] E. NELSON, *Quantum Fluctuations*, Princeton University Press.

[8] M. H. A. DAVIS, "Lectures on Stochastic Control and Nonlinear Filtering: Lectures Delivered at the Indian Institute of Science," Springer Verlag, 1984.

CONTINUATION METHODS FOR NONLINEAR EIGENVALUE PROBLEMS VIA A SINC-GALERKIN SCHEME

Jack D. Dockery * and Nancy J. Lybeck

Department of Mathematical Sciences
Montana State University
Bozeman, Montana 59717

1 Introduction

In this paper we will be looking at semilinear boundary value problems of
the form

$$\mathcal{L}u(x) \equiv u''(x) + cu(x) \;=\; f(x, u(x), \lambda) \;,\quad a < x < b \qquad (1.1)$$
$$u(a) \;=\; u(b) = 0 \;,$$

where c is a fixed constant and λ is a parameter. Here, a and b need not
be finite. This type of problem often arises as the equilibrium problem for
a scalar evolution equation. The purpose of this paper is to illustrate the
use of the Sinc-Galerkin method for one-parameter problems such as (1.1).

The Sinc method has several attractive properties, one being its exponential rate of convergence ([10, 11]). Another advantage of Sinc method is
that it handles unbounded and bounded domains equally well. There is no
need with this method for domain truncation ([5]). An attractive feature
for nonlinear problems is that the discrete approximation of (1.1) has the
property the non-linear equations are coupled only by a linear operator.
This allows for easy evaluation of the Jacobian. This should be contrasted
with high order finite element methods for which the equations are coupled not only by the linear part of the discrete operator but also through
nonlinear interactions.

In section 2, we describe the Sinc-Galerkin method for the boundary
value problem (1.1). For fixed λ, the Sinc-Galerkin approximation yields
a square matrix equation. This can be solved directly for u, unless the
Jacobian is singular. Considering λ as a free parameter, we get an underdetermined system of the form

$$F(u, \lambda) \;=\; 0, \qquad\qquad (1.2)$$
$$F : \mathbb{R}^{m+1} \;\to\; \mathbb{R}^m \;.$$

In section 3 we look at the Euler-Newton Continuation method to solve
(1.2). This method uses an Euler step to make an initial prediction. This

---

*Supported in part by NSF grant DMS–9113526.

is followed by Newton corrections, which utilize the Moore-Penrose generalized inverse of the non-square Jacobian for (1.2). Using λ as the continuation parameter, the method follows the solution curves of (1.2) as λ varies.

In section 4 numerical results are presented. First the Bratu problem is solved numerically. Second, the d-homotopy method ([1]) is used in addition to the Euler-Newton method to follow a bifurcation from infinity. The final example is a boundary value problem on the half line. This example shows off the power off the Sinc-Galerkin method for unbounded domains. In particular, the system is essentially the same as the discretization for a bounded domain.

2 The Sinc-Galerkin method

The development presented here will be for the two-point boundary value problem (1.1) where the interval (a, b) is $(0, 1)$ or $(0, \infty)$. The Sinc-Galerkin method for problems of this type is discussed in detail in [8] and [11]. The following discussion contains only the information necessary to build the discrete system.

A Galerkin method for the approximation of (1.1) may be briefly summarized as follows. Select a set of basis functions

$$\{S_j(x)\}$$

and define the approximate solution by

$$u_m(x) = \sum_{i=-M}^{N} u_i S_i(x) \quad , \quad m = M + N + 1 \quad . \tag{2.1}$$

The unknown coefficients $\{u_i\}$ in (2.1) are determined by orthogonalizing the residual with respect to the basis functions. This yields the discrete Galerkin system

$$(\mathcal{L}u_m - f(\cdot, u_m, \lambda), S_j) = 0 \quad , \quad -M \leq j \leq N \quad . \tag{2.2}$$

The choice of the inner product that is used in (2.2) along with the choice of basis functions determines the properties of the approximating method.

The inner product that is used for the Sinc-Galerkin method is defined by

$$(u, v) = \int_a^b u(x)v(x)w(x)dx \tag{2.3}$$

where $w(x)$ is, for the moment, an arbitrary weight function. The choice of $w(x)$ will be discussed at the end of this section.

In the Sinc-Galerkin method, the basis functions are derived from the Whittaker cardinal (Sinc) function

$$\text{sinc}(x) = \frac{\sin(\pi x)}{\pi x} \quad , \quad -\infty < x < \infty$$

and its translates

$$S(j, h)(x) = \text{sinc}\left(\frac{x - jh}{h}\right) \quad , \quad h > 0 \quad . \tag{2.4}$$

In order to have the Sinc translates defined on the interval $(0, 1)$, consider the conformal map

$$\phi(z) = \ell n \left(\frac{z}{1 - z}\right) \quad .$$

This map carries the eye-shaped region

$$D_E = \left\{ z = x + iy : |\arg\left(\frac{z}{1 - z}\right)| < d \leq \frac{\pi}{2} \right\}$$

onto the infinite strip

$$D_S = \left\{ \xi = \zeta + i\eta : |\eta| < d \leq \frac{\pi}{2} \right\} \quad .$$

We will also need the Sinc translates defined on the interval $(0, \infty)$. We use conformal map

$$\phi(w) = ln(\sinh(w))$$

which maps

$$D_B \equiv \{ w \in \mathbb{C} : |arg(sinh(w))| < d \leq \pi/2 \}$$

onto the infinite strip D_S.

The basis functions used in (2.1) are then defined by

$$S_j(x) = (g(x)/\phi'(x))S(j, h) \circ \phi(x), \tag{2.5}$$

where the function g plays the role of a weight adjustment that may be altered according to various properties of the sought for solution. For most examples g is taken to be the identity function but if one is not only interested in approximating the solution u but also its derivative u' then other choices of g are appropriate. This will be illustrated in section 4.

The integrals in (2.2) are approximated by the sinc quadrature rule ([8, 10]). To describe this quadrature rule we need to define the following function space, $B(D)$.

Definition 1 *Let D be a domain in the $w = u + iv$ plane with boundary points $a \neq b$. Let $z = \phi(w)$ be a one-to-one conformal map of D onto the infinite strip*

$$D_S \equiv \{z \in \mathbb{C} : z = x + iy, \quad |y| < d\}$$

where $\phi(a) = -\infty$ and $\phi(b) = +\infty$. Denote by $w = \psi(z)$ the inverse of the mapping ϕ and let

$$\Gamma \equiv \{w \in \mathbb{C} : w = \psi(x), \quad x \in \mathbb{R}\} = \psi(\mathbb{R}).$$

Let $B(D)$ denote the class of functions analytic in D which satisfy for some constant a with $0 \leq a < 1$,

$$\int_{\psi(x+L)} |F(w)dw| = \mathcal{O}(|x|^a), \quad x \to \pm\infty$$

where $L = \{iy : |y| < d\}$ and for γ a simple closed contour in D

$$N(F, D) \equiv \lim_{\gamma \to \partial D} \int_\gamma |F(w)dw| < \infty.$$

Further, for $h > 0$, define the nodes

$$w_k = \psi(kh), \quad k = 0, \pm 1 \pm 2, \dots.$$

The sinc quadrature rule is contained in the following result.

Theorem 2.1 *Let $F \in B(D)$ and $h > 0$. Let ϕ be a one-to-one conformal map of the domain D onto D_S. Let $\psi = \phi^{-1}$, $x_k = \psi(kh)$, and $\Gamma = \psi(\mathbb{R})$. Further assume that there are positive constants α, β, and C so that*

$$\left| \frac{F(\xi)}{\phi'(\xi)} \right| \leq C \left\{ \begin{array}{ll} \exp\left(-\alpha|\phi(\xi)|\right), & \xi \in \Gamma_a \\ \exp\left(-\beta|\phi(\xi)|\right), & \xi \in \Gamma_b \end{array} \right.$$

where $\Gamma_a \equiv \{\xi \in \Gamma : \phi(\xi) = x \in (-\infty, 0)\}$ and $\Gamma_b \equiv \{\xi \in \Gamma : \phi(\xi) = x \in [0, \infty)\}$. Make the selections

$$N = \left[\left| \left|\frac{\alpha}{\beta}\right| M + 1 \right| \right]$$

and

$$h = \left(\frac{2\pi d}{\alpha M} \right)^{1/2} \leq \frac{2\pi d}{\ell n 2} \tag{2.6}$$

then

$$\int_\Gamma F(\xi)d\xi = h \sum_{k=-M}^{N} \frac{F(x_k)}{\phi'(x_k)} + \mathcal{O}\left(\exp\left(-(2\pi d\alpha M)^{1/2}\right) \right). \tag{2.7}$$

The sinc quadrature rule is the replacement of the integral in (2.7) by the sum on the right hand side of (2.7).

To facilitate the recording of the sinc quadrature rule applied to (2.2) it is convenient to introduce the notation

$$\delta_{jk}^{(p)} \equiv h^p \frac{d^p}{d\phi^p} \left[S(j,h) \circ \phi(x) \right] \Big|_{x=x_k} \quad , \quad p = 0,1,2,\ldots$$

where $x_k = \phi^{-1}(kh)$. Explicitly, these quantities for $p = 0,1,2$ are given by

$$\delta_{jk}^{(0)} = \left\{ \begin{array}{ll} 1, & j = k \\ 0, & j \neq k \end{array} \right. ,$$

$$\delta_{jk}^{(1)} = \left\{ \begin{array}{ll} 0, & j = k \\ \frac{(-1)^{k-j}}{k-j}, & j \neq k \end{array} \right.$$

and

$$\delta_{jk}^{(2)} = \left\{ \begin{array}{ll} \frac{-\pi^2}{3}, & j = k \\ \frac{-2(-1)^{k-j}}{(k-j)^2}, & j \neq k \end{array} \right. .$$

We will let $I^{(p)}$ denote the $m \times m$ matrix (m = M+N+1) with j,k entries $\delta_{jk}^{(p)}$ for $j,k = -M,\ldots N$ and $p = 0,1$ and 2. For a function given h on (a,b) we will let $\vec{h} = [h(x_{-M})\ldots h(x_N)]^T$, and $\mathcal{D}(h) = diag(\vec{h})$ where the x_ks are as in Theorem 2.1.

The method of approximating the integrals in (2.2) begins by integrating by parts to remove all derivatives from u_m onto S_j. Assuming that the weights g and w are such that the boundary terms vanish and then applying the sinc quadrature rule yields the following discrete system for the unknowns $\{u_j\}$:

$$A_h D(\frac{g}{\phi'}) \, \vec{u} = D(\frac{\tilde{w}}{\phi'}) \, \vec{f}. \tag{2.8}$$

Here we have set

$$\tilde{w}(x) = \frac{g(x)w(x)}{\phi'(x)}, \tag{2.9}$$

$$\vec{f} = (f(x_{-M}, u_{-M}, \lambda), \ldots, f(x_N, u_N, \lambda))^T, \tag{2.10}$$

and A_h is the $m \times m$ matrix given by

$$A_h \equiv \left\{ \frac{1}{h^2} \, I^{(2)} D(\phi' \tilde{w}) \; + \; \frac{1}{h} \, I^{(1)} D \left(\frac{\phi''}{\phi'} \, \tilde{w} + 2\tilde{w}' \right) \right.$$

$$\tag{2.11}$$

$$\left. + \; I^{(0)} D \left(\frac{\tilde{w}'' + c\tilde{w}}{\phi'} \right) \right\}.$$

One important observation to make about the nonlinear system (2.8) is that the nonlinear terms are decoupled. This makes for easy evaluation of the Jacobian of (2.8). In particular the Jacobian of (2.8) with respect to u_m has the form

$$A_h D(\frac{g}{\phi'}) - D(\frac{\tilde{w}}{\phi'})D(\vec{f'}),$$

where

$$\vec{f'} = (f_u(x_{-M}, u_{-M}, \lambda), \ldots, f_u(x_N, u_N, \lambda))^T.$$

It follows that the Jacobian is just a diagonal perturbation of the linear part of (2.8).

In Table 2.1 below, some of the relevant quantities to fill the matrices in (2.8)-(2.11) for the various maps ϕ under consideration here are provided.

(a, b)	ϕ	$\frac{\phi''}{(\phi')^2}$	$\frac{1}{\phi'}$	$-\frac{1}{\phi'}\left(\frac{1}{\phi'}\right)''$
$(0,1)$	$\ln\left(\frac{x}{1-x}\right)$	$2x - 1$	$x(1-x)$	$2x(1-x)$
$(0,\infty)$	$\ln(\sinh(x))$	$-\text{sech}^2(x)$	$\tanh(x)$	$\frac{2\tanh^2(x)}{\cosh^2(x)}$

Table 2.1 Components for the discrete system in (2.8).

3 Continuation Methods

If we consider the parameter λ in (1.1) as unknown, then the Sinc-Galerkin system (2.8) gives rise to a under-determined system which we will denote by

$$F(w) = 0.$$

Here $F : \mathbb{R}^{m+1} \to \mathbb{R}^m$ and $w = (\vec{u}, \lambda) \in \mathbb{R}^{m+1}$. There are several methods ([1, 7]) for following solution curves to $F(w) = 0$. In this section we will discuss the Euler-Newton predictor-corrector method. Other continuation methods can be found in [1, 3, 7].

To describe the method we need to define the notion of a tangent vector to the curve $F(w) = 0$.

Definition 2 *Let A be an $(m) \times (m + 1)$ matrix with rank m. Let $A^* = \vec{A}^T$. Then the unique vector $\vec{t}(A) \in \mathbb{R}^{m+1}$ satisfying :*

1.) $A\vec{t} = 0$

2.) $\|\vec{t}\| = 1$

3.) $\det\left(\begin{array}{c} A \\ \vec{t}^* \end{array}\right) > 0$

is called the tangent vector induced by A.

If 0 is a regular value of F then it follows that the Jacobian of F, $F'(w_0)$, has a well defined tangent vector for each $w_0 \in F^{-1}(0)$. For an $(m) \times (m+1)$ matrix A with rank m we let $A^\dagger$ denote the pseudo-inverse of A:

$$A^\dagger = A^*(AA^*)^{-1}.$$

The basic Euler-Newton predictor-corrector method is as follows:
Euler-Newton algorithm:

1.) Input w such that $F(w) = 0$. Pick a stepsize ds > 0.

2.) Repeat:

    a.) $v = u + \text{ds } \vec{t}(F'(u))$.

    b.) Repeat:

$$\begin{aligned}
w &= v - (F'(v))^\dagger F(v) \\
v &= w
\end{aligned}$$

      Until convergence.

For more details on this algorithm we refer the reader to [1].

4 Numerical Results

In this section we present three numerical examples. The first example is an often used model problem for continuation methods. The second example illustrates the 'd-homotopy' method ([1]). This is one of many methods used to find a nontrivial solution on a branch of solutions. Both the first and second examples are boundary value problems on the unit interval. The last example comes from a nonlinear Schroedinger equation that arises as a simplification to the Maxwell-Bloch equations. This example illustrates how with little effort the Sinc method can be used to solve problems posed on unbounded domains.

Example 4.1 Consider the semilinear boundary value problem:

$$\begin{aligned}
u''(x) &= -\lambda e^u \quad , \; \lambda > 0 \\
u(0) &= u(1) = 0 \quad .
\end{aligned} \tag{4.1}$$

This problem provides a simplified model of nonlinear diffusion phenomena in combustion and semiconductors. It has been considered by Glowinski, Keller and Rhienhardt ([4]), as well as a number of other investigators and is known as the Bratu problem.

Since the interval is $(0,1)$ we take $\phi(x) = \ln(x/(1-x))$. With this choice of map the nodes x_k are given by

$$x_k = \frac{e^{kh}}{1+e^{kh}}$$

To complete the system (2.8) we need to choose the weights g and w. With zero boundary data the choice $g(x) \equiv \phi(x)$ is standard ([8]). In particular the expansion (2.1) takes the form

$$u_m(x) = \sum_{i=-M}^{N} u_i S(i,h) \circ \phi(x) \ , \quad m = M + N + 1 \ .$$

In order to guarantee that the boundary integrals leading to (2.8) vanish the choice $w(x) = 1/\phi'(x)$ is sufficient, thus $\tilde{w}$ in (2.9) is given by

$$\tilde{w}(x) = \frac{1}{\phi'(x)}.$$

Using Table 2.1 the discrete system for (4.1) becomes

$$\begin{aligned} F(\vec{u},\lambda) &\equiv A\vec{u} + \mathcal{D}\left(\frac{1}{(\phi')^2}\right)\vec{f}(\vec{u},\lambda) \\ &= \vec{0} \ . \end{aligned} \tag{4.2}$$

where the matrix A is given by

$$A = \frac{1}{h^2}I^{(2)} - \frac{1}{h}I^{(1)}\mathcal{D}\left((-1)\frac{\phi''}{(\phi')^2}\right) + I^{(0)}\mathcal{D}\left(\left(\frac{-1}{\phi'}\right)\left(\frac{1}{\phi'}\right)''\right)$$

and the nonlinear term $\vec{f}$ is given by

$$\vec{f}(\vec{u},\lambda) = (\lambda\exp(u_{-M}),\ldots,\lambda\exp(u_N))^T$$

The Jacobian of (4.2) with respect to $\vec{u}$ is given by

$$J_u(\vec{u}) = A + \mathcal{D}\left(\frac{1}{(\phi')^2}\right)D(\lambda\exp(\vec{u})).$$

We take h as in (2.6) with $\alpha = \beta = 1$ and $d = \pi/2$.

At $\lambda = 0$ it is clear that (4.2) has $\vec{u} \equiv 0$ as a solution. It is known that (4.1) has a one parameter family of positive solutions. We have used the Euler-Newton algorithm to solve (4.2) for $(\vec{u},\lambda)$, the results are shown in Figure 4.1. Here we have plotted $\|u\|_\infty$ of the approximate solution verses

$M = N$	$|\lambda(u_{\max}) - \lambda_A(u_{\max})|$	asy. error	$\frac{|\lambda_T - \lambda_A|}{\text{asy.error}}$
2	$1.6783e - 01$	$1.7286e - 01$	$9.7093e - 01$
4	$6.2863e - 02$	$1.8819e - 01$	$3.3404e - 01$
8	$1.3019e - 02$	$1.1952e - 01$	$1.0893e - 01$
16	$1.1937e - 03$	$3.5416e - 02$	$3.3706e - 02$
32	$1.2461e - 04$	$3.5710e - 03$	$3.4894e - 02$

Table 4.1: Error in Lambda for Example 4.1.

λ. Note, each point on the curve corresponds to a solution of the boundary value problem (4.1).

While it is difficult to compute a explicit expression for the true solution of (4.1), a simple phase plane analysis shows that for each value of $u_{\max} \in [0, \infty)$ there is only one value of $\lambda = \lambda(u_{\max})$ for which the boundary value problem (4.1) has a solution u with $||u||_\infty = u_{\max}$. One can show that $u_{\max}$ and $\lambda(u_{\max})$ are related as follows: let $k = u'(0) \geq 0$. Then

$$||u||_\infty \quad \equiv \quad u_{max} = \ln\left(\frac{k^2 + 2}{2}\right) \tag{4.3}$$

$$\lambda(u_{\max}) \quad = \quad \frac{4(\ln(k^2 + 1 + k\sqrt{k^2 + 2}))^2}{k^2 + 2} \quad .$$

The equations (4.3) provide a means for comparison of the approximate solution to the true solution. The results are given in Table 4.1. The third column of this table is given by

$$\text{error} = \max|\lambda_T(u_{\max}) - \lambda_A(u_{\max})| \, ,$$

where $u_{\max}$ is the maximum value of the approximate solution, $\lambda_T(u_{\max})$ is exact value of λ for the calculated value of $||u||_\infty$, and λ_A is the calculated value of λ. The asymptotic error for the sinc method is given by $M^2 e^{-\pi\sqrt{(\frac{M}{2})}}$ (see [8, 10, 11]). This is displayed in column four.

Example 4.2 Our next example is similar to Example 4.1 above. We include it to introduce the "d-homotopy" ([1]) method. As above there is a one parameter family of positive solutions for the boundary value problem. The difficulty is that $u \equiv 0$ is a solution for all values of the parameter

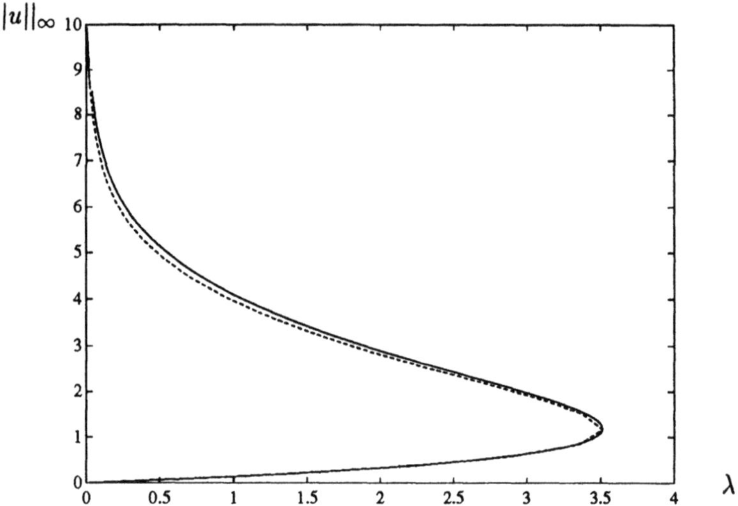

Figure 4.1: True (——) and Approximate Solution Branches for Example 4.1.

λ and there is no bifurcation from the trivial branch. To find a starting solution on the non-trivial branch we use the d-homotopy method.

Consider:

$$\begin{aligned} u'' &= \lambda u(1-u) \ , \ \lambda > 0 \\ u(0) &= u(1) = 0 \ . \end{aligned} \tag{4.4}$$

We see that $u \equiv 0$ is a solution for all values of λ. A phase analysis of (4.4) shows that there is a one parameter family of positive solutions for (4.4). As $\lambda \to \infty$, $||u||_\infty \to \infty$ and as $\lambda \to \infty$, $||u||_\infty \to 3/2$. Further, one can show that λ as a function of $||u||_\infty$ is monotone decreasing. The numerical computed bifurcation diagram is shown in Figure 4.2. This type of bifurcation diagram is often referred to as a bifurcation from ∞.

The difficulty here is that we need to "jump" from the zero solution branch to the nonzero solution branch. To achieve this we will use the d-homotopy method as follows:

Starting with $\gamma = 0$ and $\lambda_0 > 0$, use the Euler-Newton Continuation method to follow solutions of

$$u'' = \lambda_0 u(1-u) + \gamma, \tag{4.5}$$

with γ the free parameter and λ_0 fixed. The curve of solutions for (4.5) hits the $\gamma = 0$ plane at a non-trivial solution. This provides a starting solution for solutions of (4.4).

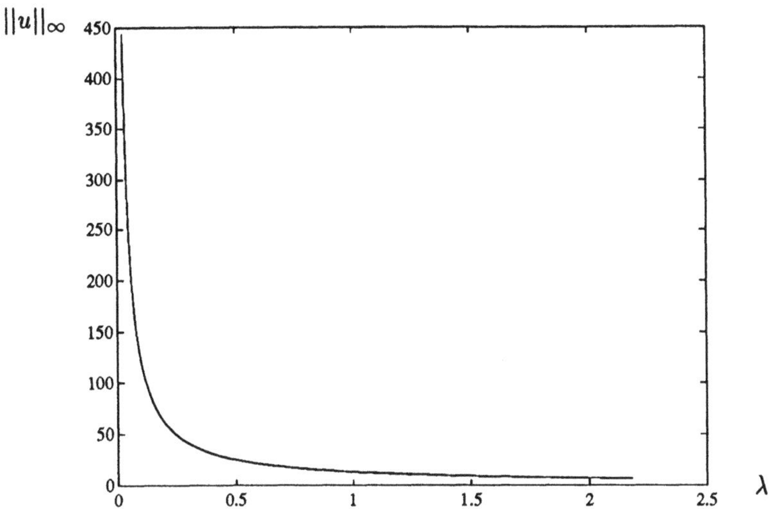

Figure 4.2: Approximate Solution Branchs via AUTO (- - -) and the Sinc Method for Example 4.2.

We take g, w and h as in Example 4.1 above. The discrete system for the d-homotopy method then becomes

$$F(\vec{u}, \lambda, \gamma) \equiv A\vec{u} + \mathcal{D}\left(\frac{1}{(\phi')^2}\right)[\vec{f}(\vec{u}, \lambda) + \gamma\vec{1}] = 0, \qquad (4.6)$$

where

$$\vec{f}(\vec{u}, \lambda) = (\lambda u_{-M}, \dots, \lambda u_N)^T.$$

We first solve (4.6) with λ fixed using γ as the continuation parameter until γ returns to zero at a non–trivial solution. This gives us a starting solution on the branch of positive solutions. Once this is done we continue the solution of (4.6) with $\gamma \equiv 0$ and λ the continuation parameter. The results of the λ continuation are shown in Figure 4.2. We have also done the same calculations using the continuation code AUTO ([3]). The results of the AUTO run using 120 collocation points are also shown in Figure 4.2. We see that there is good agreement between the two methods (there are two plots shown).

Example 4.3 In the final example we consider a problem on the half line namely

$$u''(x) - u(x) = -\frac{u(x)^3}{x^2} \quad \text{on } (0, \infty) \qquad (4.7)$$

$$u(0) = 0 \quad u \to 0 \text{ as } x \to \infty \ . \qquad (4.8)$$

J. DOCKERY AND N. LYBECK

It is known (see e.g., [6, 9]) that the solutions of (4.7)-(4.8) can be characterized by their nodal properties. In particular, there is a discrete set of numbers

$$\{a_j|\ldots < -a_2 < -a_1 < 0 < a_1 < a_2 < \ldots\}$$

such if $u'(0) \neq \pm a_j$ for some j then there is no bounded solution of (4.7). Further, if $u'(0) = \pm a_j$ then there is a solution of (4.7)-(4.8) having $j - 1$ zeros on the interval $(0, \infty)$. For example, the solution with $u'(0) = a_1$ is positive on the half line.

Here the challenge is to not only produce the solutions $u(x)$ but also the "eigenvalues" a_j. In this example we indicate how that Sinc-Galerkin method in concert with the d-homotopy method can be used to find approximate solutions of (4.7)-(4.8).

Since the problem is on the half line we take the map $\phi(x) = ln(\sinh(x))$. The nodes for this map are given by

$$x_k = ln(e^{kh} + \sqrt{e^{2kh} + 1}).$$

Since we are interested in computing not only an approximation of u but also u' we take $g(x) \equiv 1$. One reason for this is that

$$\frac{d}{dx} S(k, h) \circ \phi(x)$$

is unbounded as $x \to 0$ or ∞. For more details on this choice we refer the reader to [10, 8]. For the inner product weight we take $w(x) = \phi'(x)$. It follows from (2.3) and (2.5) that with these choices of weights we a using the Petrov-Galerkin method ([11]). In particular we are orthogonalizing the residual against the unweighted Sinc functions. With these choices of weights, $\tilde{w}(x) \equiv 1$.

There is one complication, with the above choice of weight g, the approximate solution would have zero derivative at $x = 0$, thus we could only approximate the zero solution. To overcome this we add a boundary base to approximate the derivative at the origin. In particular we look for an approximation in the form

$$u_m(x) = u_{-M-1}\, \omega_0(x) + \sum_{k=-M}^{N} u_k \frac{1}{\phi'(x)} S(k, h) \circ \phi(x), \qquad (4.9)$$

where we take $\omega_0(x) = x \exp(-x)$.

It follows that the coefficient u_{-M-1} is an approximation of the derivative of the solution at $x = 0$.

Because of the added boundary base, the discrete system is now bordered. We are going to use the d-homotopy method, thus we add γ to the right hand side of (4.7) and use the Sinc quadrature rule to approximate

$$\int_0^\infty \left(u_m'' - u_m + \frac{u_m^3}{x^2} - \gamma \right) S(j,k) \circ \phi \ \ dx = 0 \quad \text{for } j = -M-1, \ldots, N$$

with u_m given by (4.9). This yields the discrete system

$$\mathcal{A}\vec{u} = \vec{b},$$

where

$$\vec{b}_i = \left(-\frac{u_m(x_i)^3}{x_i^2} + \gamma \right) \frac{1}{\phi'(x_i)}$$

and

$$\mathcal{A} = [\vec{c} \mid A_{hb}].$$

The vector $\vec{c}$ is given by

$$\vec{c}_j = \left(\omega_0''(x_j) - \omega_0(x_j) \right) / \phi'(x_j)$$

and

$$A_{hb} = A_{hns} D\left(\frac{1}{\phi'} \right). \tag{4.10}$$

In (4.10), A_{hns} is a non-square $m+1 \times m$ version of A_h. It is obtained by using the formula (2.8) for A_h with the proviso that the matrices $I^{(l)}$, $l = 0, 1$, and 2 in (2.8) are $m+1 \times m$ and all diagonal matrices are $m \times m$. We take h as in (2.6) with $\alpha - \beta = 1$ and $d = 1.0$.

A portion of the d-homotopy path is shown in Figure 4.3. We see that there are at least to nontrival solutions. We have continued this path further, the numerics indicate that the solution path intersects the $\gamma = 0$ axies infinitely often.

In Table 4.2 we compare our estimate for a_1 to the result given by Ryder ([9]). In figure 4.4 we present the approximate solutions for a_1 and a_2. Notice, the solution corresponding to a_1 is positive on $(0, \infty)$ where as the solution at $u'(0) = a_2 \approx 14.1036$ has one internal zero.

References

[1] E. L. ALLGOWER and K. GEORG, *Numerical Continuation Methods: An Introduction*, Springer-Verlag, New York, 1990.

[2] J. CHAUVETT and F. STENGER, "The Approximate Solution of the Nonlinear Equation $\Delta u = u - u^3$," *J. Math. Anal. Appl.*, v. 51, 1975, pp. 229-242.

$M = N$	$\left|a_{1_T} - a_{1_{approx}}\right|$	asy. error	$\dfrac{\left|a_{1_T} - a_{1_{approx}}\right|}{\text{asy.error.}}$
4	$8.6210e - 02$	$4.61941e - 01$	$1.8663e - 01$
8	$1.9149e - 02$	$4.25549e - 01$	$4.4999e - 02$
16	$3.9640e - 03$	$2.13389e - 01$	$1.8576e - 02$
32	$2.5757e - 04$	$4.52731e - 02$	$5.6893e - 03$
64	$1.2129e - 05$	$2.84593e - 03$	$4.2618e - 03$

Table 4.2: Error in a_1 for Example 4.3.

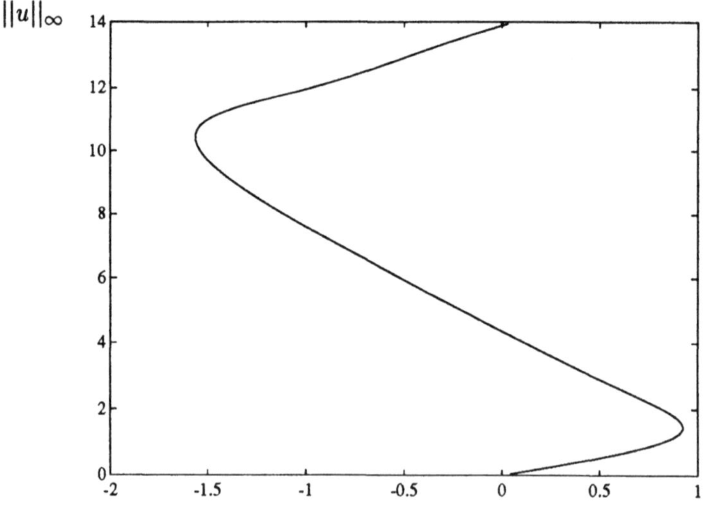

Figure 4.3: The Approximate d-Homotopy Path for Example 4.3.

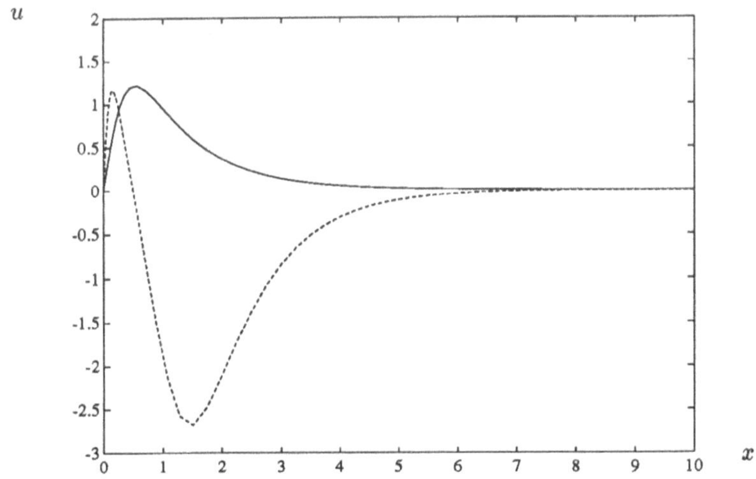

Figure 4.4: The First Two Approximate Solutions for Example 4.3.

[3] E. J. DOEDEL, " AUTO: A Program for the Automatic Bifurcation Analysis of Autonomous Systems," *Cong. Num.*, v. 30, 1981, pp. 265-284. (Proc. 10th Manitoba Conf. on Num. Math. and Comp., Univ. of Manitoba, Winnipeg, Canada, 1980).

[4] R. GLOWINSKI, H. B. KELLER and L. RHEINHART, "Continuation Conjuate Gradient Methods for the Least-Squares Solution of Nonlinear Boundary Value Problems," *SIAM J. Sci. Stat. Comput.*, v. 6, 1985, pp. 793-832.

[5] T. M. HAGSTROM and H. B. KELLER, "Asymptotic Boundary Conditions and Numerical Methods for Nonlinear Elliptic Problems on Unbounded Domains," *Math. Comp.*, v. 48, 1987, pp.449-470.

[6] C. K. R. T. JONES and T. KUEPPER, "On the Infinitely Many Solutions of a Semilinear Elliptic Equation," *SIAM J. Math. Anal.* , v. 17, 1986, pp.803-835.

[7] H. B. KELLER, "Numerical Solution of Bifurcation and Nonlinear Eigenvalue Problems," In: *Applications of Bifurcation Theory*, ed. P.H. Rabinowitz, Academic Press, New York, 1977, pp. 359-384.

[8] J. LUND and K. L. BOWERS, *Sinc Methods for Quadrature and Differential Equations.*, Siam, Philadelphia, PA, 1992.

[9] G.H. RYDER, *Boundary value problems for a class of nonlinear differential equation*," *Pacific J. Math.*, v. 22, 1967, pp. 477-503.

[10] F. STENGER, "Numerical Methods Based on Whittaker Cardinal, or Sinc Functions," SIAM Review, v. 23 No.2, April, 1981.

[11] F. STENGER, "A Sinc-Galerkin Method of Solution of Boundary Value Problems," *Math. Comp.*, v. 33, 1979, pp. 85-109.

ON THE KALMAN-YACUBOVICH-POPOV LEMMA FOR NONLINEAR SYSTEMS

Kenneth A. Doll and Christopher I. Byrnes[*]

*Department of Systems Science and Mathematics
Washington University,
St. Louis, Missouri*

1   Introduction

This paper has its origin in the confluence of two themes which are central in the use of optimization methods to generate feedback laws meeting certain desired design criteria.

On the one hand, there has been substantial historical interest in using convex variational problems to achieve feedback stabilization. For example, infinite time optimal control problems with a positive definite or convex Lagrangian provide a standard method for generating both a candidate Lyapunov function, viz. the "cost-to-go," and a stabilizing feedback law, viz. an optimal closed loop strategy. This has been no doubt most successful in the form of LQ design theory for finite-dimensional linear time-invariant systems, but has also been used with success for linear distributed parameter systems. For nonlinear systems, similar widespread application has been limited by the fact that, in general, the cost-to-go may not be smooth, or even continuous, as a function of the initial state. However, this can be established locally in the generic (i.e., hyperbolic) case using invariant manifold methods from nonlinear dynamics (see [1], [2], [9]).

On the other hand, optimization problems involving nonconvex Lagrangians can also generate stabilizing feedback laws. For example, infinite time horizon H-infinity control also results in stabilizing feedback strategies generated in a differential game theoretic setting in terms of an infinite time optimization problem which, however, involve a nonconvex Lagrangian. In this case, smoothness of the value function of the differentail game can also be establsihed using invariant manifold methods (see [6], [10]).

The stabilizability properties of certain nonconvex optimization problems is not surprising, especially in the light of the fact that the inverse problem of classical optimal control has a solution, for linear systems, derived in terms of positive real, or passive, systems. And, passive systems are themselves associated with a nonconvex optimization problem, whose solution is embodied in the classical and important Kalman-Yacubovich-Popov Lemma. Some time ago ([5]), the solution of the inverse problem of

---

[*]Supported in part by grants from AFOSR and NSF

optimal control was extended to nonlinear systems under the assumption that the value function of the associated nonconvex problem is smooth, resulting also in the derivation of the Kalman-Yacubovich-Popov Lemma for such nonlinear systems.

In section two of this paper, we recall the concept of passivity for nonlinear systems and apply invariant manifold methods to derive the Kalman-Yacubovich-Popov Lemma, for the class of proper, nonlinear systems which are affine with respect to the control variable. In section three, we illustrate the use of the nonlinear Kalman-Yacubovich-Popov Lemma in a problem of nonlinear feedback stabilization.

2 Kalman-Yacubovich-Popov Lemma for a Class of Nonlinear Systems

In this section, we will show that for a certain class of nonlinear dynamical systems, we can obtain explicitly the available storage and required supply functions as two specific solutions to a Hamilton-Jacobi equation. As a result of the available storage V_a (see eqn. (2.11)) and required supply V_r (see eqn. (2.12)) being computed this way, we also discover certain smoothness properties associated with these functions. This property allows V_a and V_r to be used in many situations where smoothness was assumed. The functions V_a and V_r also play a role in determining the optimally stabilizing and optimally destabilizing solutions to the Hamilton-Jacobi equation. They also are solutions to the nonlinear analogue of the Kalman-Yacubovich-Popov Lemma.

The class of nonlinear dynamical systems (Σ) under study have the form

$$\begin{aligned} \dot{x} &= f(x) + G(x)u \\ y &= h(x) + K(x)u \end{aligned} \qquad (2.1)$$

Admissible control inputs are those that are locally square integrable. The functions f, G, h, $K \in C^r$ ($r \geq 2$), with $f(0) = 0$, and $h(0) = 0$. So for any $x(t_0) \in \mathbb{R}^n$, there exists a unique solution to Σ such that y is also locally square integrable. The input $u(t)$ and the output $y(t)$ both take on values in $\mathbb{R}^m$ with $m \leq n$. The linearization of the above system (Σ_L) about the equilibrium point $x = 0$ is expressed as follows

$$\begin{aligned} \dot{x}_L &= Ax_L + Bu_L \\ y_L &= Cx_L + Du_L \end{aligned} \qquad (2.2)$$

where $A = \frac{\partial f}{\partial x}\big|_{x=0}$, $B = G(0)$, $C = \frac{\partial h}{\partial x}\big|_{x=0}$, and $D = K(0)$. The system Σ will also be assumed to satisfy certain dissipation conditions.

Definition 2.1 *A system Σ with supply rate $w(u,y)$ is said to be* <u>dissipative</u> *if there exists a nonnegative function V mapping from $\mathbb{R}^n$ to $\mathbb{R}$, with $V(0) = 0$, (called a storage function) such that for all admissable controls u, and initial conditions $x(t_0)$, with $t_1 \geq t_0$*

$$V(x(t_1)) - V(x(t_0)) \leq \int_{t_0}^{t_1} w(u,y)dt. \qquad (2.3)$$

It is assumed that the supply rate $w(u,y)$ satisfies for any admissable u, $\int_0^T |w(u,y)|dt < \infty$ for all $T \geq 0$.

Definition 2.2 *A system Σ is said to be* <u>passive</u> *if it is dissipative with respect to supply rate $w(u,y) = \langle u,y \rangle$ and* <u>strongly passive</u> *if it is dissipative with respect to $w(u,y) = \langle u,y \rangle - \frac{\gamma}{2}\langle u,u \rangle$ for some $\gamma > 0$.*

Similar to passivity is the property of being positive real.

Definition 2.3 *A system is said to be* <u>positive real</u> *if for all admissable u, and $T \geq 0$, $\int_0^T \langle u,y \rangle \, dt \geq 0$ whenever $x(0) = 0$. Similarly, a system is* <u>strongly positive real</u> *under the same assumptions when $\int_0^T (\langle u,y \rangle - \frac{\gamma}{2}\langle u,u \rangle)dt \geq 0$.*

The relationship between a system being (strongly) passive and (strongly) positive real involves reachability of the state space from the point of minimum storage ($x = 0$).

Definition 2.4 *A system Σ is said to be* <u>reachable from 0</u> *if for all x_1 in the state space, there exists an admissible control u and a time $t_1 \geq t_0$ such that the state can be driven from $x(t_0) = 0$ to $x(t_1) = x_1$.*

The relationship between passivity and positive realness can be summa rized in the following lemma from [3].

Lemma 2.5 *[3] A passive system is positive real. A positive real system that is reachable from the origin is passive.*

This concept can be broadened to include strongly passive and strongly positive real.

One more definition is needed before we can state the central theorem of this paper.

Definition 2.6 *A system Σ is zero-state observable if, for any trajectory such that $u(t) \equiv 0$, $y(t) \equiv 0$ implies $x(t) \equiv 0$.*

Theorem 2.7 *Suppose the system Σ is strongly positive real, reachable from the origin, zero state observable, and satisfies $(K(x) + K^T(x))$ non-singular for all x. Suppose also that its linearization Σ_L is controllable and has its spectrum in the open left half plane. Then V_a and $V_r \in C^r (r \geq 2)$.*

Proof: To prove this result, we will need several intermediate lemmas.

Lemma 2.8 *Suppose the system Σ satisfies $\int_0^T w(u, y)dt \geq 0$, with $T \geq 0$, $x(0) = 0$ for*

$$w(u, y) = \frac{1}{2}(u^T Ru + 2u^T Sy + y^T Qy), \tag{2.4}$$

where $R = R^T$, $Q = Q^T$. Then the linearized system Σ_L satisfies $\int_0^T w(u_L, y_L)dt \geq 0$, with $T \geq 0$, $x_L(0) = 0$ for

$$w(u_L, y_L) = \frac{1}{2}(u_L^T Ru_L + 2u_L^T Sy_L + y_L^T Qy_L), \tag{2.5}$$

Proof: Using an approach similar to [9], consider an input $u_L(\cdot)$ to Σ_L and define $u(t, \varepsilon) = \varepsilon u_L(t)$, ε small, as a one parameter family of inputs to Σ. Denoting the resulting family of states and outputs to Σ as $x(t, \varepsilon)$ and $y(t, \varepsilon)$ respectively, we first need to show that $x_L(t) = \left[\frac{\partial x(t, \varepsilon)}{\partial \varepsilon}\right]_{\varepsilon=0}$, and $y_L(t) = \left[\frac{\partial y(t, \varepsilon)}{\partial \varepsilon}\right]_{\varepsilon=0}$. Differentiate Σ with respect to ε and evaluate at $\varepsilon = 0$. Noting that $u(t, 0) = 0$, $x(0, \varepsilon) = 0$, and $f(0) = 0$ we get $x(t, 0) = 0$. We are left with:

$$\frac{d}{dt}\left[\frac{\partial x(t, \varepsilon)}{\partial \varepsilon}\right]_{\varepsilon=0} = A\left[\frac{\partial x(t, \varepsilon)}{\partial \varepsilon}\right]_{\varepsilon=0} + Bu_L(t) \tag{2.6}$$

$$\left[\frac{\partial y(t, \varepsilon)}{\partial \varepsilon}\right]_{\varepsilon=0} = C\left[\frac{\partial x(t, \varepsilon)}{\partial \varepsilon}\right]_{\varepsilon=0} + Du_L(t) \tag{2.7}$$

Subtracting these equations from Σ_L and letting $z(t) = x_L(t) - \left[\frac{\partial x(t, \varepsilon)}{\partial \varepsilon}\right]_{\varepsilon=0}$ we obtain $\dot{z}(t) = Az(t)$ with $z(0) = 0$. This generates a unique solution $z(t) = 0$ or $x_L(t) = \left[\frac{\partial x(t, \varepsilon)}{\partial \varepsilon}\right]_{\varepsilon=0}$. This immediately implies $y_L(t) = \left[\frac{\partial y(t, \varepsilon)}{\partial \varepsilon}\right]_{\varepsilon=0}$. Now considering the original problem, if Σ satisfies $\int_0^T w(u, y)dt \geq 0$, $x(0) = 0$, then differentiating twice with respect to ε, and setting $\varepsilon = 0$, yields $\int_0^T w(u_L, y_L)dt \geq 0$, $x_L(0) = 0$.

This lemma showed that if the system Σ is (strongly) positive real, then so is the system Σ_L. Since (A, B) is controllable, the system Σ_L will also be (strongly) passive.

Lemma 2.9 *Suppose the system Σ_L is strongly positive real. Suppose also that $\sigma(A) \subset \mathbb{C}^-$, (A, B) is controllable, and $(D + D^T)$ is invertible. Then the algebraic Riccati equation (ARE)*

$$A^T P + PA + (PB - C^T)(D + D^T)^{-1}(B^T P - C) = 0 \qquad (2.8)$$

has exactly one real symmetric solution P^- with the additional property that $\sigma(A^-) \subset \mathbb{C}^-$ where $A^- = A + B(D + D^T)^{-1}(B^T P^- - C)$, and exactly one real symmetric solution P^+ with the additional property that $\sigma(A^+) \subset \mathbb{C}^+$ where $A^+ = A + B(D + D^T)^{-1}(B^T P^+ - C)$. In addition, $P^- < P^+, 0 < P^+$, and every real symmetric solution P to the (ARE) satisfies $P^- \leq P \leq P^+$.

Proof: The assumption that Σ_L is strongly positive real gives us the following frequency domain inequality

$$H(s^*, s) = G(s) + G^T(s^*) \geq \gamma I \quad \text{for all } Re\ s \geq 0 \qquad (2.9)$$

where $G(s) = C(Is - A)^{-1}B + D$. This can easily be seen by considering a new system $\tilde{\Sigma}_L$ where $y = Cx + Du$ is replaced by $\tilde{y} = Cx + (D - \frac{\gamma}{2}I)u$ $\tilde{y} = y - \frac{\gamma}{2}u$)

The following are now equivalent:

(i) Σ_L strongly positive real

(ii) $\tilde{\Sigma}_L$ positive real

(iii) $\tilde{G}(s) + \tilde{G}^T(s^*) \geq 0$ for all $Re\ s \geq 0,\ s \notin \sigma(A)$

(iv) $G(s) + G^T(s^*) \geq \gamma I$ for all $Re\ s \geq 0,\ s \notin \sigma(A)$

Next, we need to find an $\varepsilon > 0$ such that the strong frequency domain inequality (SFDI) holds

$$H(s^*, s) \geq \varepsilon(s + s^*)B^T(Is^* - A^T)^{-1}(Is - A)^{-1}B \text{ for all } Re\ s \geq 0 \qquad (2.10)$$

Choosing $\varepsilon \leq \gamma/M$ where

$$M = \max_{s \in \overline{\mathbb{C}^+}}(s + s^*)B^T(Is^* - A^T)^{-1}(Is - A)^{-1}B < \infty$$

since $\sigma(A) \subset \mathbb{C}^-$, we have the desired inequality. Since (SFDI) holds, a theorem by [11] tells us that there exist real symmetric solutions P^- and P^+ to the ARE, such that all the desired conditions are satisfied.

Returning to the nonlinear problem, we have a system Σ that is positive real and reachable from the origin. This means that the available storage

$$V_a(x) = -\inf_u \int_0^{t_1} \langle u, y \rangle dt,\ t_1 \geq 0,\ x(0) = x \qquad (2.11)$$

and the required supply

$$V_r(x) = \inf_u \int_{t_{-1}}^0 \langle u, y \rangle dt, \quad t_{-1} \le 0, \quad x(0) = x \tag{2.12}$$

are well defined (i.e. finite).

Since $w(u, y) = \langle u, y \rangle$ is such that for all y, there exists u such that $w(u, y) \le 0$ we can restate V_a and V_r as

$$V_a(x) = -\lim_{t_1 \to \infty} \inf_u \int_0^{t_1} \langle u, y \rangle dt, \qquad x(0) = x \tag{2.13}$$

and

$$V_r(x) = \lim_{t_{-1} \to -\infty} \inf_u \int_{-t_{-1}}^0 \langle u, y \rangle dt, \qquad x(0) = x \tag{2.14}$$

with the same constraints noted above.

Consider the optimal control problem where Σ is the system dynamics and we are trying to minimize the functional $J(u, x_0) = \int_0^\infty \langle u, y \rangle dt$ over all admissable controls u with the initial condition $x(0) = x_0$. The associated Hamiltonian for this problem is

$$H(x, p, u) = \langle p, f(x) + G(x)u \rangle - \langle u, h(x) + K(x)u \rangle \tag{2.15}$$

and the problem is to maximize $H(x, p, u)$ over all admissable u.

Solving for the optimal control u_*

$$\frac{\partial H}{\partial u} = G^T(x)p - h(x) - (K(x) + K^T(x))u = 0 \tag{2.16}$$
$$u_* = (K(x) + K^T(x))^{-1}(G^T(x)p - h(x))$$

The optimal Hamiltonian can be rewritten as

$$H_*(x, p) = p^T f(x) + \frac{1}{2}(p^T G(x) - h^T(x))(K(x) + K^T(x))^{-1}(G^T(x)p - h(x)) \tag{2.17}$$

where its associated Hamiltonian vector field (X_H) is

$$\begin{bmatrix} \dot{x} \\ \dot{p} \end{bmatrix} = \begin{bmatrix} \frac{\partial H}{\partial p} \\ -\frac{\partial H}{\partial x} \end{bmatrix} \tag{2.18}$$

The linearization of the Hamiltonian vector field around the point $(x, p) = (0, 0)$ is given by the following costate Hamiltonian differential equations

$$\begin{bmatrix} \dot{x}_L \\ \dot{p}_L \end{bmatrix} = DX_H \begin{bmatrix} x_L \\ p_L \end{bmatrix} \tag{2.19}$$

where

$$DX_H = \begin{bmatrix} A - B(D + D^T)^{-1}C & B(D + D^T)^{-1}B^T \\ -C(D + D^T)^{-1}C & -(A - B(D + D^T)^{-1}C)^T \end{bmatrix} \qquad (2.20)$$

To proceed, we need to determine if the linearized Hamiltonian vector field, DX_H is hyperbolic, that is, it has no imaginary axis eigenvalues.

Consider a subspace of the form $\begin{bmatrix} I \\ P- \end{bmatrix}$. It satisfies $DX_H \begin{bmatrix} I \\ P- \end{bmatrix} = \begin{bmatrix} I \\ P- \end{bmatrix} A^-$ where A^- is defined as in Lemma 2.3. This means that $\begin{bmatrix} I \\ P- \end{bmatrix}$ spans the stable eigenspace of DX_H and has dimension n. By the structure of a Hamiltonian system, we know that the other n eigenvalues are in the open right half plane. In fact $\begin{bmatrix} I \\ P+ \end{bmatrix}$ spans the unstable eigenspace and satisfies $DX_H \begin{bmatrix} I \\ P+ \end{bmatrix} = \begin{bmatrix} I \\ P+ \end{bmatrix} A^+$ with A^+ defined as in Lemma 2.3.

Hyperbolicity of the origin allows us to conclude using invariant manifold techniques that there exists a stable manifold N^- and an unstable manifold N^+ defined in some neighborhood of the origin, $(x, p) = (0, 0)$, which are tangent at the origin to the stable eigenspace, $\mathrm{span}\{\begin{bmatrix} I \\ P- \end{bmatrix}\}$, and unstable eigenspace, $\mathrm{span}\{\begin{bmatrix} I \\ P+ \end{bmatrix}\}$, respectively.

Since N^- and N^+ are tangent at the origin to subspaces of the form $\mathrm{span} \{\begin{bmatrix} I \\ P \end{bmatrix}\}$, there exists a neighborhood W of $x = 0$ such that the submanifolds N^- and N^+ are projectable (i.e. $N^- = \{(x, \pi^-(x)) | x \in W\}$ and $N^+ = \{(x, \pi^+(x)) | x \in W\}$).

Hyperbolicity of the Hamiltonian vector field also implies that the N^- and N^+ are Lagrangian submanifolds; that is

$$d\pi|_N = \sum_{i=1}^{n} d\pi_i \wedge dx_i|_N = 0 \qquad (2.21)$$

where π, N represent either the stable or unstable manifolds.

Now that both N^- and N^+ are Lagrangian submanifolds and projectable on a neighborhood W of $x = 0$, we can deduce that $N^- =$ graph dV^- for some C^{r-1} function V^-, and similarly that $N^+ =$ graph dV^+ for some C^{r-1} function V^+.

Now the stable and unstable manifolds can be represented as

$$N^- = \{(x, \frac{\partial V^-(x)}{\partial x}) | x \in W\}, \ N^+ = \{(x, \frac{\partial V^+}{\partial x}(x)) | x \in W\} \qquad (2.22)$$

Remark 2.10 *If our initial system is C^r, $r \geq 2$, then the Hamiltonian system will be C^{r-1}. The stable and unstable manifolds will also be C^{r-1}. Since $N^- =$ graph dV^- and $N^+ =$ graph dV^+, this implies that V^- and V^+ are C^r. For $r = 2$, V^- and V^+ are twice differentiable. So on the neighborhood W, the Hamilton-Jacobi equation (HJE) is satisfied for $p^T = \frac{\partial V}{\partial x}$*

$$\frac{\partial V}{\partial x}f(x)+\frac{1}{2}\left(\frac{\partial V}{\partial x}G(x) - h^T(x)\right)(K(x)+K^T(x))^{-1}\left(G^T(x)\frac{\partial^T V}{\partial x} - h(x)\right) = 0 \tag{2.23}$$

where $V = V^-$ or $V = V^+$.

The optimal control may be rewritten as

$$u_* = (K(x) + K^T(x))^{-1}\left(G^T(x)\frac{\partial^T V}{\partial x} - h(x)\right) \tag{2.24}$$

Note also that any V that satisfies the HJE also satisfies the dissipation inequality. This can be seen by noting first that the HJE may be rewritten as

$$\frac{\partial V}{\partial x}f(x) + \frac{1}{2}u_*^T(K(x) + K^T(x))u_* = 0 \tag{2.25}$$

Substituting the identity

$$\frac{1}{2}u_*^T(K(x)+K^T(x))u_* = \frac{\partial V}{\partial x}G(x)u-\langle u, y\rangle+\frac{1}{2}(u-u_*)^T(K(x)+K^T(x))(u-u_*) \tag{2.26}$$

gives

$$\frac{\partial V}{\partial x}(f(x)+G(x)u)- \langle u, y\rangle+\frac{1}{2}(u-u_*)^T(K(x)+K^T(x))(u-u_*)=0 \tag{2.27}$$

which is maximized at $u = u_*$.

The dissipation inequality for all admissable u comes from noting that

$$\frac{\partial V}{\partial x}(f(x) + G(x)u) - \langle u, y\rangle \leq 0 \text{ for all } u$$

and integrating along solutions of $\dot{x} = f(x) + G(x)u$ to get

$$V(x(t_1)) - V(x(t_0)) \leq \int_{t_0}^{t_1} \langle u, y\rangle dt.$$

Therefore V^- and V^- satisfy the dissipation inequality. In [5], it is shown that if a system is dissipative and zero-state observable, all functions V satisfying the dissipation inequality must be positive definite. So $V^- > 0$ and $V^+ > 0$.

In the linearized discussion, we found that the maximizing solutions to the ARE satisfied $P^- < P^+$ and $P^+ > 0$. Since V^- and V^+ are solutions to the (HJE) it follows that $P^- = \frac{\partial^2 V^-}{\partial x^2}|_{x=0}$ and $P^+ = \frac{\partial^2 V^+}{\partial x^2}|_{x=0}$. Since $V^- > 0$, $\frac{\partial^2 V^-}{\partial x^2}|_{x=0} \geq 0$ so the following relationship holds for solutions to the (ARE): $0 \leq P^- < P^+$ and any solution P satisfies $P^- \leq P \leq P^+$.

Similarly, since $P^- < P^+$, we have for all $x \in W$ that $V^-(x) < V^+(x)$.

Lemma 2.11 *Suppose that $V_a(V_r)$ is well defined, and that there exists a global C^r solution $V^- \geq 0(V^+ \geq 0)$ to the Hamilton-Jacobi equation that generates the stable (unstable) manifold as defined above. We can now show that $V_a = V^-(V_r = V^+)$ (i.e. V_a and V_r are C^r functions).*

Proof: Using the solution V^-, we can rewrite the Hamilton-Jacobi equation as

$$\frac{\partial V^-}{\partial x}(f(x) + G(x)u_*) = \langle u_*, h(x)\rangle + \frac{1}{2}\langle u_*, (K(x) + K^T(x))u_*\rangle \quad (2.28)$$

where $u_* = (K(x) + K^T(x))^{-1}(g^T(x)\frac{\partial^T V^-}{\partial x} - h(x))$. By integrating along trajectories of the form,

$$\dot{x} = f(x) + G(x)u_* \quad (2.29)$$

(i.e. on the stable manifold), the (HJE) reduces to

$$V^-(x(t_1)) - V^-(x(t_0)) = \int_{t_0}^{t_1}\{\langle u_*, h(x)\rangle + \frac{1}{2}\langle u_*, (K(x) + K^T(x))u_*\rangle\}dt \quad (2.30)$$

for all $t_1 \geq t_0$. Since the integral dissipation inequality is satisfied with $V = V_a$ for every u, take $u = u_*$, which gives us

$$V_a(x(t_1)) - V_a(x(t_0)) \leq \int_{t_0}^{t_1}\{\langle u_*, h(x)\rangle + \frac{1}{2}\langle u_*, (K(x) + K^T(x))u_*\rangle\}dt \quad (2.31)$$

Subtracting (2.30) from (2.31), we obtain

$$V_a(x(t_1)) - V^-(x(t_1)) \leq V_a(x(t_0)) - V^-(x(t_0)) \quad (2.32)$$

Assuming that $x(t_0) \in W$, let $t_1 \to \infty$, and using the local exponential stability of (2.29) we obtain

$$V_a(x) \geq V^-(x) \text{ for all } x \in W. \quad (2.33)$$

Since $V^-(x) \geq 0$ and satisfies the (HJE), we also have that $V_a(x) \leq V^-(x)$ for all $x \in W$. We now have $V_a(x) = V^-(x)$ for all $x \in W$. $V_r(x) = V^+(x)$ for all $x \in W$ follows similarly if you let $t_0 \to -\infty$. ∎

The final point is that storage functions $V = V_a$ and $V = V_r$ satisfy the Kalman-Yacubovich-Popov lemma. For systems of this type, it is nothing more than the differential version of the dissipation inequality.

$$\frac{\partial V}{\partial x}f(x) + \frac{\partial V}{\partial x}G(x)u \leq h^T(x)u + \frac{1}{2}u^T(K(x) + K^T(x))u. \qquad (2.34)$$

for all u and x.

3 Applications to Stability

A very common interest in nonlinear control theory is which feedbacks will stabilize a plant that is positive real. In the linear, single input-single output, and strictly proper $(D = 0)$ case, there exist theorems which state that if a system is positive real, its state space is reachable from the point of minimum storage, and if it is completely observable, then any continuous feedback which is a I-III sector function of the output is asymptotically stabilizing. A I–III sector function $s(y)$ is one which lies only in the I-III sectors, i.e. $y > 0 \Rightarrow s(y) > 0$, $y < 0 \Rightarrow s(y) < 0$, and $s(0) = 0$. The following lemma will generalize this theory to the proper (but not strictly), nonlinear multi-input multi-output case.

Lemma 3.1 *Given a system Σ that satisfies the conditions of Theorem 2.1, any continuous feedback of the form $u = -s(y)$, where $\langle s(y), y \rangle > 0$ for all $y \neq 0$, will be locally asymptotically stabilizing.*

Proof: From the above theorem, there exist functions $V(x)$ with $0 < V_a(x) \leq V(x) \leq V_r(x)$ defined in a neighborhood W of $x = 0$ which satisfy the dissipation inequality

$$V(x(t)) - V(x(t_0)) \leq \int_{t_0}^{t} \langle u(\tau), y(\tau) \rangle d\tau \qquad (3.1)$$

for all $t \geq 0$ and all locally square integrable inputs u such that $x(t)$ stays in W.

Choose a candidate $V(x)$ which is continuous. This is possible since we just proved that at least V_a and V_r are twice differentiable. Applying the feedback $u = -s(y)$, we obtain

$$V(x(t)) - V(x(t_0)) \leq -\int_{t_0}^{t} \langle s(y(\tau)), y(\tau) \rangle d\tau \qquad (3.2)$$

which is nonpositive for all values of $t \geq 0$. Since $V(x) > 0$ is finite and continuous it may be used as a Lyapunov function to determine stability.

Choose a neighborhood $W' \subset W$ such that for all motions starting at $x(t_0) \in W'$, the subsequent motion $x(t)$ stays in W. Equation (3.2) shows that $V(x(t))$ is monotonically nonincreasing along trajectories of the closed loop system. This shows that the system

$$\begin{aligned}
\dot{x} &= f(x) - G(x)s(y) \\
y &= h(x) - K(x)s(y)
\end{aligned} \qquad (3.3)$$

is Lyapunov stable.

In addition, the system Σ is zero state observable (i.e. $y(t) \equiv 0 \ \forall t \geq t_0$ implies that $x(t) \equiv 0 \ \forall t \geq t_0$). From equation (3.2) and the assumptions on $s(y)$, $V(x(t))$ will be strictly decreasing (i.e. nonconstant) on all nontrivial trajectories of the state $x(t)$ by zero-state observability. This satisfies the conditions of LaSalle's theorem for local asymptotic stability.

4 Conclusion

In this paper, we recall the concept of passivity for nonlinear systems and apply invariant manifold methods to derive the Kalman-Yacubovich-Popov Lemma, for the class of proper, nonlinear systems which are affine with respect to the control variable. We also illustrate the use of the nonlinear Kalman-Yacubovich-Popov Lemma in a problem of nonlinear feedback stabilization.

References

[1] P. Brunovsky, "On optimal stabilization of nonlinear systems, *Mathematical Theory of Control*, A.V. Balakrishnan and Lucien W. Neustadt, eds., Academic Press, New York and London (1967).

[2] C.I. Byrnes, "New methods in nonlinear optimal control, Proceedings of 1st ECC, Grenoble, 1991.

[3] C.I. Byrnes, A. Isidori, J.C. Willems, "Passivity, Feedback Equivalence and the Global Stabilization of Minimum Phase Nonlinear Systems", to appear in *Proc. of the Conf. on Control of Dynamical Systems*, Lyon, June 1990.

[4] J.C. Doyle, K. Glover, P.P. Khargonekar, B.A. Francis, "State space solutions to standard H_2 and H_∞ control problems," *IEEE Trans. Aut. Contr.*, AC-34, (1990), pp. 831–846.

[5] D.J. Hill, P.J. Moylan, "The stability of nonlinear dissipative systems", *IEEE Trans. Aut. Contr.*, AC-21, No. 4 (1976), pp. 708–711.

[6] A. Isidori, "Feedback control of nonlinear systems, In *Proc. of 1st European Control Conf.*, Grenoble, France, July 1991.

[7] D.L. Lukes, "Optimal regulation of nonlinear dynamical systems", *SIAM J. Contr.*, vol. 7, no. 1, February 1969, pp. 75–100.

[8] P.J. Moylan, "Implications of passivity in a class of nonlinear systems", *IEEE Trans. Aut. Contr.*, AC-19, No. 4, August, 1974, pp. 373–381.

[9] A.J. Van der Schaft, "L_2-gain analysis of nonlinear systems and nonlinear H_∞ control", Memorandum No. 969, University of Twente, May 1991.

[10] A.J. Van der Schaft, "A state-space approach to nonlinear H_∞ control", *Syst. and Contr. Lett.* 16 (1991), pp. 1–8.

[11] J.C. Willems, "Least squares stationary optimal control and the algebraic riccati equation", *IEEE Trans. Aut. Contr.*, AC-16 (1971), pp. 621–634.

[12] J.C. Willems, "Dissipative dynamical systems Part I: General theory", *Arch. Rational Mech. and Analysis*, 45, 1972, pp. 312–351.

[13] J.C. Willems, 'Dissipative dynamical systems Part II: Linear systems with quadratic supply rates", *Arch. Rational Mech. and Anal.* 45, 1972, pp. 352–393.

[14] P.J. Moylan, B.D.O. Anderson, "Nonlinear regulator theory and an inverse optimal control problem," *IEEE Trans. on Aut. Contr.*, AC-18, no. 5, October 1973, pp. 460–465.

ROBUST CONTROL OF DISTRIBUTED PARAMETER SYSTEMS WITH STRUCTURED UNCERTAINTY

Richard H. Fabiano,* Andrew J. Kurdila,[†] and Thomas Strganac[†]

Texas A&M University
College Station, TX 77843

1 Introduction

Consider the following abstract Cauchy problem on a Hilbert space H:

$$\dot{x}(t) = Ax(t) + Bu(t) \tag{1.1}$$
$$x(0) = x_0.$$

Here U is a Hilbert space, $B \in L(U, H)$ and A is the infinitesimal generator of a strongly continuous semigroup $T(t)$ on H. Associated with (1.1) is the cost functional

$$J_0(u) = \int_0^\infty \left\{ |Cx(t)|_H^2 + \langle Ru, u \rangle \right\} dt, \tag{1.2}$$

where $C \in L(H, H)$ and R is a positive definite selfadjoint bounded operator on U. If we set $\bar{U} = L^2(0, \infty; U)$, then the usual linear quadratic regulator (LQR) problem is

$$\min_{u \in \bar{U}} J_0(u) \tag{1.3}$$

subject to dynamics governed by (1.1).

Now suppose that there may be some uncertainty in the model, so that instead of (1.1) we consider

$$\dot{x}(t) = (A + \Delta A)x(t) + (B + \Delta B)u(t) \tag{1.4}$$
$$x(0) = x_0,$$

where $\Delta A \in L(H, H)$ and $\Delta B \in L(U, H)$. An interesting problem is to determine control laws which are robust with respect to the uncertainties ΔA and ΔB. Recent work on these and related problems can be found

---

* Department of Mathematics. The research of the first author was supported in part by the Institute for Scientific Computing, Texas A&M University.

[†] Department of Aerospace Engineering. The research of the second and third authors was supported in part by AFOSR grant F49620-92-J-0450.

in [3], [9], [11]-[14]. A standard approach is to assume that there is some known structure to the uncertainty. That is, assume that ΔA and ΔB can be factored as

$$\Delta A = D L_A E \tag{1.5}$$
$$\Delta B = F L_B G$$

where $D, E \in L(H,H)$, $F \in L(U,H)$, $G \in L(U,U)$ are known, and $L_A \in L(H,H)$, $L_B \in L(U,U)$ are unknown. In a recent paper [15] Rhee and Speyer have shown that by treating the uncertainty terms as disturbances and then applying results from differential game theory, a robust feedback control can be constructed from the solution of a nonstandard game theoretic algebraic Riccati equation. However this is a finite dimensional result, and in this paper we extend these ideas to the infinite dimensional setting under consideration here. We also point out that in certain cases there is a connection between the game theoretic algebraic Riccati equation and the standard LQR algebraic Riccati equation (as observed in [2]). In these cases known LQR convergence results can be applied to approximation schemes for this problem. An example is presented in which a control is designed to be robust with respect to possible actuator failure.

2 Linear system with structured uncertainty

In this section we follow the ideas in [15] and show how the factorization (1.5) of the uncertainty can be used view the problem in a differential game framework. A robust stability result is then given. To proceed, set

$$w_1 = L_A\, y_1 \tag{2.1}$$
$$w_2 = L_B\, y_2$$

where

$$y_1 = Ex$$
$$y_2 = Gu.$$

Set $w = (w_1, w_2) \in H \times H$ and $y = (y_1, y_2) \in H \times U$. Then

$$w = \begin{bmatrix} L_A & 0 \\ 0 & L_B \end{bmatrix} y \tag{2.2}$$

and the 'fictitious' disturbance w can be viewed as a feedback signal of y amplified by an unknown gain. Thus, the uncertain system (1.4) can be written as

$$\dot{x}(t) = Ax(t) + Bu(t) + \Phi w(t) \tag{2.3}$$
$$x(0) = x_0.$$

Here $\Phi \in L(H \times H, H)$ is defined by $\Phi(w_1, w_2) = Dw_1 + Fw_2$. Then (2.3) can be treated in the differential game framework by considering the cost

$$
\begin{aligned}
J_1(u, w) &= \int_0^\infty \left\{ |Cx|_H^2 + \langle Ru, u \rangle_U + |y|_{H \times U}^2 - \frac{1}{\gamma^2} |w|_{H \times H}^2 \right\} dt \\
&= \int_0^\infty \left\{ \langle Qx, x \rangle_H + \langle \hat{R}u, u \rangle_U - \frac{1}{\gamma^2} |w|_W^2 \right\} dt,
\end{aligned} \tag{2.4}
$$

where $Q = C^*C + E^*E$, $\hat{R} = R + G^*G$, and $W = H \times H$. Associated with this cost is the following algebraic Riccati equation

$$
A^*\Pi + \Pi A + Q - \Pi(B\hat{R}^{-1}B^* - \gamma^2 \Phi\Phi^*)\Pi = 0. \tag{2.5}
$$

In order to prove a robust stability result, we require the following lemma which is a modification of a result by Zabczyk [16, Lemma 3]. Recall that if $C \in L(H, V)$ then the pair (C, A) is detectable if and only if there exists an operator $S \in L(V, H)$ such that $A - SC$ is stable (that is, $A - SC$ generates a semigroup $T(t)$ satisfying $|T(t)| \leq \mathrm{Me}^{-\alpha t}$ for some $\alpha > 0$).

Lemma 2.1 *Suppose that $K \in L(H)$ and (C, A) is detectable. If there is an operator $\Pi \geq 0$ satisfying the Liapunov equation*

$$
\langle \Pi x, Ay \rangle + \langle Ax, \Pi y \rangle = -\langle Cx, Cy \rangle - \langle Kx, Ky \rangle \qquad \forall x, y \in \mathrm{dom}A \tag{2.6}
$$

then A is stable.

Proof: The proof follows an argument very similar to that used in [17]. Let $T(t)$ be the semigroup generated by A. By a lemma due to Datko (see [4] or [10, page 116]) it is sufficient to show that

$$
\int_0^\infty |T(t)x|^2 \, dt < \infty \qquad \text{for all} \quad x \in H.
$$

For $x \in \mathrm{dom}A$, set $x(t) = T(t)x$. Then $\dot{x}(t) = Ax(t)$, so that

$$
\begin{aligned}
\frac{d}{dt} \langle \Pi x(t), x(t) \rangle &= \langle \Pi x(t), Ax(t) \rangle + \langle Ax(t), \Pi x(t) \rangle \\
&= -|Cx(t)|^2 - |Kx(t)|^2. \tag{2.7}
\end{aligned}
$$

Integrating from 0 to T gives

$$
\langle \Pi x(T), x(T) \rangle + \int_0^T |Cx(s)|^2 \, ds + \int_0^T |Kx(s)|^2 \, ds = \langle \Pi x, x \rangle \tag{2.8}
$$

for all $x \in \mathrm{dom}A$. A straightforward limit argument shows that (2.8) holds for all $x \in H$. It follows since all terms on the left hand side are nonnegative and the right hand side is independent of T that $\int_0^\infty |C\,T(s)x|^2\,ds < \infty$ for all $x \in H$. Next, since (C, A) is detectable, there exists an operator $S \in L(V, H)$ such that $A - SC$ generates a semigroup $\widetilde{T}(t)$ satisfying $|\widetilde{T}(t)| \le Me^{-\alpha t}$ for some $\alpha > 0$. Observe that $A = A - SC + SC$, so that

$$T(t)x = \widetilde{T}(t)x + \int_0^t \widetilde{T}(t - \tau)SC\,T(\tau)x\,d\tau$$

and it follows that

$$|x(t)| \le |S| \int_0^t |\widetilde{T}(t - \tau)|\,|C\,T(\tau)x|\,d\tau + |\widetilde{T}(t)x|.$$

Applying Youngs inequality yields

$$\left(\int_0^\infty |x(t)|^2\,dt\right)^{\frac{1}{2}} \le |S| \int_0^\infty |\widetilde{T}(t)|\,dt \left(\int_0^\infty |C\,T(t)x|^2\,dt\right)^{\frac{1}{2}}$$

$$+ \left(\int_0^\infty |\widetilde{T}(t)|^2\,dt\right)^{\frac{1}{2}} |x|$$

$$< \infty$$

The proof is complete. This lemma allows us to prove the following robust stability result.

Theorem 2.2 *Suppose that (C, A) is detectable. If Π is a nonnegative definite solution of (2.5), then the feedback control $u(t) = -\hat{R}^{-1}B^*\Pi x(t)$ stabilizes (1.4) for all $|L_A| < \gamma$, $|L_B| < \gamma$.*

Proof: The proof extends the ideas in [15] to Hilbert space. It must be shown that the closed loop operator $A_c = A + \Delta A - (B + \Delta B)\hat{R}^{-1}B^*\Pi$ is stable. Note that equation (2.5) can be written as

$$\langle \Pi x, Ay \rangle + \langle Ax, \Pi y \rangle + \langle Qx, y \rangle - \langle (B\hat{R}^{-1}B^* - \gamma^2[DD^* + FF^*]\Pi x, \Pi y \rangle = 0$$

for all $x, y \in \mathrm{dom}A$. It follows that

$$\langle \Pi x, A_c y \rangle + \langle A_c x, \Pi y \rangle =$$

$$\underline{\langle \Pi x, Ay \rangle} + \langle \Pi x, \Delta Ay \rangle - \langle \Pi x, (B + \Delta B)\hat{R}^{-1}B^*\Pi y \rangle + \underline{\langle Ax, \Pi y \rangle}$$

$$+ \langle \Delta Ax, \Pi y \rangle - \underline{\langle (B\hat{R}^{-1}B^* - \gamma^2[DD^* + FF^*])\Pi x, \Pi y \rangle} + \underline{\langle Qx, y \rangle}$$

$$- \gamma^2 \langle (DD^* + FF^*)\Pi x, \Pi y \rangle - \langle Qx, y \rangle - \langle \Delta B\hat{R}^{-1}B^*\Pi x, \Pi y \rangle$$

The underlined terms cancel to zero, so that

$$\langle \Pi x, A_c y \rangle + \langle A_c x, \Pi y \rangle =$$

$$- \langle Cx, Cy \rangle - \gamma^2 \langle D^*\Pi x, D^*\Pi y \rangle - \gamma^2 \langle F^*\Pi x, F^*\Pi y \rangle$$

$$- \langle B^*\Pi x, \hat{R}^{-1}B^*\Pi y \rangle + \langle \Pi x, DL_AEy \rangle + \langle DL_AEx, \Pi y \rangle$$

$$- \langle Ex, Ey \rangle - \langle \Pi x, FL_BG\hat{R}^{-1}B^*\Pi y \rangle - \langle FL_BG\hat{R}^{-1}B^*\Pi x, \Pi y \rangle.$$

Continuing, we have

$$\langle \Pi x, A_c y \rangle + \langle A_c x, \Pi y \rangle =$$

$$- \langle Cx, Cy \rangle - \gamma^2 \langle (D^*\Pi - \frac{1}{\gamma^2} L_A E)x, (D^*\Pi - \frac{1}{\gamma^2} L_A E)y \rangle$$

$$- \frac{1}{\gamma^2} \langle (\gamma^2 F^* + L_B G \hat{R}^{-1} B^*)\Pi x, (\gamma^2 F^* + L_B G \hat{R}^{-1} B^*)\Pi y \rangle$$

$$- \langle R \hat{R}^{-1} B^* \Pi x, \hat{R}^{-1} B^* \Pi y \rangle - \langle Ex, (I - \frac{1}{\gamma^2} L_A^* L_A) Ey \rangle$$

$$- \langle G \hat{R}^{-1} B^* \Pi x, (I - \frac{1}{\gamma^2} L_B^* L_B) G \hat{R}^{-1} B^* \Pi y \rangle$$

Now if $|L_A| < \gamma$ and $|L_B| < \gamma$, then the right hand side can be written as $-\langle Cx, Cy \rangle - \langle Kx, Ky \rangle$, so that the result follows from Lemma 2.1.

<u>Remark</u> The result in [15] requires that (Q, A) be detectable, while our result only requires that (C, A) be detectable.

Theorem 2.1 says that a robust control can be constructed from a solution of the algebraic Riccati equation (2.5). In order to devise a numerical method to compute solutions, one can introduce finite dimensional approximations to the system. We point out that under the condition that

$$B \hat{R}^{-1} B^* - \gamma^2 [DD^* + FF^*] \geq 0 \tag{2.9}$$

equation (2.5) can be written as

$$\langle \Pi x, Ay \rangle_H + \langle Ax, \Pi y \rangle_H + \langle C^* C x, y \rangle_H - \langle \Pi \Omega^2 \Pi x, y \rangle_H = 0 \tag{2.10}$$

for all $x, y \in \text{dom } A$, where

$$\Omega = \left[B \hat{R}^{-1} B^* - \gamma^2 [DD^* + FF^*] \right]^{\frac{1}{2}}. \tag{2.11}$$

As was observed in [2], (2.10) is the algebraic Riccati equation for a standard LQR problem. Thus, if (2.9) holds, standard convergence results for approximation of LQR problems in Hilbert space can be applied. We do not give details here; see [5]. For known LQR approximation results, see [1], [6]-[8]. We use this observation to justify our numerical approximations for the example considered in the next section.

3 Example: flexible beam with actuator damage

In this section we consider a model for a situation in which one of two actuators attached to a flexible beam may suffer damage or failure, and it is desired to construct a feedback control law robust with respect this uncertainty. We describe a numerical experiment in which a control is computed first with the standard LQR approach, and then with the method described in this paper. The performance of the two controllers is then compared in the presence of possible actuator failure.

To proceed, consider the following equation for a cantilevered Euler-Bernoulli beam with Kelvin-Voigt damping.

$$\frac{\partial^2 y}{\partial t^2} + \frac{\partial^2}{\partial x^2} EI(x) \frac{\partial^2 y}{\partial x^2} + c_1 \frac{\partial y}{\partial t} + c_2 \frac{\partial^5 y}{\partial x^4 \partial t}$$
$$= [b_1(x) + \epsilon_b \Delta_b(x)] u_1(t) + b_2(x) u_2(t) \qquad (3.1)$$

The boundary conditions are given by

$$y(t,0) = y_x(t,0) = 0$$
$$[EI(x) y_{xx}(t,x) + c_2 y_{xxt}(t,x)]_{x=1} = 0$$
$$\frac{\partial}{\partial x}[EI(x) y_{xx}(t,x) + c_2 y_{xxt}(t,x)]_{x=1} = 0$$

and initial conditions by

$$y(0,x) = y_0(x)$$
$$y_t(0,x) = y_1(x).$$

In this equation $y(t,x)$ represents the tranverse displacement of the beam at time t and position x along the beam of length 1. The parameter $EI(x)$ represents the stiffness, c_1 is an external viscous damping coefficient, c_2 is the internal Kelvin-Voigt damping coefficient. The functions $b_1(x)$ and $b_2(x)$ are given by

$$b_1(x) = \begin{cases} 10, & \text{if } .38 < x < .42 \\ 0, & \text{otherwise,} \end{cases} \qquad b_2(x) = \begin{cases} 10, & \text{if } .88 < x < .92 \\ 0, & \text{otherwise.} \end{cases}$$

These functions approximate two actuators placed at positions .4 and .9 along the beam. Also included in the equation is structured uncertainty in the control corresponding to the first actuator, and this is represented by the term $\epsilon_b \Delta_b(x)$. For this model, $\Delta_b(x) = b_1(x)$, so that $\epsilon_b = -1$ corresponds to complete failure of the actuator at .4.

This system may be formulated in a Hilbert space setting (that is, like equation (1.4)) in a fairly straightforward fashion. Briefly, set $U = \mathbb{R}^2$ and $H = H_L^2(0,1) \times L^2(0,1)$, where $H_L^2(0,1) = \{f \in H^2(0,1) : f(0) = 0 = f'(0)\}$ with norm

$$|(f,g)|_H^2 = \int_0^1 |f''(x)|^2 \, dx + \int_0^1 |f(x)|^2 \, dx.$$

ROBUST CONTROL WITH STRUCTURED UNCERTAINTY

By setting $z(t) = (y(t, \cdot), y_t(t, \cdot))$ the system (3.1) can be reformulated as

$$\dot{z}(t) = Az(t) + (B + \Delta B)u(t) \tag{3.2}$$
$$z(0) = z_0.$$

As above, $\Delta B = FL_B G$, and the appropriate operators are given by

$$A(f, g) = \left(g, -\frac{d}{dx^2}[EI(x)f''] - c_1 g - c_2 \frac{d^4}{dx^4} g \right)$$
$$B(u_1, u_2) = (0, b_1(x)u_1 + b_2(x)u_2)$$
$$Fu = (0, \Delta_b(x)u_1) \qquad G(u_1, u_2) = (u_1, 0) \qquad L_B = \epsilon_b I.$$

Note that there is no ΔA term, since in this example we only consider uncertainty in the control. If there was uncertainty in the stiffness or damping terms (for example, to account for damage to the structure) then this would be modeled by an operator ΔA. However we note that to treat uncertainty in the stiffness would require a modification of our theory, as it would lead to an unbounded D operator (recall the factorization (1.5)).

Continuing with our example, we first compute the optimal feedback gain K_0 for a standard LQR problem for the cost functional

$$J_0(u) = \int_0^\infty \left\{ q^2 |z(t)|_H^2 + |u(t)|^2 \right\} dt.$$

Hence in the notation used earlier, $C = qI$ and $R = I$. From standard LQR theory, $K_0 = -R^{-1}B^*\Pi_0$, where Π_0 is a solution to the algebraic Riccati equation

$$A^*\Pi + \Pi A + C^*C - \Pi B R^{-1} B^*\Pi = 0. \tag{3.3}$$

Next we compute the gain $K_1 = -\hat{R}^{-1}B^*\Pi_1$ where Π_1 is a solution to (2.5). Then we look at the closed loop eigenvalues so as to compare the performance of both the LQR controller and the game theory controller in the presence of the uncertainty ΔB. A cubic spline-based finite element scheme is used for the approximation. Ten elements on $[0, 1]$ are used for approximation. The following data is used: $EI = 40$, $c_1 = .02$, $c_2 = .002$, $\epsilon_b = -.4$, $q^2 = 5$ and $\gamma = .4$. In the tables below we list the eigenvalues for the first 3 modes. The controllers have very little influence on the third and higher modes, which are already quite damped due to the Kelvin-Voigt damping. In Table 3.1 we list the open loop eigenvalues of A.

$$-.0223 \pm 22.237i$$
$$-.4955 \pm 139.362i$$
$$-3.8188 \pm 390.308i$$

Table 3.1: Open loop eigenvalues

Table 3.2 shows the closed loop eigenvalues with no uncertainty. That is, we list the eigenvalues of $A + BK_0$ and $A + BK_1$.

$A + BK_0$	$A + BK_1$
$-.2609 \pm 22.235i$	$-.7751 \pm 22.223i$
$-.5259 \pm 139.362i$	$-.5216 \pm 139.361i$
$-3.8198 \pm 390.308i$	$-3.8196 \pm 390.308i$

Table 3.2: Closed loop eigenvalues without actuator damage

Table 3.3 shows the closed loop eigenvalues for the case in which the actuator at .4 has suffered damage. The standard LQR feedback is not designed to account for this, and its performance is poor relative to the robust feedback.

$A + (B + \Delta B)K_0$	$A + (B + \Delta B)K_1$
$-.1655 \pm 22.236i$	$-.4740 \pm 22.232i$
$-.5137 \pm 139.362i$	$-.5112 \pm 139.362i$
$-3.8194 \pm 390.308i$	$-3.8193 \pm 390.308i$

Table 3.3: Closed loop eigenvalues with actuator damage

These preliminary numerical results indicate that the control law designed using the method described in this paper outperforms the standard LQR optimal feedback control in the presence of possible actuator damage.

References

1. Banks, H.T. and K. Ito, Approximation in LQR problems for infinite dimensional systems with unbounded input operators, to appear.

2. Bensoussan, A., Saddle points of convex concave functionals, in *Differential Games and Related Topics*, H.W. Kuhn and G.P. Szego, eds., North-Holland, Amsterdam, 1971, 177-200.

3. Curtain, R., H_∞-control for distributed parameter systems, Proc. 29th IEEE Conf. Dec. and Control, December 1990, 22-26.

4. Datko, R., Extending a theorem of A.M. Liapunov to Hilbert space, *J. Math. Analysis Applic.*, 32, 1970, 610-616.

5. Fabiano, R.H., Kurdila, A.J. and X. Li, Approximation of a min-max control problem in Hilbert space, submitted.

6. Gibson, J.S., Linear-quadratic optimal control of hereditary differential systems: infinite dimensional Riccati equations and numerical approximations, *SIAM J. Control and Optimization*, 21, 1983, 95-139.

7. Ito, K., Strong convergence and convergence rates of approximating solutions for algebraic Riccati equations in Hilbert spaces, in *Distributed Parameter Systems*, Springer Lec. Notes in Control and Inf. Sci., 102, 1987, 153-166.

8. Kappel, F. and D. Salamon, An approximation theorem for the algebraic Riccati equation, *SIAM J. Control and Optimization*, 28, 1990, 1136-1147.

9. Pandolphi, L., Stability of perturbed linear distributed parameter systems: a Lyapunov equation approach, Proc. IFAC Symposium on DPS, Perpignan, June, 1989, Eds. El Jai and Amaroux, 57-62.

10. Pazy, A., *Semigroups of Linear Operators and Applications to Partial Differential Equations*, Applied Mathematical Sciences Vol. 44, Springer Verlag, New York, 1983.

11. Peterson, I.R., Some new results on algebraic Riccati equations arising in linear quadratic differential games and the stabilization of uncertain systems, *Systems and Control Letters*, 10, 1988, 341-348.

12. Peterson, I.R., Disturbance attenuation and H_∞-optimization: a design method based on the algebraic Riccati equation, *IEEE Transactions Automatic Control*, 32, 1987, 427-429.

13. Pritchard, A.J. and S. Townley, Robustness optimization for abstract, uncertain control systems: unbounded inputs and perturbations, Proc. IFAC Symposium on DPS, Perpignan, June, 1989, Eds. El Jai and Amaroux, 117-121.

14. Pritchard, A.J. and S. Townley, Robustness of linear systems, *J. Differential Equations*, 77, 1989, 254-286.

15. Rhee, I. and J. Speyer, A game theoretic controller for a linear time-invariant system with parameter uncertainty and its application to the space station, AIAA paper number AIAA-90-3220-CP, 1990.

16. Zabczyk, J., Remarks on the algebraic Riccati equation in Hilbert space, *Applied Mathematics and Optimization*, Vol. 2, No. 3, 1976, 251-258.

ON THE PHASE PORTRAIT OF THE KARMARKAR'S FLOW

Leonid E. Faybusovich,

Department of Mathematics
University of Notre Dame
Notre Dame, Indiana 46556

1 Introduction

A so-called projective scaling vector is defined on the interior of a given polyhedron P and depends on a cost function. These vector fields (under some additional assumptions) [2] can be considered as continuous versions of the Karmarkar's algorithm for solving linear programming problems. For the case where P is a simplex

$$P = \{p \in R^n : p_1 + p_2 + \ldots + p_n = 1, p_i \geq 0, i = 1, 2 \ldots n\}, \qquad (1.1)$$

the Karmarkar's flow $K_c(p)$ has the form:

$$\dot{p}_i = c_i p_i^2 - p_i \sum_{j=1}^n c_j p_j^2, i = 1, 2, \cdots n. \qquad (1.2)$$

Let us observe that more general dynamical systems

$$\dot{p} = D(p)(Ap - e < p, Ap >), \qquad (1.3)$$

where $D(p) = diag(p_1, \cdots p_n), e = (1, \cdots 1)$ and A is an arbitrary n by n matrix were studied in biomathematics [3].

In the present paper, we are interested in the special case of dynamical systems (1.3), where $A = A^T$ is symmetric. Indeed, consider the interior of the simplex

$$int(P) = \{p \in P : p_i > 0, i = 1, 2 \cdots n\}$$

as a Riemannian submanifold of the positive orthant $R_+^n = \{p \in R^n : p_i > 0, i = 1, 2 \cdots n\}$ endowed with the Riemannian metric

$$g(p; \xi, \mu) = \sum_{i=1}^n \frac{\xi_i \mu_i}{p_i} \qquad (1.4)$$

If

$$f(p) = \frac{1}{2} < p, Ap >,$$

then the gradient of f relative to (4) projected onto $int(P)$ will have the form

$$\dot{p} = D(p)(Ap - \frac{<p, Ap>}{<e, p>}e) \tag{1.5}$$

which coincides with (1.3) on the simplex P. The systems of this type solve nonlinear optimization problems of the form

$$<p, Ap> \to max, p \in P$$

and were studied in detail in [1]. Here we apply the technique developed in [1] to completely describe the phase portrait of (1.2).

2 The analysis of stationary points

Here we describe the results of [1] for the particular case where P is a simplex (1.1). Observe that in [1] we actually analyze the situation where P is an arbitrary polyhedron.

Suppose that f is a smooth function on R^n. Consider the following optimization problem:

$$f(p) \to max, p \in P. \tag{2.1}$$

Let $\nabla f(p)$ be the gradient of f relative to the standard Euclidean metric. Consider a dynamical system

$$\dot{p} = D(p)(\nabla f(p) - <p, \nabla f(p)> e), p \in P. \tag{2.2}$$

The main idea of [1] is to consider the problem

$$\tilde{f}(x) = f(\phi(x)) \to max, \tag{2.3}$$

$$x \in S^{n-1}$$

instead of the problem (2.1). Here $\phi(x_1, \ldots x_n) = (x_1^2, \ldots x_n^2)$ and S^{n-1} is the unit sphere in R^n. It was shown that $x \in S^{n-1}$ is a stationary point of the vector field

$$\dot{x} = \nabla \tilde{f}(x) - <\nabla \tilde{f}(x), x> x \tag{2.4}$$

if and only if $\Phi(x)$ is a stationary point of the vector field (2.2). The indices of stationary points of (2.4) were described in terms of (2.1) in the following way. Given $I \subset [1, n]$ denote by $P(I) = \{p \in P : p_i = 0, i \in I, p_i > 0, i \bar{\in} I\}$. Let $X(I) = \{\xi \in R^n :<e, \xi> = 0, \xi_i = 0, i \in I\}$.

Proposition 2.1 *Suppose that* $x \in S^{n-1}, p = \Phi(x) \in P(I)$ *for some* $I \subset [1, n]$. *Then* x *is a stationary point of the vector field (2.4) if and only if*

$$< \nabla f(p), \xi > = 0$$

for any $\xi \in X(I)$.

Given $I \subset [1, n], I \neq [1, n]$, introduce vectors $\mu_s = e_s - e_i, s \in I$ for some $i \bar\in I$

Proposition 2.2 *Suppose that x is a stationary point of (2.4), $p = \phi(x) \in P(I)$. Then the index of x (the number of eigenvalues with negative real parts of the linearization of (2.4) at x) is given by the formula:*

$$ind(x) = card\{s \in I :< \nabla f(p), \mu_s >< 0\} + ind(D^2 f | X(I)) \qquad (2.5)$$

Here by $ind(D^2 f | X(I))$ we denoted the index of the quadratic form $D^2 f$ restricted to $X(I)$. Here $D^2 f$ denoted the second Frechet derivative of f.

Proposition 2.3 *Suppose that $x(\cdot)$ is an integral curve of the vector field (2.4) on S^{n-1}. Then $t \to \Phi(x(\frac{t}{4}))$ is an integral curve of the vector field (2.2).*

Propositions 2.1,2.2,2.3 (for an arbitrary polyhedron P) are proved in [1].

Proposition 2.4 *Suppose that $x \in S^{n-1}$ is such that $p = \Phi(x) \in P(I)$ for some $I \subset [1, n]$. Then x is a stationary point of (2.4) if and only if*

$$< \nabla f(p), e_i >= \alpha, i \bar\in I, \qquad (2.6)$$

for some α.

Proof: By Proposition 2.1 it is sufficient to verify that (2.6) is equivalent to:

$$< \nabla f(p), \xi >= 0 \qquad (2.7)$$

for any $\xi \in X(I)$.

If $\xi \in X(I)$, then $\xi_i = 0, i \in I$ and

$$\sum_{i \bar\in I} \xi_i = 0.$$

If (2.6) holds, then $< \nabla f(p), \xi >= \alpha \sum \{\xi_i : i \bar\in I\} = 0$. On the other hand if (2.7) holds, then $< \nabla f(p), e_i - e_j >= 0$ for any $i, j \bar\in I$ ∎

Proposition 2.5 *Under assumptions of Proposition 2.4*

$$ind(x) = ind(D^2 f(p)|X(I)) + card\{s \in I :< \nabla f(p), e_s >< \alpha\}.$$

Proof: This is an immediate corollary of Propositions 2.2,2.4 ∎

Consider the special case of Karmarkar's flow, where $f(p) = \frac{1}{2}(c_1 p_1^2 + \ldots + c_n p_n^2)$. Suppose that $c_i \neq 0$ for all i. Introduce $J_+ = \{i \in [1, n] : c_i > 0\}, J_- = [1, n] \setminus J_+$.

Proposition 2.6 *Suppose that $x \in S^{n-1}$ is a stationary point of (2.4), $\Phi(x) = p \in P(I)$. Then either $J = [1, n] \setminus I \subset J_+$ or $J \subset J_-$.*

Proof: Indeed, by Proposition 2.4 $c_i p_i = \alpha, i \widetilde{\in} I$. If $\alpha > 0$, then $c_i > 0, i \widetilde{\in} I$. If $\alpha < 0$, then $c_i < 0, i \widetilde{\in} I$. ∎

Proposition 2.7 *For any nonempty J such that $J \subset J_+$ or $J \subset J_-$, there exists a unique stationary point p such that $p \in P([1, n] \setminus J)$.*

Proof: Let, say, $J \subset J_+$. Set

$$p_i = \frac{\frac{1}{c_i}}{\sum_{i \in J} \frac{1}{c_i}}, i \in J,$$

$$p_i = 0, i \widetilde{\in} J.$$

We immediately verify by Proposition 2.4, that p is a stationary point of (2.2). By the same proposition, this is the only choice if $p \in P([1, n] \setminus J)$. ∎

Let Γ be the set of all nonempty subsets J of [1,n] such that $J \subset J_+$ or $J \subset J_-$. By Propositions 2.6,2.7 there exists a one-to-one correspondence between subsets $J \in \Gamma$ and stationary points of (2.2) (for the case $f(p) = \frac{1}{2}(c_1 p_1^2 + \ldots + c_n p_n^2)$.) Given $J \in \Gamma$, denote the corresponding stationary point by $p(J)$.

Proposition 2.8 *We have*

$$indp(J) = n - cardJ \tag{2.8}$$

for $J \subset J_+$,

$$indp(J) = cardJ - 1 \tag{2.9}$$

for $J \subset J_-$.

Proof: The result immediately follows by Proposition 2.5 and the observation that $< \nabla f(p), e_s > = p_s c_s = 0, s \in [1, n] \setminus J$. ∎

Proposition 2.9 *For any $i, j \in [1, n]$ the hyperplane $V_{ij} = \{p \in R^n : c_i p_i = c_j p_j\}$ is an invariant manifold for the dynamical system (1.2).*

Proof: We have

$$\frac{d}{dt}(c_i p_i - c_j p_j) =$$

$$c_i^2 p_i^2 - c_j^2 p_j^2 - (c_i p_i - c_j p_j)\left(\sum_{k=1}^{n} c_k p_k^2\right), =$$

$$= (c_i p_i - c_j p_j)\left(c_i p_i + c_j p_j - \sum_{k=1}^{n} c_k p_k^2\right), \tag{2.10}$$

where the derivative is taken along solutions to (1.2). ∎

Remark 2.10 It is, of course, follows from the Proposition 2.9, that the corresponding half-spaces $V_{ij}^{\pm}$ are invariant manifolds ,too.

We are now in position to describe the phase portrait of (1.2). Let $p^* \in int(P) = \{p \in P : p_i > 0, i = 1, 2, \ldots n\}$. Suppose that

$$\alpha = max\{c_i p_i^* : i \in [1, n]\}, \beta = min\{c_i p_i^* : i \in [1, n]\}.$$

Let, further, $J_\alpha = \{i \in [1, n] : c_i p_i^* = \alpha\}, J_\beta = \{i \in [1, n] : c_i p_i^* = \beta\}$.
We distinguish three cases:
a). $J_+ \neq \phi, J^- \neq \phi$
b). $J_+ = [1, n]$
c). $J_- = [1, n]$

Theorem 2.11 *For any $p(0) \in int(P)$, the corresponding integral curve $p(t)$ of (1.2) such that $p(0) = p^*$ has the following properties. In the case a).*

$$\lim p(t) = p(J_\alpha), l \to +\omega, \tag{2.11}$$

$$\lim p(t) = p(J_\beta), t \to -\infty. \tag{2.12}$$

In the case b).

$$\lim p(t) = p(J_\alpha), t \to +\infty, \tag{2.13}$$

$$\lim p(t) = p([1, n]), t \to -\infty. \tag{2.14}$$

In the case c).

$$\lim p(t) = p([1, n]), t \to +\infty, \tag{2.15}$$

$$\lim p(t) = p(J_\beta), t \to -\infty. \tag{2.16}$$

Proof: In the case b) the function $f(p) = \frac{1}{2}(c_1 p_1^2 + \ldots + c_n p_n^2)$ is strictly convex. and in the case c) $f(p)$ is strictly concave. Thus (2.14), (2.15) follow from more general results of [1]. An easy calculation shows that

$$< \nabla f(p), K_c(p) >= \sum_{i=1}^{n} p_i (c_i p_i - \sum_{j=1}^{n} c_j p_j^2)^2 \geq 0. \tag{2.17}$$

where K_c is defined in (1.2). Since by Propositions 2.6,2.7, K_c has only finitely many stationary points on P, we can conclude from (2.17) that any integral curve of K_c converges to one of the stationary points when $t \to \pm\infty$. Consider now the case a). Let $\lim p(t) = p(J), t \to +\infty$. By Proposition 2.9 there should be $J_\alpha \subset J$. Indeed, if $c_i p_i^* > c_j p_j^*$, then $c_i p_i(t) > c_j p_j(t)$ for any i, j and $t \in R$. Hence $c_i p_i(J) \geq c_j p_j(J)$. If $c_i p_i^* = c_j p_j^*$, then $c_i p_i(t) = c_j p_j(t)$ for any t and hence $c_i p_i(J) = c_j p_j(J)$. It is then clear that $\alpha(J) > 0$ where $\alpha(J) = c_j p_j(J), j \in J$. Suppose that $j \in J \setminus J_\alpha$. If $i \in J_\alpha$, we will clearly have:

$$c_i p_i(t) + c_j p_j(t) - \sum_{k=1}^{n} c_k p_k^2(t) \to \alpha(J) > 0, t \to +\infty.$$

Then by (2.10) $\frac{d}{dt}(c_i p_i(t) - c_j p_j(t)) > 0$ for sufficiently large positive t. But this means that $c_i p_i(t) - c_j p_j(t)$ is strictly monotonically increasing for sufficiently large t. Since $c_i p_i(t) - c_j p_j(t)) > 0$ for any t we can conclude that $c_i p_i(J) > c_j p_j(J)$. Contradicts the choice of j. Thus $J = J_\alpha$ which completes the proof of (2.11). The proof of (2.12), (2.13) and (2.16) is completely similar. $\blacksquare$

Remark 2.12 Theorem 1 yields a complete description of the phase portrait of the vector field K_c for the case where all $c_i \neq 0$. Indeed, each face of P is the simplex again. Since each face F is invariant under K_c and $K_c|F$ has exactly the same form as K_c on P (with fewer variables) we can apply Theorem 2.11 to describe the asymptotic behavior of integral curve of K_c lying in $int(F)$.

Let $C = \{c \in R^n : c_i \neq c_j, c_i \neq 0 \text{ for any } i, j\}$.

Theorem 2.13 *The point* $p^* = (e_1 + e_2 + \ldots + e_n)/n$ *is the only point in* P *with the following property. For any* $c \in C$ *the integral curve* $p_c(t)$ *of the vector field* K_c *such that* $p_c(0) = p^*$, *we have:*

$$\lim p_c(t) = p(J_1), t \to +\infty, \lim p_c(t) = p(J_2), t \to -\infty, \tag{2.18}$$

where

$$f_c(p(J_1)) = \max\{f_c(p) : p \in P\}, f_c(p(J_2)) = \min\{f_c(p) : p \in P\}.$$

Here

$$f_c(p) = \sum_{i=1}^{n} c_i p_i^2.$$

Proof: Suppose that $c \in C$ and $i_1, i_2 \in [1, n]$ are such that $c_{i_1} > c_i > c_{i_2}, i \neq i_1, i_2$. If $c_{i_1} > 0$, then

$$f(p) \leq \sum_{k:c_k>0} c_k p_k^2 \leq \sum_{k:c_k>0} c_k p_k \leq c_{i_1}.$$

on the other hand, $f_c(e_{i_1}) = c_{i_1}$. Now $c_{i_1} p_{i_1}^* > c_i p_i^*$ for any $i \neq i_1$. By theorem 2.11 $\lim p_c(t) = e_{i_1}, t \to +\infty$. Similarly, if $c_{i_2} < 0$, then

$$f(p) \geq \sum_{k:c_k<0} c_k p_k^2 \geq \sum_{k:c_k<0} c_k p_k \geq c_{i_2}.$$

Thus, $\min\{f_c(p) : p \in P\} = f(e_{i_2})$. On the other hand, $c_{i_2} p_{i_2}^* < c_i p_i^*$ for any $i \neq i_2$. By theorem 2.11 $\lim p_c(t) = e_{i_2}, t \to -\infty$. If $c_{i_1} < 0$, then the function f_c is strictly concave on P. Hence

$$f_c(p([1, n]) = \max\{f_c(p) : p \in P\}.$$

On the other hand, by theorem 2.11, $p_c(t) \to p([1, n]), t \to +\infty$. The case $c_{i_2} > 0$ is considered similarly. Suppose that $\tilde{p} \in P$ is such that (2.18) holds for $p_c(0) = \tilde{p}$. It is clear that $\tilde{p} \in int(P)$. If $\tilde{p}$ belongs to some proper face F of P, then there exists an extreme point (say, e_i) of P which does not belong to F. We can always find $c \in C$ such that $f_c(e_i) = \max\{f(p) : p \in P\}$. Since F is an invariant manifold for K_c, it is clear that (2.18) does not hold in this case. Suppose that $\tilde{p}$ is such that $\tilde{p}_i \neq \tilde{p}_j$ for some $i, j \in [1, n]$. Take $c_i = 1/\tilde{p}_i, c_j = 1/\tilde{p}_j, c_s = -s, s \neq i, j$. By theorem 2.11 we have $\lim p_c(t) = p(J), J = \{i, j\}, t \to +\infty$. But since $c_i \neq c_j$, we clearly have:

$$\max\{f_c(p) : p \in P\} = \max\{c_i, c_j\} > f(p(J)).$$

■

3 A Lax representation

We now show that the equation (2.4) admits a Lax representation in the double-bracket form. Although this is again a very particular case of [1], we outline an independent proof here.

Proposition 3.1 *Let* $X = xx^T, x \in R^n$. *If* x *evolves according to (2.4), then*

$$\dot{X} = 2[[D((c_1, \cdots, c_n)^T)diag(X), X], X]. \tag{3.1}$$

Here $diag(X)$ *means the diagonal part of the matrix* X.

L. FAYBUSOVICH

Proof: Let $x \in S^{n-1}, v \in R^n$. An easy calculation shows that

$$[xx^T, [xx^T, D(v)]] = x(D(v)x)^T + (D(v)x)x^T - 2xx^T < D(v)x, x > . \quad (3.2)$$

If

$$\dot{x} = r(x) - < x, r(x) > x, r(x) \in R^n, x \in S^{n-1}, \quad (3.3)$$

then

$$\frac{d}{dt}(xx^T) = r(x)x^T + xr(x)^T - 2 < x, r(x) > xx^T. \quad (3.4)$$

We see that (3.2) coincides with (3.4) if $r(x) = D(v)x$. In the case of (1.4) we have $r(x) = 2D((c_1 x_1^2, \cdots, c_n x_n^2)^T)x$. We finally observe that for $X = xx^T$ we have $diag(X) = D((x_1^2, \cdots, x_n^2)^T)$. ∎

4 Concluding remarks

Karmarkar's equations have many interesting properties [2]. In this paper we completely described the phase portrait of this equation. Our main observation is that this equation is a very particular case of dynamical systems studied in [1]. In particular, it solves in general a quadratic programming problem rather than linear one. From our perspective the Theorem 2.13 is really remarkable. It shows that for a certain set C of cost functions there exists a unique universal point e/n such that solutions of Karmarkar's equations with the initial condition e/n converge to the optimal (global) solution of the corresponding quadratic programming problem (including an open subset in C which correspond to problems with many local optimal solutions). This phenomenon is new to our knowledge and opens interesting opportunities for the analysis of nonconvex optimization problems with the help of dynamical system theory.

References

[1] L.E. FAYBUSOVICH, "Dynamical Systems Which Solve Optimization Problems with Linear Constraints," *IMA Journal of Mathematical Control and Information*, 1991, v. 8, pp. 135-149.

[2] N. KARMARKAR, "Riemannian Geometry Underlying Interior-point Methods for Linear Programming," *Contemporary Mathematics*, v. 114, pp. 51-76.

[3] E.C. ZEEMAN, "Population Dynamics from Game Theory," *Lecture notes in Mathematics*, v. 819, pp. 471-497.

THE REDUCED BASIS METHOD IN
CONTROL PROBLEMS

Max D. Gunzburger* and Janet S. Peterson†

*Department of Mathematics and
Interdisciplinary Center for Applied Mathematics
Virginia Tech
Blacksburg, VA 24061-0531*

1 Introduction

The standard methods for determining solutions of partial differential equations are only loosely tied to the particular problem being solved. For example, in the context of finite element methods, one uses the same type of local basis functions, e.g., piecewise polynomials, for solving Poisson's equation, the heat equation, the Navier-Stokes equations, etc. What if one were to use basis functions that still belong to one's favorite finite element space and are *global* in nature, but which are intimately related to the solution of the particular partial differential equation of interest? This is the rationale behind the *reduced basis method*.

The reduced basis method was developed for structural engineering applications, and has met with considerable success in that setting. See, e.g., [1], [2], [6]-[10]. Its potential usefulness for fluid calculations has also been demonstrated; see [11]. Applications to other mathematical problems as well as some analyses of the reduced basis method may be found in [3], [5], [12], and [13]. In this paper we examine the possibilities for its application to the computational resolution of control problems. Although our focus is on flow control problems, it is clear that our discussion can also be applied to other settings having partial differential equations playing the role of state contraints.

2 Reduced basis methods for nonlinear PDEs

We describe one class of reduced basis methods, known as the *Taylor basis* method. Other classes of reduced basis methods can be defined

---

* Supported in part by the Air Force Office of Scientific Research under grant number AFOSR-90-0179, and in part by the Office of Naval Research under grant number N00014-91-J-1493.

† Supported in part by the Office of Naval Research under grant number N00014-91-J-1493.

based on interpolation or least squares approximation. On purpose, in order to clarify the exposition, we set our discussion here in a rather vague environment.

We denote our nonlinear partial differential equation by

$$F(\phi, g) = 0,\qquad(2.1)$$

where ϕ denotes the state and g parameters appearing in the problem, e.g., the Reynolds number in fluid dynamics problems or the load parameter in structural mechanics problems. Now, let us denote by ϕ_0 the solution of (2.1) for some value g_0 of the parameters. Next, denote by ϕ_j the j-th derivative of the state ϕ with respect to the parameters g, evaluated at g_0. Then, for example, ϕ_1 and ϕ_2 may be determined as solutions of

$$F_\phi(\phi_0, g_0)\phi_1 = -F_g(\phi_0, g_0)\qquad(2.2)$$

and

$$F_\phi(\phi_0, g_0)\phi_2 = -\phi_1 F_{\phi\phi}(\phi_0, g_0)\phi_1 - 2F_{\phi g}(\phi_0, g_0)\phi_1 - F_{gg}(\phi_0, g_0)\,,\quad(2.3)$$

respectively, where F_ϕ, $F_{\phi\phi}$, F_g, etc., denote the obvious Frechet derivatives of F. Note that ϕ_1 is nothing else but the sensitivity derivative of the state with respect to the design parameter g. Equations for ϕ_j, $j > 2$, the higher derivatives of ϕ, may be defined in a similar manner. Note that (2.2) and (2.3), and indeed, all of the equations that determine ϕ_j, $j = 1, 2, \ldots$, have the *same* left-hand side and are *linear* partial differential equations.

Next, let us denote by ϕ_j^h an approximation to ϕ_j found by a standard finite element, finite difference, or other method. For the sake of concreteness, we will focus on finite element methods. We assume, as would be done in practice, that the same finite element space V^h is used to determine all the ϕ_j^h, $j \geq 0$. Note that for $j \geq 1$, all the ϕ_j^h may be determined by solving linear systems with the same coefficient matrices, but with different right-hand sides. Also, recall that all of the ϕ_j^h are found by approximating differential equations with the parameter g set to g_0.

Now we seek an approximate solution of the discretization of (2.1) at a different value of g, say $g = g_1$, and which is of the form

$$\phi^h = \phi^h(g_1) = \sum_{j=0}^{J} \alpha_j \phi_j^h\,.\qquad(2.4)$$

Note that since $\phi_j^h \in V^h$ for $j = 0, \ldots, J$, we have that $\phi^h \in V^h$ as well. Thus, instead of finding an approximate solution, belonging to the space

V^h, of (2.1) for $g = g_1$ that is a combination of the usual local finite element basis functions, we instead look for a solution in the subspace

$$V_J^h = \text{span}\{\phi_0^h, \phi_1^h, \ldots, \phi_J^h\} \subset V^h. \tag{2.5}$$

The set $\{\phi_0^h, \phi_1^h, \ldots, \phi_J^h\}$ of $J+1$ functions is called the *reduced Taylor basis*, and a solution of the form (2.4) is called a *reduced basis solution*.

We still haven't indicated how one chooses the coefficients α_j in (2.4). One could, of course, simply choose them to be the Taylor coefficients so that, if we are dealing with a scalar parameter g, $\alpha_j = (g_1 - g_0)^j/j!$. In this case (2.4) may be expressed as

$$\phi^h = \phi^h(g_1) = \sum_{j=0}^{J} \frac{(g_1 - g_0)^j}{j!}\, \phi_j^h. \tag{2.6}$$

However, it has been found that one can obtain better approximations by choosing the coefficients α_j so that (2.4) satisfies the discrete form of (2.1) in the subspace V_J^h, e.g., we apply Galerkin's method in that subspace. We now elaborate a little on this approach.

A Galerkin method for (2.1) is usually based on a weak formulation of that system. The latter is obtained by first multiplying (2.1) by a test function ψ, then integrating the result over a domain Ω, and then perhaps applying an integration by parts procedure to some terms in the result and applying some of the boundary conditions given in the problem specification. A typical result of this process takes the following form. For a given choice of g, one seeks a $\phi \in V$ such that

$$A\big(G(\phi, g), \psi\big) = 0 \quad \forall \psi \in V, \tag{2.7}$$

where $A(\cdot, \cdot)$ is a bilinear form, $G(\cdot, \cdot)$ is a nonlinear mapping, and V a suitable function space. A discretization of (2.7) may be defined by choosing a finite dimensional subspace $V^h \subset V$ and then requiring that $\phi^h \in V^h$ satisfy

$$A\big(G(\phi^h, g), \psi^h\big) = 0 \quad \forall \psi^h \in V^h. \tag{2.8}$$

In fact, ϕ_0^h, the first of the reduced basis functions, can be found from (2.8) with $g = g_0$. Of course, the other reduced basis function can be defined as solution of derivatives of (2.8) with respect to g, evaluated at $g = g_0$.

Now, we turn to finding an approximation to the solution of (2.7) for $g = g_1$. We could, of course, define such an approximation to be the solution of (2.8) with $g = g_1$; this would yield the usual Galerkin approximation in the subspace V^h in terms of the local basis functions for V^h. We could also use the Taylor series (2.6). However, instead, we discretize (2.7) with respect to the reduced basis space (2.5), i.e., we seek a $\phi^h \in V_J^h$ such that

$$A\big(G(\phi^h, g_1), \psi^h\big) = 0 \quad \forall \psi^h \in V_J^h. \tag{2.9}$$

Thus, α_j, $j = 0, \ldots, J$, in (2.4) are now determined from

$$A\left(G(\sum_{j=0}^{J} \alpha_j \phi_j^h, g_1), \phi_k^h\right) = 0 \quad \text{for } k = 0, \ldots, J. \tag{2.10}$$

It seems that one pays a high price if one uses the reduced basis function instead of the usual finite element basis functions since the former are *nonlocal* in character. In fact, due to the local character of the usual finite element basis function, the nonlinear system (2.8) for $g = g_1$ is sparse. On the other hand, due to the nonlocal character of the reduced basis functions, the nonlinear system (2.9) or (2.10) is dense. However, it is found that for a large range of parameter values, that accurate approximations may be obtained with J in the range of 5 to 10. Thus, although the discrete problems are dense, they are very small compared to the sparse but huge systems (of size in the hundreds or thousands) that would result from using more standard basis functions.

We now briefly examine the question of why so few reduced basis functions are needed in order to well-approximate the solution of (2.7). In fluid flow applications, at least, one reason is that, at least away from singularities such as bifurcation points, the reduced basis functions decrease in size dramatically with j; see [4]. In fact, one has that, in a suitable norm,

$$\|\phi_j^h\| \propto \frac{1}{g^{j-1}}$$

where here $g =$ Reynolds number. Thus, even for moderate values of the Reynolds number and for relative small values of j, the basis functions ϕ_j are very small in size, at least relative to ϕ_0. This can be used to explain the observation that even the Taylor series (2.6) yields a good approximation to the solution of (2.7).

A more important reason why the reduced basis functions are so effective in determining approximations of (2.7) is that it seems that as J increases, the reduced basis set $\{\phi_0^h, \phi_1^h, \ldots, \phi_J^h\}$ becomes more "linearly dependent" in the following sense. We let θ_J denote the angle between ϕ_{J+1} and V_J^h. Then, the experimental evidence for both structural mechanics and fluid dynamic problems suggests that $\theta_J \to 0$ rapidly with increasing J. Thus, adding more reduced basis functions does little to improve the ability of these functions to approximate. Note that this property cannot be exploited if one uses the Taylor series (2.6) since there the coefficients α_j are fixed independent of ϕ_j^h. However, this property of the reduced basis functions can clearly be exploited by using a projection method, such as a Galerkin method.

3 Reduced basis methods for control problems

There are many ways to exploit reduced basis ideas in the context of the approximate resolution of control and optimization problems. All of them are based on the observation that the most costly element of any algorithm that determines approximations to optimal states and controls is the calculation of states and sensitivities, or states and adjoint states. Here, we take the view that the parmeters g appearing in (2.1) are the control or design parameters of the problem.

First, let us consider algorithms based on optimization methods and sensitivity derivatives. Here, we presume that one is using an optimization algorithm for updating the guesses for the controls or design parameters that requires knowledge of the state and the sensitivities. Thus, given a guess for the controls g_k, we have that the new gues for the controls is obtained from

$$g_{k+1} = Q\bigg(g_k, \phi_0^h(g_k), \phi_1^h(g_k)\bigg) \qquad (3.1)$$

for some implicitly or explicitly defined function Q.

In the present notation, we could employ an algorithm having the following outline. Note that the algorithm incorporates the possibility that the reduced basis functions may have to be updated from time to time if the current values of the control parameters become to distant from the values at which the reduced basis functions were computed.

Given an initial guess $g = g_0$ for the design parameters or controls,

   I. determine the reduced basis functions ϕ_j^h, $j = 1, \ldots, J$, by solving discretizations of (2.1)-(2.3) and the analogous equations for the higher-order derivatives of the state ϕ with respect to the controls g;

  II. use $\phi_0^h \approx \phi(g_0)$ and $\phi_1^h \approx (\partial\phi/\partial g)(g_0)$ to determine a new guess for the controls from a formula such as (3.1), i.e., from $g_1 = Q(g_0, \phi_0^h, \phi_1^h)$;

 III. for $k = 1, 2, \ldots, K$,

     i. by solving a discretization of (2.7) in the reduced space $V_j^h = \mathrm{span}\{\phi_0^h, \phi_1^h, \ldots, \phi_j^h\}$, determine an approximation $\phi^{(k)}$ to $\phi(g_k)$ of the form (2.4), i.e., of the form $\phi^{(k)} = \sum_{j=0}^J \alpha_j^{(k)} \phi_j^h$;

ii. determine an approximation $\phi'^{(k)}$ to $\phi'(g_k)$ by differentiating the result of (III.i), i.e., $\phi'^{(k)} = \sum_{j=1}^{J} \alpha_j^{(k)} \phi_{j+1}^h$;

iii. determine a new guess for the control from (3.1), i.e., $g_{k+1} = Q(g_k, \phi^{(k)}, \phi'^{(k)})$;

IV.

   i. if the current approximations g_K and $\phi^{(K)}$ for the optimal controls and state, respectively, are sufficiently converged, go on to step (V);

   ii. if the current approximations g_K and $\phi^{(K)}$ are not sufficiently converged and if the reduced basis functions are not adequate for approximating states beyond g_K, set $g_0 = g_K$ and return to step (I) to update the reduced basis functions;

   iii. if the the current approximations g_K and $\phi^{(K)}$ are not sufficiently converged and if the reduced basis functions are adequate for approximating states beyond g_K, set $g_1 = g_K$ and return to step (III);

V. if the converged reduced basis solutions are not sufficiently accurate, compute new approximations to the optimal state using the full finite element discretization of (2.7); if necessary, return to step (I) to improve the accuracy of the approximation to the optimal control.

Reduced basis methods can also be used to advantage in the context of adjoint methods for solving flow control problems. Briefly, if we wish to solve the problem

$$\min_{\phi, g} \mathcal{J}(\phi, g)$$

subject to the constraint

$$F(\phi, g) = 0 \tag{3.2}$$

via adjoint, or co-state, or Lagrange multiplier methods, we are required to approximately solve systems of the type (3.2) coupled with

$$F_\phi^*(\phi, g)\xi - \mathcal{J}_\phi(\phi, g) = 0 \tag{3.3}$$

and

$$F_g^*(\phi, g)\xi - \mathcal{J}_g(\phi, g) = 0 \tag{3.4}$$

for the optimal state, control, and adjoint state ϕ, g, and ξ, respectively. In (3.3) and (3.4), $(\cdot)_g$ and $(\cdot)_\phi$ denote the obvious Frechet derivatives and $F_\phi^*(\phi, g)$ denotes the adjoint of the Frechet derivative F_ϕ evaluated at (ϕ, g). A similar definition holds for $F_g^*(\phi, g)$.

Usually, for control problems involving state equations that are partial differential equations, (3.2)-(3.4) is a formidable system to approximate. For example, if the constraint equations (3.2) are the Navier-Stokes equations, then (3.2)-(3.4) is a system roughly twice the size of the Navier-Stokes equations! If one were to solve (3.2)-(3.4) by standard finite element or finite difference methods, the size of the resulting discrete problems would be prohibitively large. However, by using reduced bases, one could easily solve (3.2)-(3.4).

A possible algorithm is outlined as follows. Using (3.2) and its derivatives with respect to g, we determine the reduced basis functions $\phi_j = \partial^j \phi / \partial g^j$, $j = 1, \ldots, J$, at an initial guess g_0 for the controls. We linearize (3.2)-(3.4) by some standard iterative method, e.g., Newton's method. We then solve for the Newton iterates in the reduced basis space based on the Newton linearization of (3.2)-(3.4). Once convergence is achieved, we could, if necessary, correct the solution by solving (3.2) in the full finite element space.

References

[1] B. ALMROTH, P. STERN and F. BROGAN, "Automatic choice of global shape functions in structural analysis," *AIAA J.*, v. 16, 1978, pp. 525–528.

[2] A. CHAN and K. HSIAO, "Nonlinear analysis using a reduced number of variables," *Comput. Meth. Appl. Mech. Engrg.*, v. 52, 1985, pp. 899-913.

[3] J. FINK and W. RHEINBOLDT, "On the error behavior of the reduced basis technique for nonlinear finite element approximations," *ZAMM*, v. 63, 1983, pp. 21–28.

[4] M. GUNZBURGER and J. PETERSON, "Predictor and steplength selection in continuation methods for the Navier-Stokes equations," *Computers Math. Applic.*, v. 22, 1991, pp. 73–81.

[5] M. LEE, "The reduced basis method for differential algebraic equation systems," *ICMA Technical Report ICMA-85-85*, U. of Pittsburgh, 1985.

[6] D. NAGY, "Modal representation of geometrically nonlinear behavior by the finite element method," *Computers & Struct.*, v. 10, 1979, pp. 683–688.

[7] A. NOOR, "Reecent advances in reduction methods for nonlinear problems," *Comput. & Struc.*, v. 13, 1981, pp. 31-44.

[8] A. NOOR, C. ANDERSEN, and J. PETERS, "Reduced basis technique for collapse analysis of shells," *AIAA J.*, v. 19, 1981, pp. 393–397.

[9] A. NOOR and J. PETERS, "Reduced basis technique for nonlinear analysis of structures," *AIAA J.*, v. 18, 1980, pp. 455–462.

[10] A. NOOR and J. PETERS, "Tracing post-limit-point paths with reduced basis technique," *Comput. Meth. Appl. Mech. Engrg.*, v. 28, 1981, pp. 217-240.

[11] J. PETERSON, "The reduced basis method for incompressible viscous flow calculations," *SIAM J. Sci. Stat. Comput.*, v. 10, 1989, pp. 777–786.

[12] T. PORSCHING, "Estimation of the error in the reduced basis method solution of nonlinear equations," *Math. Comp.*, v. 45, 1985, pp. 487-496.

[13] T. PORSCHING and M. LEE, "The reduced basis method for initial value problems," *ICMA Technical Report ICMA-86-95*, U. of Pittsburgh, 1986.

NUMERICAL TREATMENT OF OSCILLATING INTEGRALS APPEARING IN HEAT CONDUCTION PROBLEMS

Sven-Åke Gustafson

Høgskolesenteret i Rogaland
Box 2557 Ullandhaug
N-4004 Stavanger, NORWAY

1 Introduction

We shall discuss the numerical treatment of Fourier integrals of the general form

$$F(t) = \int_{-\infty}^{\infty} e^{ist} f(s)ds, \tag{1.1}$$

when the parameter t is moderate to large and f decays slowly. We will illustrate the efficiency of the methods to be presented on the three examples

$$f_1(s) = \frac{1}{\sqrt{1+s^2}}, \tag{1.2}$$

$$f_2(s) = \exp\left(-\sqrt[4]{1+s^2}\right) \tag{1.3}$$

and

$$f_3(s) = \frac{1}{\cosh\sqrt{is}}. \tag{1.4}$$

f_3 is of interest, since it is related to the functions appearing in [2] All of these functions decay slowly. Thus

$$f_1(s) = \mathcal{O}(|s|^{-1}), \ |s| \to \infty$$

and

$$f_3(s) = \mathcal{O}(e^{-\sqrt{|s|}}), \ |s| \to \infty$$

We give some sample values on the values of $|f_3|$:

$$\begin{aligned}
|f_3(0)| &= 0.5, \ |f_3(10)| = 0.35, \ |f_3(100)| = 0.029, \\
|f_3(1000)| &= 1.4 \cdot 10^{-5}, \ |f(2000)| = 1.4 \cdot 10^{-7}
\end{aligned}$$

It is therefore not advantageous to approximate (1.1) with an integral over a bounded interval, since the latter would have to be chosen very large and many values of the integrand would be required. Instead we shall use methods based on convergence acceleration. We will discuss two main strategies:

i) Approximate (1.1) with an infinite trapezoidal sum, which then is evaluated using convergence acceleration. A class of formulas which are suitable for this purpose will be described in Section 2. Our treatment will generalize that in [6]. We will show that the calculated sum may be expressed in two algebraically equivalent ways: either as a rational expression with t in (1.1) as an argument or as a linear combination of the first few terms which are used for determining the approximate sum.

ii) Approximate the integrand of (1.1) for moderate to large values of s with an exponential sum which is integrated exactly over the corresponding (infinite) interval. Here we build on the treatment given in [1] An account is given in Section 3.

The methods to be described in this paper generate a rapidly converging sequence of approximations to a desired limit value. However, the calculated values are influenced by round-offs which limits the obtainable accuracy. The choice of a good approximation from this sequence of calculated values is governed by stopping rules analogous to those described in [4] and [6].

2 Linear convergence acceleration schemes derived from quadrature and interpolation

Consider the general power series

$$\mathcal{F}(z) = \sum_{r=0}^{\infty} c_r z^r, \tag{2.1}$$

where the function $\mathcal{F}$ is defined for those z which are such that the series (2.1) converges. It is often advantageous to extend the definition of $\mathcal{F}$ also to such areas in the complex plane where the series is divergent. We next introduce:

Definition 2.1 *Let n be a positive integer, $\eta_0(z), \ldots, \eta_{n-1}(z)$ n numbers, which as indicated may depend on z. We call*

$$\mathcal{F}_n(z) = \sum_{r=0}^{n-1} \eta_r(z) c_r, \tag{2.2}$$

a linear transformation of the power series (2.1).

It is desirable that $\mathcal{F}_n(z)$ should give an approximation to $\mathcal{F}(z)$ and often one seeks to construct sequences of transformations $\mathcal{F}_1(z)$, $\mathcal{F}_2(z), \ldots$ such that

$$\lim_{n \to \infty} \mathcal{F}_n(z) = \mathcal{F}(z).$$

Definition 2.2 *Associated with the transformation $\mathcal{F}_n(z)$ we introduce its generating polynomial $Q_n(z;\cdot)$ with argument t given by*

$$Q_n(z;t) = \sum_{r=0}^{n-1} \eta_r(z)t^r. \tag{2.3}$$

Sometimes it is easiest to define a linear transformation in terms of its generating polynomial.

Example 2.1 *Let w be a real or complex number and consider the polynomial*

$$\frac{1}{1-zw} \sum_{r=0}^{n-1} \left(\frac{zt-zw}{1-zw} \right)^r, \tag{2.4}$$

obtained by expanding $(1-tz)^{-1}$ in a Taylor series around $t = w$ and retaining the first n terms. For each w (2.4) is a generating polynomial of a linear transformation. We mention the special cases:

- *$w = 0$: term by term summation*

- *$w = 1$: generalized Euler transformation*

- *$w = 1/2$: optimal Taylor acceleration (See [4])*

Thus if the generating polynomial of a linear transformation is given, one may determine the transformation itself by expanding the polynomial in power form. Sometimes it is helpful to work with operators to facilitate the calculations. We introduce

Definition 2.3 *Let c_0, $c_1,\ldots$, be a sequence of numbers. We define*

$$c_{r+1} = Ec_r, \quad c_{r+1} - c_r = \Delta c_r, \quad Ic_r = c_r. \tag{2.5}$$

E and Δ are called the shift and difference operators, I the identity operator.

Remark 2.1 *We may form powers and polynomials of shift and difference operators. They are linear operators and obey familiar laws for multiplication, e.g.:*

$$\Delta^0 = E^0 = I,$$
$$E^{m+n} = E^m \cdot E^n,$$
$$E^m \cdot \Delta = E^{m+1} - E^m.$$

Lemma 2.4 *Let now the linear transformation (2.2) have the generating polynomial (2.3). Then we may write*

$$\mathcal{F}_n(z) = \sum_{r=0}^{n-1} \eta_r(z)E^r c_0 = Q_n(z;E)c_0 \tag{2.6}$$

Thus we replace the argument t in the polynomial with the shift operator E. Using the laws mentioned above we may now derive the well-known recursion formulæ associated with the Euler transformation. See e.g. ([4]). We next show how to derive rational expressions approximating the sum (2.1) by fitting linear combinations of geometric series to this power series. Let namely

$$t_1, t_2, \ldots, t_n, \tag{2.7}$$

be n fixed numbers. Next determine the unique solution

$$x_1, x_2, \ldots, x_n, \tag{2.8}$$

to the linear system

$$\sum_{i=1}^{n} x_i t_i^r = c_r, \quad r = 0, 1, \ldots, n-1. \tag{2.9}$$

Next put

$$\hat{c}_r = \sum_{i=1}^{n} x_i t_i^r, \quad r = 0, 1, \ldots, \tag{2.10}$$

and set

$$\hat{\mathcal{F}}(z) = \sum_{r=0}^{\infty} \hat{c}_r z^r. \tag{2.11}$$

We note that

$$\hat{c}_r = c_r, \quad r = 0, 1, \ldots, n-1. \tag{2.12}$$

We next show that $\hat{\mathcal{F}}(z)$ can equivalently be expressed either as a rational expression with z as variable or as a linear transformation of the power series (2.1). We find namely

$$\hat{\mathcal{F}}(z) = \sum_{r=0}^{\infty} \hat{c}_r z^r = \sum_{r=0}^{\infty} \sum_{i=1}^{n} x_i t_i^r z^r = \sum_{i=1}^{n} x_i \sum_{r=0}^{\infty} z^r t_i^r = \sum_{i=1}^{n} \frac{x_i}{1 - z t_i}. \tag{2.13}$$

Next we demonstrate that $\hat{\mathcal{F}}(z)$ may be expressed as a linear transformation of the form (2.2).

Theorem 2.5 *Let $t_1, \ldots, t_n$ be as before and determine $\eta_0(z), \ldots, \eta_{n-1}(z)$ as the solution of the linear system*

$$\sum_{r=0}^{n-1} \eta_r(z) t_i^r = \frac{1}{1 - t_i z}, \quad i = 1, 2, \ldots, n. \tag{2.14}$$

Thus we get from the above

$$\hat{\mathcal{F}}(z) = \sum_{i=1}^{n} \frac{x_i}{1 - z t_i} = \sum_{i=1}^{n} x_i \sum_{r=0}^{n-1} \eta_r(z) t_i^r = \sum_{r=0}^{n-1} \eta_r(z) c_r. \tag{2.15}$$

Remark 2.2 *We note that we determined the numbers x_i by solving a linear system of the same type as that encountered when one seeks to determine the weights of a mechanical quadrature rule while $\eta_r(z)$ were obtained after determining an interpolating polynomial. Since these systems are of Vandermonde type they may be treated by using e.g. the codes in ([3]). The derivations in the present section are valid for any sequence. In order to obtain bounds on $|\hat{\mathcal{F}}(z) - \mathcal{F}((z)|$ we need to introduce assumptions on the coefficients $c_0, c_1, \ldots$. This topic is dealt with in [4] and [7]*

3 Approximation by means of trapezoidal sums

We shall require that the function f in (1.1) satisfies the following
General assumptions There is a constant $D > 0$ such that f is analytic in the strip $\Im(z) \leq D$. Further f shall have the representation

$$f(s) = u(s) + iv(s),$$

where u and v are real for s real and satisfy on the real line:

$$u(s) = u(-s), \quad v(s) = -v(-s)$$

It is easily verified that all of the functions f_1, f_2, f_3 in Section 1 satisfy these general assumptions. We note that we may write (1.1) in the form

$$F(t) = 2 \cdot \Re(\int_0^\infty e^{ist} f(s)ds.) \tag{3.1}$$

If we now select a step-size $h > 0$ we may define the trapezoidal approximation $F(t; h)$ to $F(t)$ by

$$F(t; h) = h \cdot \sum_{n=-\infty}^{\infty} e^{inht} f(nh) \tag{3.2}$$

which alternatively may be written

$$F(t; h) = h \cdot \Re[f(0) + e^{iht} \sum_{n=0}^{\infty} e^{inht} f((n+1)h)] \tag{3.3}$$

From the theory in [9] and [12] now follows that the trapezoidal approximation converges rapidly to the true value of the integral when $h \to 0$. Using the classical residue theorem we may evaluate (1.1) exactly for f_3 and find the result

$$2\pi^2 \sum_{n=0}^{\infty} (-1)^n (1+2n) e^{-\pi^2 t(1+2n)^2/4}, \tag{3.4}$$

step-size	# of func.-values	absolute error
$2 \cdot \pi/t$	30	$.18 \cdot 10^{7}$
π/t	25	$.12 \cdot 10^{-1}$
$0.5\pi/t$	14	$.87 \cdot 10^{-4}$
$0.25\pi/t$	48	$.45 \cdot 10^{-8}$
$0.125\pi/t$	80	$.32 \cdot 10^{-12}$

Table 3.1: Results of evaluating (1.1) for $f = f_3$ for $t = 1$. ($F(1) \approx 1.674$)

which may be compared to the values which are obtained numerically. The theory and methods of [6] apply and we give some samples of the computational results for $f = f_3$. Several other values of t were tested and the general conclusion was that the accuracy deteriorated when t was much smaller than 1 even if smaller step-sizes h and more functional values were used. Absolute error remained small for t-values greater than 10 but the relative accuracy became worse. Similar observations were made for the other two test-functions f_1 and f_2.

step-size	# of func.-values	absolute error
$2. \cdot \pi/t$	30	$.10 \cdot 10^6$
π/t	17	$.48 \cdot 10^{-13}$
$0.5\pi/t$	25	$.58 \cdot 10^{-15}$
$0.25\pi/t$	46	$.17 \cdot 10^{-14}$

Table 3.2: Results of evaluating (1.1) for $f = f_3$ for $t = 10$. $(F(10) \approx 3.80 \cdot 10^{-10})$

4 Methods based on integrating an exponential approximation of f

(2.10) may be used to define $\hat{c}_r$ for non-integer values of r as well as for $r > n - 1$. Therefore (2.9), (2.10) define a fitting of a sum of exponentials with coefficients x_i to the values c_r which in turn form an equidistant table of the integrand in (1.1). This exponential expression defines a linear combination of geometrical series which is entered into the trapezoidal sum which is evaluated by means of the convergence acceleration algorithms. Therefore, the discretization error caused by replacing the integral (1.1) with a trapezoidal sum remains, even if f should happen to coincide exactly with the exponential sum obtained from (2.9). An alternative approach would be to enter this exponential expression into (1.1) and evaluate the integral exactly. This idea has been used with success for Fourier integrals over of the type of (3.1) when the function f is completely monotonic or may be written as a difference of two such functions. See [1] and [8]. The error analysis in these papers is based on exploiting the special character which is assumed for f but the construction of the formula can be carried out for any function. Therefore this strategy was tested on the three functions f_1, f_2 and f_3 from Section 1. The numbers t_i in (2.9) were taken to be the zeroes of the shifted Chebyshev polynomials both in [6] and in the calculations reported in Section 3 but this choice did not give good results for our three test-function. This observation agrees with the findings in [1]. Instead the zeroes of the shifted Jacobi-polynomials with parameters $\alpha = 0$ and $\beta = 2 \cdot t$ were used in (2.9). The step-size $h = 0.125 \cdot \pi/t$ turned out to be a good choice and the results in Table 4.1 emerged. As in the preceding experiments the absolute accuracy was good for all $t > 1$ but

the relatived accuracy become worse for larger values of t. For small t the
accuracy was less satisfactory. These observations agree qualitatively with
the conclusions in [1] for a more restricted class of functions. The following
theoretical result explains our observations from the experiments

Theorem 4.1 *Let f satisfy the General assumptions in the beginning of
Section 3. In addition we require f to be bounded on the strip $\Im(z) \leq D$
and be such that*

$$\lim_{x \to \infty} \int_{x-iD}^{x+iD} f(z)dz = 0 \tag{4.1}$$

and also such that

$$\int_L |f(z)|\,|dz| < \infty,$$

*where L is a straight line parallel with the real axis and lying in the strip.
Denote by $E(s)$ the exponential approximation obtained from (2.9) and assume that*

$$|f(s) - E(s)| \leq \delta, \ |z| \leq R$$

where R is a fixed number.
*   Then*

$$|\int_0^\infty e^{st}(f(s) - E(s)ds| \leq Ae^{-bt} + 2R\delta. \tag{4.2}$$

where the constants A, b are independent of t.
Proof. *The numerical method defined by (2.9) consists of replacing f in
(3.1) with the exponential sum $E(s)$. This function is bounded in the part
of the strip corresponding to $\Re(s) \geq 0$. Hence the left hand side of (4.2) is
the error we need to estimate. Due to (4.1) we may replace the integral of
(3.1) with an integral along the real axis to a point P at x_1, a positive real
number, from thence to $x_1 + iy_1$, where $y_1 > 0$ and from there to infinity
along a straight line parallel to the real axis. Here x_1, y_1 are chosen such
that*

$$|x_1 + iy_1| \leq R.$$

The desired result now follows from direct calculations.

The form of the error bound indicates that it is important to approximate
f accurately in a neighbourhood of the origin and that the error bound
decreases in absolute terms when t becomes large. This is consistent with
the results from the computed examples.

5   Conclusions

Trapezoidal approximation combined with convergence acceleration is an
efficient method for large classes of functions f. Since the optimal step-size

t	$F(t)$	# of func.-values	absolute error
0.1	9.202	14	$.95 \cdot 10^{-2}$
1.0	1.674	14	$.33 \cdot 10^{-5}$
3.0	$0.120 \cdot 10^{-1}$	17	$.45 \cdot 10^{-6}$
10.0	$0.380 \cdot 10^{-9}$	17	$.69 \cdot 10^{-9}$

Table 4.1: Results of evaluating (1.1) for $f = f_3$ using exponential approximation

is inversely proportional to the value of t, it is possible to save computational effort, if $F(t)$ is determined for a sequence of t-values of the form $t_{j+1} = .5 \cdot t_j$, $j = 0, 1, \ldots$. Another possibility could be to construct rational approximations for $F(t)$ using (2.10). Methods based on exponential approximation of f were described in Section 4. They use a smaller number of functional values and the construction of a rational approximation for $F(t)$ is straight-forward. However, the accuracy is not as good as for the approach earlier mentioned. There seems to be a need to search for a stabilization devise similar to the batching used in summation of series. See [6].

References

[1] G. DAHLQUIST and S.-Å. GUSTAFSON, "On the computation of slowly convergent Fourier integrals", *Methoden und Verfahren der Mathematischen Physik*, v. 6, 1972, pp. 93–112.

[2] D. S. GILLIAM, J. R. LUND, B. A. MAIR and C. F. MARTIN, "Regularization for inverse heat conduction problems", K. Bowers and J. Lund (Eds) *Computation and Control II*, Birkhäuser, Boston, Basel, Berlin, 1991, pp. 135–150

[3] S.-Å. GUSTAFSON,"Rapid computation of general interpolation formulas and mechanical quadrature rules", *CACM* v. 14, 1971, pp. 797–801.

[4] S.- Å. GUSTAFSON, "Convergence acceleration on a general class of power series," *Computing*, v. 21, 1978, pp. 53–69.

[5] S.-Å. GUSTAFSON, "Two computer codes for convergence acceleration", *Computing*, v.21. 1978, pp. 87–91.

[6] S.- Å. GUSTAFSON, "Computing inverse Laplace transforms using convergence acceleration", K. Bowers and J. Lund (Eds) *Computation and Control II*, Birkhäuser, Boston, Basel, Berlin, 1991, pp. 151–163

[7] S.- Å. GUSTAFSON, "Numerical inversion of Laplace transforms using integration and convergence acceleration", *Techn. Rep.*91-18 Swedish Nuclear Fuel and Waste Management Co., Stockholm, Sweden, 1991.

[8] S.- Å. GUSTAFSON and I. MELINDER "Computing Fourier integrals by means of near-optimal rules of the Lagrangian type", *Computing*, v. 11, 1978, pp. 21–26.

[9] J. LUND and K. L. BOWERS, *Sinc Methods for Differential Equations*, SIAM, Philadelphia, 1992

[10] I. MELINDER, "Accurate approximation in weighted maximum norm by interpolation", *J. Approx. Theory*, v. 22, 1978, pp. 33-45.

[11] F. STENGER, "Numerical Methods Based on Whittaker Cardinal, or Sinc Functions," *SIAM Rev.*, v. 23, 1981, pp. 165–224.

[12] F. STENGER, *Numerical Methods Based on Sinc and Analytic Functions*, Springer-Verlag, to appear 1993.

[13] D. V. WIDDER, *The Laplace Transform*, Princeton University Press, Princeton, N. J., 1971.

[14] J. WIMP, *Sequence Transformations and Their Applications* , Academic Press, New York, 1981.

ROOT LOCUS FOR CONTROL SYSTEMS WITH COMPLETELY SEPARATED BOUNDARY CONDITIONS

Jianqiu He [*]
Department of Mathematics
Texas Tech University
Lubbock, Texas 79409

1 Introduction

Recently, Byrnes and Gilliam [2, 3] initiated the study of a root locus methodology for distributed parameter control systems and established a quite complete analog of the finite dimensional root locus theory for an important class of distributed parameter control systems. As a continuation of their work, He [5] studied a class of so-called "Birkhoff regular systems", which is a generalization of the systems in [2, 3]. Many of the results for the systems studied in [2, 3] have been generalized to Birkhoff regular systems. In particular, for these systems, the quantities that specify and characterize the asymptotic behavior of the closed loop poles were analyzed in detail and the various behaviors of the root locus were described and illustrated.

In this paper, we consider another class of distributed parameter control systems, which is related to the systems in [2, 3], but unlike the systems in [5]. These systems are almost never self-adjoint or even Birkhoff regular. We focus on the root locus of the systems and show that the return difference equation for these systems can be written in the same form as for Birkhoff regular systems. Thus, many of the results obtained for Birkhoff regular systems are also true in the present case. Indeed, the conclusions for our systems are parallel to what has been established for the systems in [2, 3].

2 The System

First we describe a general control system which contains all the classes mentioned above as special cases. By a "general control system" we mean a system of the form:

$$
\begin{aligned}
\dot{w}(x,t) &= Aw(x,t), \quad x \in (0,1) \\
Bw(t) &= u(t) \\
w(x,0) &= \phi(x) \in \mathrm{L}^2(0,1) \\
y(t) &= Cw(t)
\end{aligned}
\tag{2.1}
$$

[*]Supported in part by grants from AFOSR and Texas Advanced research Program.

where

$$A = (-1)^{(\mu-1)} D^n + \sum_{j=2}^{n} p_j(x) D^{n-j}, \quad D = \frac{d}{dx}$$

with n even, $n = 2\mu$, $p_j(x) \in L^2(0,1), j = 2, \cdots, n$. The operators B, C and the domain of A are defined in terms of a set of boundary operators $\{\mathcal{W}_j\}_{j=0}^{n}$:

$$B = \mathcal{W}_1, \quad C = \mathcal{W}_0$$

$$D(A) = \{f \in H^{(n)}(0,1) : \mathcal{W}_i(f) = 0, \ i = 2, \cdots, n\}$$

where $H^{(n)}$ is the usual Sobolev space and

$$\mathcal{W}_i(f) \equiv \mathcal{W}_{i0}(f) + \mathcal{W}_{i1}(f), \qquad i = 0, 1, \cdots, n \tag{2.2}$$

with

$$\mathcal{W}_{i0}(f) \equiv \alpha_i f^{(m_i)}(0) + \sum_{j=0}^{m_i-1} \alpha_{ij} f^{(j)}(0)$$

$$\mathcal{W}_{i1}(f) \equiv \beta_i f^{(m_i)}(1) + \sum_{j=0}^{m_i-1} \beta_{ij} f^{(j)}(1)$$

$\alpha_i, \alpha_{ij}, \beta_i, \beta_{ij} \in \mathbb{R}, \ i = 0, 1, \cdots, n, \ 0 \leq m_0, m_1, \cdots, m_n \leq n - 1.$

Thus the input and output are assumed to occur through the boundary and without loss of generality it is assumed that they appear in $\mathcal{W}_1$ and $\mathcal{W}_0$ respectively. It is also assumed that the difference of the orders of the input and output is positive, i.e.

$$\ell \equiv m_1 - m_0 > 0. \tag{2.3}$$

A closed loop system is obtained through a simple scalar boundary output feedback law

$$u(t) = -ky(t) + v(t). \tag{2.4}$$

The corresponding closed loop system can be expressed in terms of a family of operators depending on the parameter k as

$$\dot{w}(x,t) = A_k w(x,t)$$
$$w(x,0) = \phi(x) \in L^2(0,1) \tag{2.5}$$

where

$$A_k f = A f,$$

$$D(A_k) = \{f \in H^{(n)}(0,1) \ : \ \mathcal{W}_i(f) = 0, \ i = 2, \cdots, n,$$
$$\mathcal{W}_1(f) + k\,\mathcal{W}_0(f) = 0\}$$

By "separated" we mean the boundary conditions (2.2) can be written as

$$\mathcal{W}_i(f) = \alpha_i f^{(m_i)}(0) + \sum_{j=0}^{m_i-1}\left(\alpha_{ij} f^{(j)}(0) + \beta_{ij} f^{(j)}(1)\right), \ i = 0, \cdots, \nu \quad (2.6)$$

$$\mathcal{W}_i(f) = \beta_i f^{(m_i)}(1) + \sum_{j=0}^{m_i-1}\left(\alpha_{ij} f^{(j)}(0) + \beta_{ij} f^{(j)}(1)\right), \ i = \nu+1, \cdots, n$$

where $m_1 > m_2 > \cdots > m_\nu$, $m_{\nu+1} > m_{\nu+2} > \cdots > m_n$ for some $1 \le \nu \le n-1$.

Systems (2.5) with separated boundary conditions (2.6) when $\nu = \mu$ are those studied in [2, 3] and [4]. Systems (2.5) when the boundary conditions $\{\mathcal{W}_j(f)\}_{j=1}^{n}$ and $\{\mathcal{W}_j(f)\}_{j=0,2}^{n}$ are both regular (cf. [1], [6]) are called "Birkhoff regular".

In this paper we investigate systems (2.5) with "completely separated" boundary conditions

$$\mathcal{W}_i(f) = \alpha_i f^{(m_i)}(0) + \sum_{j=0}^{m_i-1} \alpha_{ij} f^{(j)}(0), \ i = 0, 1, \cdots, \nu \quad (2.7)$$

$$\mathcal{W}_i(f) = \beta_i f^{(m_i)}(1) + \sum_{j=0}^{m_i-1} \beta_{ij} f^{(j)}(1), \ i = \nu+1, \cdots, n$$

where $m_1 > m_2 > \cdots > m_\nu$, $m_{\nu+1} > m_{\nu+2} > \cdots > m_n$ for some $1 \le \nu \le n-1$.

Remark 2.1 *Systems with separated boundary conditions when $\nu \ne \mu$ can never be either Birkhoff regular or self-adjoint while they are always Birkhoff regular when $\nu = \mu$. Furthermore, systems with separated but not completely separated boundary conditions are quite different from those with completely separated boundary conditions, from the point of view of root locus, unless $\nu = \mu$ (cf. Example 3.1). On the other hand, if $\nu = \mu$, the arguments and results presented below remain true for systems with boundary conditions separated only.*

Remark 2.2 *Among systems with completely separated boundary conditions, those with colocated controls and sensors present very different behavior from those with noncolocated controls and sensors (cf. Example 3.2).*

The result in [2]-[5] regarding the open and closed loop transfer function is also true for the general control systems. In particular, we can copy the following proposition from [2]-[5] for the systems we are considering.

Proposition 2.3 *The open loop transfer function* $\mathcal{G}_0 = \dfrac{\mathcal{N}(\lambda)}{\mathcal{D}(\lambda)}$ *for the system* $((2.5),(2.7))$ *is real, i.e.,*

$$\mathcal{G}_0(\bar{\lambda}) = \overline{\mathcal{G}_0(\lambda)},$$

where

$$\mathcal{N}(\lambda) \equiv \det\left(\{W_i(y_j)\}_{i=0,2,j=1}^{n,n}\right), \mathcal{D}(\lambda) \equiv \det\left(\{W_i(y_j)\}_{i=1,j=1}^{n,n}\right) \qquad (2.8)$$

are entire functions of λ *with* $\{y_j\}_{j=1}^n$ *a basis of solutions of* $Ay = \lambda y$*, which are entire functions of* λ.

The closed loop transfer function for the feedback law

$$u = -ky + v$$

is

$$\mathcal{G}_k = \frac{\mathcal{N}(\lambda)}{\mathcal{D}(\lambda) + k\,\mathcal{N}(\lambda)}. \qquad (2.9)$$

The closed loop poles are defined to be the zeros of the so-called return difference equation

$$\mathcal{D}(\lambda) + k\,\mathcal{N}(\lambda) = 0. \qquad (2.10)$$

These closed loop poles are also exactly the spectra of A_k as k varies. It is easy to see that the closed loop poles occur in conjugate pairs. To study the root locus of $((2.5),(2.7))$, we define

$$S_j = \left\{ z \;\middle|\; \frac{j\pi}{n} \le \arg(z) < \frac{(j+1)\pi}{n} \right\}. \qquad (2.11)$$

Then the entire complex λ-plane under the map $\lambda = i^n z^n$ is covered by the image of two adjacent regions S_j from (2.11).

For each region S_j we prescribe a particular ordering of the nth roots of -1, denoted by ω_k, $k = 1, 2, \cdots, n$, so that for all z in S_j, we have

$$\mathrm{Re}(z\omega_1) \le \mathrm{Re}(z\omega_2) \le \cdots \le \mathrm{Re}(z\omega_{\mu-1}) < 0$$

$$\mathrm{Re}(z\omega_\mu) \le 0 \le \mathrm{Re}(z\omega_{\mu+1}) \qquad (2.12)$$

$$0 < \mathrm{Re}(z\omega_{\mu+2}) \le \mathrm{Re}(z\omega_{\mu+3}) \le \cdots \le \mathrm{Re}(z\omega_n)$$

In particular, for S_0

$$\omega_{2j-1} = e^{(1-\frac{2j-1}{n})\pi i}, \quad \omega_{2j} = e^{(1+\frac{2j-1}{n})\pi i}, \quad j = 1, 2, \cdots, \mu \qquad (2.13)$$

for S_1

$$\omega'_{2j-1} = e^{(1+\frac{2j-3}{n})\pi i}, \quad \omega'_{2j} = e^{(1-\frac{2j+1}{n})\pi i}, \quad j = 1, 2, \cdots \mu \qquad (2.14)$$

With the notation

$$[a] = a + O\left(\frac{1}{z}\right), \qquad (2.15)$$

the main results of Birkhoff in [1] establish the existence of n linearly independent solutions

$$y_j = e^{z\omega_j x}[1], \quad j = 1, \cdots, n$$

of

$$(-1)^{\mu-1}Ay + z^n y = 0.$$

In addition, this basis of solutions satisfies

$$\mathcal{W}_i(y_j) = [\alpha_i](z\omega_j)^{m_i}, \quad i = 0, 1, \cdots, \nu$$

$$\mathcal{W}_i(y_j) = [\beta_j](z\omega_j)^{m_i} e^{z\omega_j}, \quad i = \nu+1, \cdots, n$$

for $|z|$ large. Using these asymptotic expressions in (2.8) and defining

$$U_j = \begin{pmatrix} \omega_j^{m_1} \\ \vdots \\ \omega_j^{m_\nu} \end{pmatrix}, \qquad V_j = \begin{pmatrix} \omega_j^{m_{\nu+1}} \\ \vdots \\ \omega_j^{m_n} \end{pmatrix}$$

we obtain

$$\Delta(z,0) \equiv \det\left(\mathcal{W}_i(y_j)\right)_{i=1,j=1}^{i=n,j=n}$$

$$= z^m \begin{vmatrix} \text{diag}([\alpha_1],\cdots,[\alpha_\nu])(U_1,\cdots,U_n) \\ \text{diag}([\beta_{\nu+1}],\cdots,[\beta_n])(e^{z\omega_1}V_1,\cdots,e^{z\omega_n}V_n) \end{vmatrix}$$

$$= z^m \sum_{j_1<\cdots<j_\nu} \pm \left[\prod_{j=1}^{\nu}\alpha_j \prod_{j=\nu+1}^{n}\beta_j\right] e^{-z\sum_{i=1}^{\nu}\omega_{j_i}} .$$

$$\cdot |(U_{j_1},\cdots,U_{j_\nu})| \cdot |(V_1,\cdots,\hat{V}_{j_1},\cdots,\hat{V}_{j_\nu},\cdots,V_n)|$$

where $m = \displaystyle\sum_{j=1}^{n} m_j$ and $\hat{V}_j$ means that V_j is not included in the expression.

Therefore, defining

$$D_1 = \alpha_1|(U_1,\cdots,U_{\nu-1},U_{\nu+1})| \cdot |(V_\nu,V_{\nu+2},\cdots,V_n)|,$$

$$D_{-1} = \alpha_1|(U_1, \cdots, U_{\nu-1}, U_\nu)| \cdot |(V_{\nu+1}, V_{\nu+2}, \cdots, V_n)|,$$

$$h(z) = \left[\prod_{j=2}^{\nu} \alpha_j \prod_{j=\nu+1}^{n} \beta_j\right] e^{-z\sum_{i=1}^{\nu+1}\omega_i},$$

it can be shown that

$$D_1 \neq 0, \quad D_{-1} \neq 0,$$

$$\Delta(z,0) = z^m h(z)\left\{-[D_1]e^{z\omega_\nu} + [D_{-1}]e^{z\omega_{\nu+1}}\right\},$$

and then $\Delta(z,0) = 0$ can, up to a power of z, be equivalently written as

$$\delta(z,0) \equiv -[D_1]e^{z\omega_\nu} + [D_{-1}]e^{z\omega_{\nu+1}} = 0. \tag{2.16}$$

Similarly, if we define

$$U_j^0 = \begin{pmatrix} \omega_j^{m_0} \\ \omega_j^{m_2} \\ \vdots \\ \omega_j^{m_\nu} \end{pmatrix},$$

and let

$$N_1 = \alpha_0|(U_1^0, \cdots, U_{\nu-1}^0, U_{\nu+1}^0)| \cdot |(V_\nu, V_{\nu+2}, \cdots, V_n)|,$$

$$N_{-1} = \alpha_0|(U_1^0, \cdots, U_{\nu-1}^0, U_\nu^0)| \cdot |(V_{\nu+1}, V_{\nu+2}, \cdots, V_n)|,$$

then

$$N_1 \neq 0, \quad N_{-1} \neq 0,$$

$$\Delta(z,\infty) \equiv \det\left(\mathcal{W}_i(y_j)\right)_{i=0,2,j=1}^{i=n,j=n} = z^{m_0} h(z)\left\{-[N_1]e^{z\omega_\nu} + [N_{-1}]e^{z\omega_{\nu+1}}\right\},$$

where $m^0 = \displaystyle\sum_{j=0,2}^{n} m_j$, has the equivalent form

$$\delta(z,\infty) \equiv -[N_1]e^{z\omega_\nu} + [N_{-1}]e^{z\omega_{\nu+1}}. \tag{2.17}$$

Finally, the return difference equation (2.10)

$$\Delta(z,0) + k\Delta(z,\infty) = 0$$

in the z-plane can, up to a power of z, be written as

$$\delta(z,k) \equiv z^\ell \delta(z,0) + k\delta(z,\infty) = 0, \tag{2.18}$$

where $\ell = m - m^0$.

3 Statement of Main Results

In this section we state our results about systems $((2.5),(2.7))$. For the sake of simplicity and due to the fact that most of the proofs depend heavily on the results in [5], we omit proofs and refer to [5]. Note that most of our results are similar to the corresponding ones for general Birkhoff regular systems, except that in the present case the results depend on μ and ν rather than just $\mu = n/2$.

Just as in the general Birkhoff regular case [5], to investigate the root locus of $((2.5),(2.7))$, we need only consider $z \in S_0$ when ν is odd, $z \in S_1$ when ν is even because the closed loop poles occur in conjugate pairs and the relation between λ and z is given by $\lambda = i^n z^n$.

Proposition 3.1 *We have the following:*

1. $\overline{D}_{-1} = \begin{cases} (-1)^{\mu-1} D_1, & \text{for } \nu \text{ odd} \\[2ex] (-1)^\mu D_{-1} e^{\frac{4m\pi i}{n}}, & \text{for } \nu \text{ even} \end{cases}$

   $\overline{N}_{-1} = \begin{cases} (-1)^{\mu-1} N_1, & \text{for } \nu \text{ odd} \\[2ex] (-1)^\mu N_{-1} e^{\frac{4m^0 \pi i}{n}}, & \text{for } \nu \text{ even} \end{cases}$

2. *No matter ν is odd or even*

   $$D_{-1} = D_1 e^{-\frac{2m\pi i}{n}}, \qquad N_{-1} = N_1 e^{-\frac{2m^0 \pi i}{n}}$$

3. $\arg \tau = \begin{cases} s\pi + \dfrac{\ell\pi}{n}, & \text{for } \nu \text{ odd} \\[2ex] s\pi + \dfrac{2\ell\pi}{n}, & \text{for } \nu \text{ even} \end{cases}$,

   where $\tau = \dfrac{N_{-1}}{D_{-1}}$ and $s = \dfrac{1 - \operatorname{sgn}(\alpha_0/\alpha_1)}{2} + \ell + \ell^*$ with ℓ^* the number of $\{m_j\}_{j=2}^\nu$ between m_0 and m_1.

As in [2]-[4] and [5], we introduce

Definition 3.2 *The* instantaneous gain *is defined by*

$$\mathcal{K} = (-1)^s \tag{3.1}$$

where s is defined in Proposition 3.1.

Assumption 3.3 *The sign of k is chosen from the instantaneous gain formula so that*

$$k \cdot \mathcal{K} \geq 0.$$

With this assumption, we are able to establish the following important proposition.

Proposition 3.4 *With the above assumption:*

1. *When ν is odd, all the zeros of*

$$\delta(z, k) = 0$$

   occur in the region

$$\Omega \equiv S_0 \cup S_{2n-1} \cap \{z \mid \ | \operatorname{Im} z | \leq y_0\} \tag{3.2}$$

   for some y_0.

2. *To study the asymptotic behavior of the roots of*

$$\delta(z, k) = 0$$

   we need only consider an equation in the form

$$w^\ell \left[\sin \left(w + \frac{m\pi}{n} \right) \right] + | \, g \, | \left[\sin \left(w + \frac{m^0 \pi}{n} \right) \right] = 0, \tag{3.3}$$

   where

$$w = z \sin \theta, g = k\tau \sin^\ell \theta, \ \text{for } \nu \ \text{odd}, \ z \in S_0,$$

$$w = ze^{-\frac{\pi i}{n}} \sin \theta, g = k\tau e^{-\frac{\ell \pi i}{n}} \sin^\ell \theta, \ \text{for } \nu \ \text{even}, \ z \in S_1,$$

   with $\theta = \dfrac{\nu \pi}{n}$.

We now present an example to demonstrate that if the control and sensor are not colocated, then the previous proposition is not true.

Example 3.1 Consider the system

$$\dot{w} \ = \ Aw, \quad A = -\frac{d^4}{dx^4}, \quad x \in (0, 1)$$

$$u(t) \ = \ W_1(w)(t) = w''(1, t)$$

$$y(t) \ = \ W_0(w)(t) = w'(0, t)$$

$$W_2(w)(t) = w(0, t)$$

$$W_3(w)(t) = w'(1, t)$$

$$W_4(w)(t) = w(1, t)$$

Note that the control u and sensor y are not colocated.

For $z \in S_1$, a straightforward computation and estimation provide the asymptotic formula

$$\delta(z, k) = \left\{ 4e^{\frac{\pi i}{4}} z \left(e^{-(e^{-\frac{\pi i}{4}} + e^{\frac{\pi i}{4}})z} + i \right) + \right.$$

$$\left. +2k \left(e^{e^{\frac{\pi i}{4}} z} + e^{-e^{\frac{\pi i}{4}} z} \right) \right\} [1].$$

Defining $w = ze^{-\frac{\pi i}{4}} \in S_0$. The roots of $\delta(z, k)$ are asymptotically determined by the equation

$$iw \left(e^{-(1+i)w} + i \right) + k \cos w = 0.$$

Therefore the roots of $\delta(z, 0) = 0$ and $\delta(z, \infty) = 0$ are asymptotically given by

$$p_j = i\frac{\sqrt{2}}{2} \left(2j\pi + \frac{\pi}{2} \right),$$

$$z_j = e^{\frac{\pi i}{4}} \left(j\pi + \frac{\pi}{2} \right).$$

Notice that Proposition 3.4 says that $\delta(z, k) = 0$ has no roots at all outside Ω defined in (3.2) in the ν odd case, and that the ν even case can be converted into the ν odd case. Thus, from the expressions of p_j and z_j, and the relation between λ and z, we conclude that Proposition 3.4 can not be applied to this example.

The following is our form of a proposition due to M. V. Keldysh (cf. [6]). Note that our $m = \sum_{i=1}^{n} m_j$ is the sum of the weights of the matrices of coefficients of the boundary conditions defined there.

Proposition 3.5 *The roots of* $\delta(z, 0) = 0$ *and* $\delta(z, \infty) = 0$ *are asymptotically given, respectively, by*

$$p_j = \begin{cases} \dfrac{j\pi - m\pi/n}{\sin(\nu\pi/n)} + O\left(\dfrac{1}{z}\right), & \text{for } \nu \text{ odd} \\[2ex] e^{\frac{\pi i}{n}} \dfrac{j\pi - m\pi/n}{\sin(\nu\pi/n)} + O\left(\dfrac{1}{z}\right), & \text{for } \nu \text{ even} \end{cases} \tag{3.4}$$

$$z_j = \begin{cases} \dfrac{j\pi - m^0\pi/n}{\sin(\nu\pi/n)} + O\left(\dfrac{1}{z}\right), & \text{for } \nu \text{ odd} \\[2ex] e^{\frac{\pi i}{n}} \dfrac{j\pi - m^0\pi/n}{\sin(\nu\pi/n)} + O\left(\dfrac{1}{z}\right), & \text{for } \nu \text{ even} \end{cases} \tag{3.5}$$

Now we give an example to demonstrate why we require the boundary conditions to be completely separated.

Example 3.2 Consider the system

$$\dot{w} \;=\; Aw, \quad A = -\frac{d^4}{dx^4}, \quad x \in (0,1)$$

$$u(t) \;=\; W_1(w)(t) = w'''(0,t)$$

$$y(t) \;=\; W_0(w)(t) = w''(0,t) + w'(1,t)$$

$$W_2(w)(t) = w'(0,t)$$

$$W_3(w)(t) = w(0,t)$$

$$W_4(w)(t) = w(1,t)$$

Then the boundary conditions are separated but not completely separated. For $z \in S_0$, we have

$$\Delta(z,\infty) = 2\omega_1 z^2 e^{(\omega_3+\omega_4)z}\left\{\omega_3 + 2ze^{\omega_2 z}\right\}[1].$$

Thus the roots of $\Delta(z,\infty) = 0$ are asymptotically determined by the equation

$$\omega_3 + 2ze^{\omega_2 z} = 0.$$

It is easy to check that this equation does not have any roots when z is close to the boundary of S_0 for $\mathrm{Re}\, z$ large, or more precisely, when z is in the set

$$\{z = x + iy \in S_0 \mid d(z, \partial S_0) \le 1, \ x \ \text{large}\}.$$

Hence, according to Proposition 3.5, the root locus of this system is much different from that for the systems $((2.5),(2.7))$.

Parallel to the results in [2]-[4] and [5], we have the following theorems with

$$\chi = \mu + \nu. \tag{3.6}$$

Theorem 3.6 *The open loop poles and zeros of $((2.5),(2.7))$ are asymptotically given by*

$$\lambda_j = (-1)^{\chi-1}\left(\frac{j\pi}{\sin \nu\pi/n}\right)^n\left\{1 - \frac{M\pi}{j\pi} + O\left(\frac{1}{j^2}\right)\right\}, \tag{3.7}$$

where $M = m$ for poles, $M = m^0$ for zeros. Moreover, all but a finite number of open loop poles and zeros are real and simple. The open loop poles and zeros of large modulus interlace on the positive real line when χ is odd, on the negative real line when χ is even.

Theorem 3.7 *The closed loop poles vary continuously from open loop poles to open loop zeros as $|\,k\,|$ varies from zero to infinity. Every branch of the root locus is bounded. Furthermore, the closed loop poles of large modulus are real and simple, and move to the right when χ is odd, to the left when χ is even.*

Finally, we state a direct corollary of Theorem 3.4 of [5].

Theorem 3.8 *If $k \cdot \mathcal{K} \geq 0$, then a real point is on the root locus if and if it lies to the right when χ is odd, to the left when χ is even, of an odd number of open loop poles and zeros.*

References

[1] G.D. BIRKHOFF, "Boundary value and expansion problems of ordinary linear differential equations", *Trans. Amer. Math. Soc.*, Vol. 9, (1908), 373-395.

[2] C.I. BYRNES, D.S. GILLIAM, "Asymptotic properties of root locus for distributed parameter systems," *Proc. of IEEE Conf. on Dec. and Control*, Austin, 1988.

[3] C.I. BYRNES, D.S. GILLIAM, "Boundary feedback stabilization of distributed parameter systems", *Proceedings of 9th MTNS* (1989).

[4] C.I. BYRNES, D.S. GILLIAM, J. HE, "Root locus and boundary feedback design for a class of distributed parameter systems", submitted

[5] J. HE, "Root Locus for Birkhoff regular Distributed Parameter Control Systems", preprint

[6] M.A. NAIMARK, "Linear differential operators, I," Ungar, New York, (1967).

ON THE PROBLEM OF PARAMETER IDENTIFICATION IN PERSPECTIVE SYSTEMS AND ITS APPLICATION TO MOTION ESTIMATION PROBLEMS IN COMPUTER VISION

Edward P. Loucks and Bijoy K. Ghosh[*]

Department of Systems Science and Mathematics
Washington University
St. Louis, MO 63130, U.S.A.

We consider the problem of motion and shape estimation of a moving textured surface with the aid of a monocular camera. We show that the estimation problem reduces to a specific parameter estimation of a perspective dynamical system. In this paper we analyze a planar textured surface undergoing a rigid flow and an affine flow. These two cases have been analysed for perspective and orthographic projection. The main result of this paper is to identify what parameters or function of parameters are identifiable.

1 Introduction

The problem that we consider in this paper is described as follows.

Given a textured plane which is moving in continuous time following an unknown affine vector field where we assume that both the position of the plane and the parameters of the vector field are unknown. Assume that a camera produces a perfect image of the textured surface in continuous time. The problem of interest is to estimate the position and motion parameters of the surface from the observed time varying image produced by the camera.

In this paper we shall assume that the surface being observed is constantly under focus i.e. we shall not study the effect of blurring. We shall also assume that the photometric effects on the image due to the light source and the physical properties of the surface is negligible and can be ignored. Thus on the whole the process of image formation is assumed to be such that the intensity corresponding to each pixel on the surface is transferred to the image plane unattenuated via the projection process.

The existing approaches to the estimation problem in the literature can be divided broadly into two categories depending upon what is assumed to be measured from the scene. If the data observed is assumed to be the

---

*Partially supported by DOE under grant No. DE-FG02-90ER14140.

brightness pattern which the object produces on the image plane, a well known approach in the literature is based on analysing the optical flow field (see [3], [18], [19]). On the other hand if the data observed are assumed to be the discontinuity-curves in the brightness pattern on the image plane, a well known feature based approach is to identify the correspondence of various features such as points, lines and curves between one frame and the next (see [4], [5], [6], [7], [8], [9], [11], [12]). The former approach assumes that the image intensity is a smooth function and restricts attention to only the smooth part of the image plane. The latter approach assumes that the image intensity is a piecewise smooth function and restricts attention to the region of the image plane wherein the image intensity is separated by a discontinuity curve. Ofcourse for each of the two approaches, there are various projection models that one might like to consider. The two projection models well known in the literature are known as 'orthographic' and 'perspective.'

In this paper, we consider in details the problem of estimating the unknown parameters of a planar surface undergoing an affine motion. The proposed affine motion generalizes the rigid motion already considered in the literature (see [4],[13], [14], [15], [16], [17]). While preserving the shape of the surface being observed, an affine motion adequately models many other non-rigid deformations. We also consider a generalized projection which includes as a special case both "image centered projection" and "viewer centered projection" (see [10]), together with orthographic and perspective projections.

The details about the estimation scheme proposed in this paper, are now explained as follows. As shown in Fig. 1 the estimation problem is broken up into two modules known as the 'Intensity Dynamic Module' (IDM) and the 'Shape Dynamic Module' (SDM). Data from the observed surface is first processed in the IDM in order to estimate a set of 'essential parameters.' Effectively IDM performs a spatial averaging throughout the entire image plane from either the observed sequence of features or the optical flow data. For lack of space, in this paper we concentrate only on the optical flow data.

The essential parameters are functions of motion and shape parameters. The shape dynamic module views them as an observation function corresponding to the "shape-dynamics" introduced in this paper. The shape dynamical system together with the essential parameters (viewed as an output) can be regarded as an example of a perspective system introduced in [2]. By observing the essential parameters over time, SDM obtains an estimate of the motion and shape parameters.

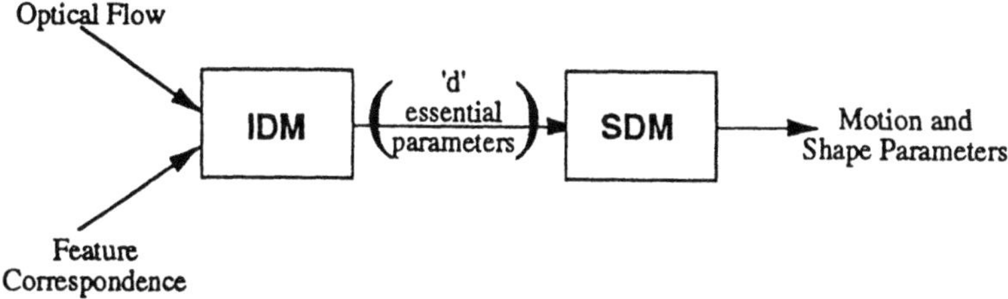

Figure 1. A two module approach to parameter estimation. IDM: Intensity Dynamic Module. SDM: Shape Dynamic Module.

2   Shape and Intensity Dynamics of a Surface Patch

Let us assume that (X, Y, Z) is the world co-ordinate frame wherein we have a surface defined by the equation

$$Z = pX + qY + r \tag{2.1}$$

where p, q, r are shape parameters that are changing in time. We assume that the motion field is given by the equation

$$\dot{\mathcal{X}} = A\mathcal{X} + b \tag{2.2}$$

where A is an arbitrary 3×3 matrix and b is a 3×1 vector. Let us define

$$A = \begin{pmatrix} a_{11} & a_{12} & a_{13} \\ a_{21} & a_{22} & a_{23} \\ a_{31} & a_{32} & a_{33} \end{pmatrix} \tag{2.3}$$

$$b = \begin{pmatrix} b_1 & b_2 & b_3 \end{pmatrix}^T. \tag{2.4}$$

The vector $\mathcal{X}$ is defined as $(X\ Y\ Z)^T$. If we homogenize the vector $(X\ Y\ Z)$ as

$$X = \frac{\bar{X}}{\bar{W}}, Y = \frac{\bar{Y}}{\bar{W}}, Z = \frac{\bar{Z}}{\bar{W}}$$

and the vector $(p\ q\ r)$ as

$$p = \frac{\bar{p}}{\bar{s}}, q = \frac{\bar{q}}{\bar{s}}, r = \frac{\bar{r}}{\bar{s}} \tag{2.5}$$

we are able to rewrite (2.1) as

$$\begin{pmatrix} \bar{p} & \bar{q} & -\bar{s} & \bar{r} \end{pmatrix} \bar{\mathcal{X}} = 0 \tag{2.6}$$

and (2.2) as

$$\dot{\bar{\mathcal{X}}} = \bar{A}\bar{\mathcal{X}} \tag{2.7}$$

where

$$\bar{\mathcal{X}} = (\ \bar{X}\ \ \bar{Y}\ \ \bar{Z}\ \ \bar{W}\ )^T \tag{2.8}$$

and

$$\bar{A} = \begin{pmatrix} A & b \\ 0 & 0 \end{pmatrix}. \tag{2.9}$$

It follows that

$$\frac{d}{dt}\begin{pmatrix} \bar{p} \\ \bar{q} \\ -\bar{s} \\ \bar{r} \end{pmatrix} = \begin{pmatrix} -A^T & 0 \\ -b^T & 0 \end{pmatrix}\begin{pmatrix} \bar{p} \\ \bar{q} \\ -\bar{s} \\ \bar{r} \end{pmatrix}. \tag{2.10}$$

The above equation (2.10), known as the 'shape dynamics,' describes how the surface (2.1) moves as points on the surface moves following the motion field (2.2). If we assume initial condition to be $\bar{s}(0) = 1$, $\bar{p}(0) = p(0)$, $\bar{q}(0) = q(0)$, $\bar{r}(0) = r(0)$, it may be concluded that the dynamical system (2.10) describes the motion of the shape parameters p, q, r. In fact the dynamics of p, q, r is quadratic and is given by

$$\begin{aligned} \dot{p} &= (a_{33} - a_{11})p - a_{21}q + a_{31} - a_{13}p^2 - a_{23}pq \\ \dot{q} &= (a_{33} - a_{22})q - a_{12}p + a_{32} - a_{13}pq - a_{23}q^2 \\ \dot{r} &= -(a_{33} + a_{23}q + a_{13}p)r + (b_3 - b_2q - b_1p). \end{aligned} \tag{2.11}$$

where $b = (b_1\ b_2\ b_3)^T$.

Assume that the surface described by (2.1) is textured i.e. the intensity $E(X, Y, Z, t)$ of a point (X, Y, Z) on the surface at time t does not change along the integral curves of (2.2). We also assume that the camera is perfectly focused on the surface i.e. intensity from the surface to the image plane is transferred unattenuated under the camera correspondence. The above two assumptions together imply that the intensity on the image plane does not change along the projection of the integral curves of (2.2). In this paper we consider the projection to be described as follows. Let (x, y) be the coordinates of the image point obtained under the projection of the point (X, Y, Z) on the surface of the object (2.1). We define

$$x = \frac{fX}{Z + \delta}, \quad y = \frac{fY}{Z + \delta} \tag{2.12}$$

where $\delta \in [0, f]$ and f is the focal length of the camera. Note that if $\delta = 0$ we obtain what is known as a viewer centered projection. If $\delta = f$ we obtain an image centered projection. These have been described in [10]. Finally note that if $\delta = f$ and $f \to \infty$, we obtain

$$x = X, \quad y = Y$$

which is known in the literature as the 'orthographic projection.' For a given fixed value of f, δ we have a new set of coordinates (x,y,Z). We now rewrite the shape equation (2.1) in the new set of coordinates as

$$1 = px + qy + r/Z \tag{2.13}$$

From (2.2), (2.12) and (2.13) we obtain

$$\dot{x} = F(x,y), \quad \dot{y} = G(x,y) \tag{2.14}$$

where

$$\begin{aligned}
F(x,y) &= (f(a_{13} + c_1)) + ((a_{11} - a_{33}) - (c_3 + pc_1))x + \\
&\quad (a_{12} - qc_1)y - (\tfrac{1}{f}(a_{31} - pc_3))x^2 - (\tfrac{1}{f}(a_{32} - qc_3))xy \\
G(x,y) &= (f(a_{23} + c_2)) + ((a_{22} - a_{33}) - (c_3 + qc_2))y + \\
&\quad (a_{21} - pc_2)x - (\tfrac{1}{f}(a_{32} - qc_3))y^2 - (\tfrac{1}{f}(a_{31} - pc_3))xy
\end{aligned} \tag{2.15}$$

and where $c_1 = (b_1 - a_{13}\delta)/(r + \delta)$, $c_2 = (b_2 - a_{23}\delta)/(r + \delta)$, $c_3 = (b_3 - a_{33}\delta)/(r + \delta)$. The integral curves of (2.14) are exactly the projection of the integral curves of (2.2) under the generalized projection (2.12). The vector field described by (2.14) has been described in the literature as 'optical flow.'

Let $e(x,y,t)$ be the intensity of a point (x,y) on the image plane at time instant t. It follows from hypothesis that

$$\frac{\partial e}{\partial t} + F(x,y)\frac{\partial e}{\partial x} + G(x,y)\frac{\partial e}{\partial y} = 0. \tag{2.16}$$

We shall call the dynamical system (2.16) as the 'intensity dynamics.'

Assume that the plane produces a time varying intensity profile on the image plane. Assume, furthermore, that the intensity function is smooth so that all its partial derivatives exist and can be computed. Let us rewrite $F(x,y)$ and $G(x,y)$ from (2.14) as follows.

$$\begin{aligned}
F(x,y) &= A + Bx + Cy + Dx^2 + Exy \\
G(x,y) &= F + Hx + Gy + Ey^2 + Dxy
\end{aligned} \tag{2.17}$$

where A, B, C, D, E, F, G and H can be defined from (2.15). Combining (2.15) and (2.16) we now write

$$v^T d = -\frac{\partial e}{\partial t} \tag{2.18}$$

where

$$v^T = (\begin{array}{cccccc} e_x & e_y & xe_x & ye_x & xe_y & ye_y \end{array}$$
$$\begin{array}{cc} x^2 e_x + xye_y & xye_x + y^2 e_y \end{array}) \tag{2.19}$$

and the vector

$$d = (\begin{array}{ccc} f(a_{13} + c_1) & f(a_{23} + c_2) & (a_{11} - a_{33}) - (c_3 + pc_1) \end{array} \tag{2.20}$$

$$\begin{array}{ccc} a_{12} - qc_1 & a_{21} - pc_2 & (a_{22} - a_{33}) - (c_3 + qc_2) \end{array}$$

$$\begin{array}{cc} -\frac{1}{f}(a_{31} - pc_3) & -\frac{1}{f}(a_{32} - qc_3) \end{array})^T$$

Let us choose $n \geq 8$ points on the image plane denoted by (x_i, y_i), $i = 1, \cdots, n$. From the observed data $e(x, y, t)$ we now form the matrix

$$V = (\begin{array}{cccc} v(x_1, y_1) & v(x_2, y_2) & \cdots & v(x_n, y_n) \end{array}) \tag{2.21}$$

and

$$u = (\begin{array}{cccc} -e_t(x_1, y_1) & -e_t(x_2, y_2) & \cdots & -e_t(x_n, y_n) \end{array})^T \tag{2.22}$$

From (2.18) it follows that

$$V^T d = u \tag{2.23}$$

If the points are chosen in such a way that rank $V = 8$, we compute

$$\hat{d} = (V^T V)^{-1} V^T u \tag{2.24}$$

as an estimate of d.

It may be remarked that the computation of an estimate of d is the goal of the 'Intensity Dynamic Module' introduced in the introduction. It can be shown (see [1]) that the vector d can also be estimated by observing the motion of various features viz. points, lines and curves on the screen. Note in particular that the estimate of the vector d above is computed via a spatial averaging over various points on the image plane. It could have been also obtained by spatial averaging over various features on the image plane. In this paper, we shall call this vector d as the vector of essential parameters.

3 Motion Estimation as a Parameter Identification Problem in Perspective Systems

Let us consider the shape dynamics (2.10) where the output function of the shape dynamics is considered to be given by (2.20). Homogenizing the vector d we now obtain the following output function for the shape dynamics.

$$
\begin{pmatrix} \bar{d}_1 \\ \bar{d}_2 \\ \bar{d}_3 \\ \bar{d}_4 \\ \bar{d}_5 \\ \bar{d}_6 \\ \bar{d}_7 \\ \bar{d}_8 \\ \bar{d}_9 \end{pmatrix} = \begin{pmatrix} 0 & 0 & -fb_1 & fa_{13} \\ 0 & 0 & -fb_2 & fa_{23} \\ -b'_1 & 0 & b_3 - \delta a_{11} & a_{11} - a_{33} \\ 0 & -b'_1 & -\delta a_{12} & a_{12} \\ -b'_2 & 0 & -\delta a_{21} & a_{21} \\ 0 & -b'_2 & b_3 - \delta a_{22} & a_{22} - a_{33} \\ -\frac{b'_3}{f} & 0 & -\frac{\delta}{f}a_{31} & \frac{a_{31}}{f} \\ 0 & -\frac{b'_3}{f} & -\frac{\delta}{f}a_{32} & \frac{a_{32}}{f} \\ 0 & 0 & -\delta & 1 \end{pmatrix} \begin{pmatrix} \bar{p} \\ \bar{q} \\ -\bar{s} \\ \bar{r} \end{pmatrix} \qquad (3.1)
$$

where p, q and r have been defined in (2.5) and where

$$
b' = (b_1 - a_{13}\delta \ \ b_2 - a_{23}\delta \ \ b_3 - a_{33}\delta) = (b'_1 \ b'_2 \ b'_3)
$$

The vector d in (2.20) is given by

$$
\begin{pmatrix} \frac{\bar{d}_1}{\bar{d}_9} & \frac{\bar{d}_2}{\bar{d}_9} & \cdot & \cdot & \cdot & \frac{\bar{d}_8}{\bar{d}_9} \end{pmatrix} \qquad (3.2)
$$

The pair (2.10), (3.1) is to be regarded as a fourth order system with outputs as points in $\mathbb{RP}^8$, the real projective space of homogeneous lines in $\mathbb{R}^9$. This is precisely an example of a perspective system introduced in [2]. The perspective identifiability problem is to identify the parameters of the 4×4 matrix (2.10) together with the initial condition on the state vector $(\bar{p} \ \bar{q} \ \bar{s} \ \bar{r})$ upto a scale factor given that the output vector $(\bar{d}_1, \cdots, \bar{d}_9)$ is observed upto a scale factor.

The main result of this section is to derive an identifiability condition for the shape dynamics (2.10) together with the observation function (3.1). Note in particular that the perspective system (2.10), (3.1) is parameterized by a set of 12 motion parameters A, b and a set of three shape parameters p, q and r. We shall show that not all the fifteen parameters are identifiable i.e. there is a nonunique chice of parameters for which the observation described by (3.1) is the same. The main theorem of this section is described as follows.

Theorem 3.1: Under a suitable generic condition on the set of fifteen parameters of the perspective system (2.10), (3.1), the following parameters or the function of the parameters are identifiable. They are a_{ij}, $i = 1, 2, 3$, $j = 1, 2, 3$; p, q, c_1, c_2, c_3, where c_1, c_2 and c_3 has been defined in section 2.

For a proof of Theorem 3.1 see [1]. The conclusion of Theorem 3.1 can be summarized as follows. The matrix A can be completely identified together with the shape parameters p and q. The parameters b_1, b_2, b_3 and r can be identified upto a line in the (b_1, b_2, b_3, r) space. This line is described by

$$c_1(r + \delta) = b_1 - a_{13}\delta$$
$$c_2(r + \delta) = b_2 - a_{23}\delta \tag{3.3}$$
$$c_3(r + \delta) = b_3 - a_{33}\delta.$$

Thus a total of 14 parameters or function of parameters out of a total of 15 free parameters are identifiable. It is of considerable theoretical interest to understand what happens to the limit of the solution expressed by Theorem 3.1 assuming $f = \delta$ and letting $f \to \infty$. This is because under this limit, the projection equation (2.12) approaches the orthographic projection. We have the following result.

Theorem 3.2: Assume $f = \delta$. Consider the perspective system (2.10), (3.1). Under the same generic condition as that expressed by Theorem 3.1, the following parameters or the function of parameters can be identified asymptotically as $f \to \infty$. They are a_{ij}, $i = 1, 2, 3$, $j = 1, 2, 3$; p, q, $b_1 + a_{13}r$, $b_2 + a_{23}r$ and $b_3 + a_{33}r$.

The proof of Theorem 3.2 follows as a corollary of Theorem 3.1. In fact as $f \to \infty$ the line (3.3) degenerates to the line

$$b_1 + a_{13}r = h_1$$
$$b_2 + a_{23}r = h_2 \tag{3.4}$$
$$b_3 + a_{33}r = h_3.$$

where h_1, h_2, h_3 are constants given by the actual values of the left hand sides of the equation (3.4).

Let us now consider orthographic projection by choosing $f = \delta$ and letting $f \to \infty$ in equation (2.12). Two of the output equations corresponding to $\bar{d}_7$ and $\bar{d}_8$ in (3.1) degenerates to zero under this limit. The homogenized output map is given by

$$
\begin{pmatrix} \bar{d}_1 \\ \bar{d}_2 \\ \bar{d}_3 \\ \bar{d}_4 \\ \bar{d}_5 \\ \bar{d}_6 \\ \bar{d}_9 \end{pmatrix}
=
\begin{pmatrix}
0 & 0 & -b_1 & a_{13} \\
0 & 0 & -b_2 & a_{23} \\
a_{13} & 0 & -a_{11} & 0 \\
0 & a_{13} & -a_{12} & 0 \\
a_{23} & 0 & -a_{21} & 0 \\
0 & a_{23} & -a_{22} & 0 \\
0 & 0 & -1 & 0
\end{pmatrix}
\begin{pmatrix} \bar{p} \\ \bar{q} \\ -\bar{s} \\ \bar{r} \end{pmatrix}
\tag{3.5}
$$

In this case we can show that the If we consider the shape dynamics (2.10) in addition to the output equation (3.5) we have the following result.

Theorem 3.3: Consider the perspective system (2.10), (3.5). Under a suitable generic condition on the parameters of the perspective system, the functions of the parameters that can be identified are given by

$$a_{11} - \pi_2 a_{13}$$
$$a_{21} - \pi_2 a_{23}$$
$$a_{12} - \pi_3 a_{13}$$
$$a_{22} - \pi_3 a_{23}$$
$$\pi_1 a_{13}$$
$$\pi_1 a_{23}$$
$$b_1 - \pi_4 a_{13}$$
$$b_2 - \pi_4 a_{23}$$
$$\frac{1}{\pi_1}(\pi_2 a_{11} + \pi_3 a_{21} + a_{31}) - \frac{\pi_2}{\pi_1}(\pi_2 a_{13} + \pi_3 a_{23} + a_{33})$$
$$\frac{1}{\pi_1}(\pi_2 a_{12} + \pi_3 a_{22} + a_{32}) - \frac{\pi_3}{\pi_1}(\pi_2 a_{13} + \pi_3 a_{23} + a_{33})$$
$$\frac{1}{\pi_1}(\pi_2 b_1 + \pi_3 b_2 + b_3) - \frac{\pi_4}{\pi_1}(\pi_2 a_{13} + \pi_3 a_{23} + a_{33})$$
$$\pi_2 a_{13} + \pi_3 a_{23} + a_{33}$$
$$\frac{1}{\pi_1}(p + \pi_2)$$
$$\frac{1}{\pi_1}(q + \pi_3)$$
$$\frac{1}{\pi_1}(r + \pi_4)$$

$$(3.6)$$

where π_1, π_2, π_3, π_4 are arbitrary real parameters with the restriction that $\pi_1 \neq 0$.

In fact the functions (3.6) describes a four parameter orbit of a perspective group action on the space of parameters (see [1] for details). Under the nominal value $\pi_1 = 1$, $\pi_2 = 0$, $\pi_3 = 0$, $\pi_4 = 0$ the above functions (3.6) are respectively the parameters a_{11}, a_{21}, a_{12}, a_{22}, a_{13}, a_{23}, b_1, b_2, a_{31}, a_{32}, b_3, a_{33}, p, q and r. Theorems 3.2 and 3.3 basically asserts that under generalized projection (2.12) parameters can be identified upto a one parameter family. As $f \rightarrow \infty$, the one parameter family of orbits approach a limiting orbit contained in the four parameter orbit described by (3.6). Thus roughly speaking "the limiting process of choosing f asymptotically large recovers more about the parameters than what can be recovered in the limit via orthographic projection."

4   Parameter estimation of a planar textured surface undergoing a rigid flow

In this section we shall assume that the motion field describes a rigid flow and consider the motion field equation (2.2) with the matrix A replaced by the matrix Ω where

$$\Omega = \begin{pmatrix} 0 & \omega_1 & \omega_2 \\ -\omega_1 & 0 & \omega_3 \\ -\omega_2 & -\omega_3 & 0 \end{pmatrix} \tag{4.1}$$

We shall analyze the perspective system (2.10), (3.1) specialized to the rigid flow dynamics and derive the following result.

Theorem 4.1: For a planar surface undergoing a rigid flow, consider the perspective system (2.10), (3.1) parameterized by a set of nine parameters. Under a suitable generic condition on the set of parameters, the vector $(c_1, c_2, c_3, \omega_1, \omega_2, \omega_3, p, q)$ is identifiable uniquely, where c_1, c_2, c_3 has been defined in section 2.

We also consider orthographic projection by choosing $f = \delta$ and letting $f \to \infty$. Two of the output equations corresponding to $\bar{d}_7$ and $\bar{d}_8$ in (3.1) degenerates to zero under this limit. Thus in this case we have a total of six equations in nine variables described by (3.5). No information is obtained about the parameter b_3 from the output equation (3.5) alone. Furthermore the parameters ω_1, ω_2, ω_3, p, q , r , b_1, b_2 are obtained upto a choice of two free parameters. If we consider the shape dynamics (2.10) in addition to the output equation (3.5) we have the following result.

Theorem 4.2: Consider the perspective system (2.10), (3.5) (with the matrix A replaced by Ω) parameterized by ω_1, ω_2, ω_3, p, q , r , b_1, b_2 and b_3. Under a suitable generic condition, the parameters or the functions of the parameters that can be identified are given by

$$\omega_1, \pm\omega_2, \pm\omega_3, \pm b_3, \mp p, \mp q, \omega_2 r + b_1, \omega_3 r + b_2. \tag{4.2}$$

It follows from the above theorem 4.2 that under orthographic projection, the parameters of the perspective system (2.10), (3.5) can be identified upto one parameter and a sign ambiguity. It is of considerable theoretical interest to understand what happens to the limit of the solution expressed by Theorem 4.1 as $f \to \infty$. We have the following result.

Theorem 4.3: For a planar surface undergoing a rigid flow, consider the perspective system (2.10), (3.1). Under the same generic

condition as that expressed by Theorem 4.1, on the set of parameters, as $f \to \infty$ the line (3.3) degenerates to the line

$$\omega_2 r + b_1 = h_1, \omega_3 r + b_2 = h_2, b_3 = h_3 \tag{4.3}$$

where h_1, h_2, h_3 are constants given by the actual values of the left hand sides of the equation (4.3).

Thus from (4.3) we infer that for large f, one can identify the parameter b_3 exactly together with the functions $\omega_2 r + b_1$ and $\omega_3 r + b_2$. Ofcourse we are also able to identify the vector $(\omega_1, \omega_2, \omega_3, p, q)$ as it can be derived directly from theorem 4.1. Note from (4.2) that this is exactly what one could identify under orthographic projection (upto a sign ambiguity).

5   Conclusion

The main results of this paper is now summarized. For a visual system with its focal length permanently fixed at infinity, parameters of a planar motion with an affine flow are recovered via orthographic projection upto a surface of dimension 4. On the other hand if the focal length can be varied, then parameters can be recovered via projection (2.12) and in this case parameters can be identifiable upto a one parameter ambiguity. For a biological system, this justifies the need for a pair of eyes with a capability to vary its focal length at will in order to decipher a non rigid environment. For a rigid motion, whether the projection is orthographic or not, parameters are always recovered upto a one parameter ambiguity even when f approaches infinity. Thus in order to decipher an cnvironment with only rigid objects, it is enough to have a pair of eyes or visual systems without the capability to vary its focal length.

References

[1] B. K. Ghosh, E. P. Loucks, A Perspective Theory for Motion and Shape Estimation in Machine Vision, Submitted for publication in Journal of Mathematical Systems Estimation and Control.

[2] B. K. Ghosh, M. Jankovic and Y. T. Wu, Perspective Problems in System Theory and its Application to Machine Vision, Accepted for publication in Journal of Mathematical Systems Estimation and Control, 1993.

[3] B. K. P. Horn, Robot Vision, The MIT Press, 1986.

[4] Y. Liu and T.S. Huang, Estimation of rigid body motion using straight line correspondence, Proc. Workshop on Motion: Representation and Analysis, IEEE Computer Society Press, 1986.

[5] Y. Liu and T.S. Huang. A linear algorithm for motion estimation using straight line correspondences, Computer Vision, Graphics, and Image Process, vol. 44, 1988.

[6] O.D. Faugeras, F. Lustman, and G. Toscani. Motion and structure from point and line matches, INRIA, BP105, Domain de Volucen, rocquencourt, 78153 Le Chesnay, France, 1987.

[7] O.D. Faugeras, On the motion of $3D$ curves and its relation to optical flow. Proc. First European Conference on Computer Vision (ed. O. Faugeras). Lecture Notes in Computer Science, vol. 427, pp. 107-117. Berlin, Heidelberg and New York: Springer Verlag, 1990.

[8] A.M. Waxman and S. Ullman, Surface structure and 3-D motion from image flow: kinematic analysis, Int'l. J. Robotics Research 4 (1985) 72-94.

[9] K.I. Kanatani. Detecting motion of a planar surface by line and surface integrals, Computer Vision, Graphics and Image Processing 29 (1985) 13-22.

[10] K. Kanatani, Group-Theoretical Methods in Image Understanding, Springer Verlag, 1990.

[11] A. Mitiche, S. Seida and J.K. Aggarwal, Line Based Computation of Structure and Motion Using Angular Invariance, Proc. Workshop on Motion: Representation and Analysis, IEEE Computer Society Press, 1986.

[12] M. E. Spetsakis and J. Aloimonos, Structure from motion using line correspondences, International J. Computer Vision, 4, pp. 171-183, 1990.

[13] H. H. Nagel, Representation of moving rigid objects based on visual observations, Comput., vol. 14, no. 8, pp. 29-39, 1981.

[14] R.Y. Tsai and T.S. Huang. Estimating three dimensional motion parameters of a rigid planar patch, IEEE Trans. on ASSP 29 (1981) 1147-1152.

[15] R.Y. Tsai and T.S. Huang and W.L. Zhu. Estimating three dimensional motion parameters of a rigid planar patch, II: Singular value decomposition, IEEE Trans. on ASSP 30 (1982) 525-534; errata, 31 (1983) 514.

[16] R.Y. Tsai and T. S. Huang. Uniqueness and estimation of three dimensional motion parameters of rigid objects with curved surfaces, IEEE Transactions on pattern analysis and machine intelligence, vol. 6, no. 1, pp. 13-26, January 1984.

[17] R.Y. Tsai, T.S. Huang. Estimating 3-dimensional motion parameters of a rigid planar patch, III: Finite point correspondences and three views problem, IEEE Trans. on ASSP 32 (1984) 213-220.

[18] G. Adiv. Determining three dimensional motion and structure from optical flow generated by several objects, IEEE Trans. on Pattern Anal. Mach. Intell. 7 (1985) 384-401.

[19] J. Aloimonos and C.M. Brown, Preception of structure from motion, I: Optical flow v.s. discrete displacements, in Proc. IEEE Conf. Computer Vision and Pattern Recognition, June, 1986.

OVER–REGULARIZATION OF ILL–POSED PROBLEMS

B. A. Mair *

*Department of Mathematics
University of Florida
Gainesville, FL 32611*

1 Introduction

The method of Tikhonov regularization is one of the most widely applied methods for solving ill-posed inverse problems which arise in a wide variety of problems in science and engineering. As is well-known, the main difficulty in applying this method occurs in the choice of a regularizing parameter, usually denoted by α, depending on the error in the data. Both apriori and aposteriori methods of choosing α have been developed, mainly for the case of finitely smoothing forward operators (cf. [7, 15, 16]). However, many problems occurring in partial differential equations (cf. [2–4, 6, 8–12, 17–19]), statistics (cf. [5]), and the inversion of the Laplace transform (cf. [1]), are characterized by linear ill-posed problems determined by infinitely smoothing forward operators.

The main aim of this paper is to outline a generalization of the "over–regularization" result of Natterer in [16] to the case of infinitely smoothing forward operators, and to demonstrate the numerical advantages of this technique. More general results, and the analytic details of the proofs will appear in [14].

The Tikhonov regularization method for solving the operator equation

$$Tf = g$$

in the Hilbert space $L^2(\mathbf{R})$, consists of minimizing the functional

$$f \to \|Tf - g_\delta\|^2 + \alpha^2 \|f\|_q^2$$

over the Sobolev space H_q in which the true solution lies.

The idea introduced by Natterer [16], is to increase the index on the Sobolev norm in the above functional without changing the assumption on the degree of smoothness of the true solution. This corresponds to regularizing over a much smaller space than that prescribed by the assumption on the true solution. It therefore seems natural to refer to this process by the term "over–regularization". In [16] it is shown that for finitely smoothing operators, the same apriori error estimates are obtained for over–regularization

---

* Supported in part by NSF Grant # DMS 9006308

as for the classical regularization, for an appropriate choice of the regularizing parameter. This idea has been utilized in obtaining an aposteriori method of choosing the regularizing parameter in [17].

Sections 2 and 3 outline a method for extending Natterer's result to a very general class of operator equations which include both finitely and infinitely smoothing operators. As in [16], these results are stated in the slightly more general context of approximation in a Hilbert scale instead of Sobolev spaces. It is interesting to point out that the method employed here is quite different from that in [16]. Natterer's proof relies heavily on interpolation inequalities which do not carry over readily to the infinitely smoothing case. The method here consists of two basic parts. The first is to obtain a general result for the classical Tikhonov regularization by using techniques found in [13, 15, 18], and then to reduce the over–regularization problem to the classical one by a careful analysis of the denseness properties of the spaces involved.

Although over–regularization has no effect on the theoretical error bounds, it seems natural to expect it to have numerical advantages due to the arbitrary smoothness of the functions over which the minimization is taking place. This numerical advantage is demonstrated in Section 4, by applying the result to the inverse heat conduction problem of recovering the temperature at one end of a rod from a discrete set of noisy temperature measurements at an interior point on the rod.

Many thanks to Scott McCullough and Murali Rao for many stimulating discussions.

2 Approximation in Hilbert scales

Let X, Y be Hilbert spaces and $T : X \to Y$ be a bounded, injective, linear operator with non-closed range $\mathcal{R}(T)$.

Consider the problem of approximating the "solution" to the "equation"

$$Tx = y_0 \tag{2.1}$$

based on inaccurate data y_δ, where y_0 is not assumed to be in $\mathcal{R}(T)$.

Assume y_0 is in the domain of the Moore–Penrose inverse $T^\dagger$, which is $\mathcal{R}(T) + \mathcal{R}(T)^\perp$.

For any such y_0, let

$$x_0 = T^\dagger(y_0). \tag{2.2}$$

This is the so–called best approximate solution to (2.1) and is the unique true solution if $y_0 \in \mathcal{R}(T)$.

The aim is to obtain an approximation to x_0 based on approximate data y_δ satisfying

$$\|Q(y_\delta - y_0)\| \le \delta. \tag{2.3}$$

where Q denotes the orthogonal projection of Y onto $\overline{\mathcal{R}(T)}$.

As in [16, 17], the stabilizing functional will be determined by a Hilbert scale $\{X_s : s \in \mathbf{R}\}$ generated by a densely defined, unbounded, self-adjoint operator L, on X, with $\|Lx\| \geq \|x\|$, for all $x \in \mathcal{D}(L)$.

As usual, $X_0 = X$, $X_s = \mathcal{D}(L^s)$, and $\|\cdot\|_p$ denotes the norm in X_p.

Then, it is well-known (cf. [2, 7, 13, 15]), that a suitable approximation $x_{p,\alpha,\delta}$ of x_0 is obtained by minimizing the quadratic functional

$$x \mapsto \|Tx - y_\delta\|^2 + \alpha^2\|x\|_p^2$$

over X_p.

Furthermore,

$$x_{p,\alpha,\delta} = (T^*T + \alpha^2 L^{2p})^{-1}T^*y_\delta. \tag{2.4}$$

To obtain an estimate on the rate of convergence of $x_{p,\alpha,\delta}$ to x_0, introduce the modulus of continuity (cf. [2, 12, 15, 16, 19]),

$$\nu(\varepsilon) = \sup\{\|x\| : x \in V, \ \|Tx\| \leq \varepsilon, \ \|x\|_p \leq 1\}.$$

The following result is well-known.

Theorem 2.1 *Suppose* $\|x_0\|_p \leq E$. *If* $\alpha(\delta) = \delta/E$ *then*

$$\|x_{p,\alpha(\delta),\delta} - x_0\| \leq \sqrt{2}\, E\, \nu(\delta/E).$$

To estimate this function ν, it is necessary to prescribe the degree of ill-posedness of the operator T. It is often assumed (cf. [2, 16, 17]) that there exist constants $a, m, M > 0$, such that

$$m\|x\|_{-a} \leq \|Tx\| \leq M\|x\|_{-a}, \ for \ all \ x \in X. \tag{2.5}$$

Then, by using the classical interpolation inequality for Hilbert scales:

$$\nu(\varepsilon) \leq m^{-p/(a+p)}\varepsilon^{p/(a+p)}. \tag{2.6}$$

This provides the classical estimate (cf. [2]):

$$\|x_{p,\alpha(\delta),\delta} - x_0\| \leq \sqrt{2}E^{a/(a+p)}m^{-p/(a+p)}\varepsilon^{p/(a+p)} \tag{2.7}$$

under the assumptions in Theorem 2.1.

However, many important ill-posed inverse problems involve operators T which do not satisfy the finitely smoothing property expressed by (2.5). To deal with these infinitely smoothing operators, introduce the following generalization of (2.5).

Assumption 2.1 *There exists a decreasing, continuous* $w : (0, \infty] \mapsto [0, \infty)$ *such that:*

(i) $w(s) = 0 \Leftrightarrow s = \infty$

(ii) the function $s \mapsto s\, w(s^{-t})$ *is convex on [0, 1] for each* $t > 0$

(iii) there exist constants $m,\ M > 0,$ *and* $0 < \gamma \leq 1$ *such that*

$$m^2 \int w\, d\mathcal{L}_{x,x} \leq \|Tx\|^2 \leq M^2 \int w^\gamma\, d\mathcal{L}_{x,x}\ \text{for all } x \in X_{2p},$$

where $\mathcal{L}$ *is the spectral measure of* L.

The rate of convergence of the regularized solution will be prescribed by the function

$$\varphi_r(\lambda) = \lambda\, w(\lambda^{-1/2r}),\ \lambda \geq 0,\ r > 0. \tag{2.8}$$

By expressing the first inequality in Assumption 2.1(iii) in terms of φ_p, and using its convexity property together with Jensen's inequality (cf. [18]), the following generalization of (2.6) is obtained in [14].

$$\nu(\varepsilon) \leq \sqrt{\varphi_p^{-1}(\varepsilon^2/m^2)}$$

Then, by using Theorem 2.1,

Theorem 2.2 *Assume* $\|x_0\|_p \leq E$. *If* $\alpha(\delta) = \delta/E$, *then*

$$\|x_{p,\alpha(\delta),\delta} - x_0\| = O\left(\sqrt{\varphi_p^{-1}(\delta^2)}\right).$$

The second inequality in Assumption 2.1(iii) will play a major role in the next section, which deals with over–regularization.

Assumption 2.1 reduces to the classical (2.5) when $w(\lambda) = \lambda^{-2a}$, $\gamma = 1$. In this case, $\varphi_p(s) = s^{1+a/p}$ so that $\varphi_p^{-1}(s) = s^{p/(a+p)}$. Thus Theorem 2.2 generalizes (2.7).

For the examples of infinitely smoothing operators associated with inverse heat conduction problems (cf. [3, 4, 7–11, 18, 19]), and inversion of the Laplace transform (cf. [1]), the appropriate choice for w is:

$$w_{a,b}(\lambda) = \exp(-b\lambda^a),\ a > 0,\ b \geq 1. \tag{2.9}$$

The corresponding φ_r given by (2.8) is

$$\varphi_{a,b,r}(s) = s \exp\left(-b\lambda^{-a/2r}\right).$$

Then, by using methods similar to those in [18],

$$\varphi_{a,b,r}^{-1}(s) = \left(\frac{b}{\ln 1/s}\right)^{2r/a} (1+o(1)), \quad as \ s \to 0+ . \qquad (2.10)$$

Thus the regularized solution $x_{p,\alpha(\delta),\delta}$ converges to x_0, as $\delta \to 0+$, at the rate of

$$(\ln 1/\delta)^{-p/a} .$$

3 Over–Regularization

This section does not assume that the sought–after solution x_0, is in the same space as that over which the minimization of the quadratic functional takes place. This idea was first introduced by Natterer in [16] for the finitely smoothing operators satisfying (2.5), and further developed for aposteriori methods in [17]. The following generalization holds under Assumption 2.1.

Theorem 3.1 *Assume (2.3) and* $\|x_0\|_q \le E$. *For any* $p \ge q$ *and* $x_{p,\alpha,\delta}$ *as in (2.4),*

$$\|x_{p,\alpha(\delta),\delta} - x_0\| = O\left(\left[\frac{\varphi_q^{-1}(\delta^{2\gamma})}{\varphi_q^{-1}(\delta^2)}\right]^{p/2q} \sqrt{\varphi_q^{-1}(\delta^2)}\right)$$

where

$$\alpha(\delta) = \frac{1+EM}{E}\delta^{\gamma} \left[\varphi_q^{-1}(\delta^2)\right]^{(p-q)/2q} .$$

The basic idea is quite natural. It is to use the denseness of X_p to obtain an approximation, x_1, to x_0. Then, use the basic Tikhonov regularization to estimate x_1 in X_p. However, these approximations have to be carefully done to maintain suitable orders of convergence.

Corollary 3.1 *Suppose there exist constants* a, b, c, m, M, *such that*

$$m^2 \int \exp(-b\lambda^a) \, d\mathcal{L}_{x,x}(\lambda) \le \|Tx\|^2 \le M^2 \int \exp(-c\lambda^a) \, d\mathcal{L}_{x,x}(\lambda)$$

for all $x \in X_{2q}$, *and* $\|x_0\|_q \le E$. *Then for any* y_δ *satisfying (2.3), and* $p \ge q$,

$$\|x_{p,\alpha(\delta),\delta} - x_0\| = O\left((\ln 1/\delta)^{-q/a}\right) ,$$

where

$$\alpha(\delta) = O\left(\delta^{\gamma}(\ln 1/\delta)^{-(p-q)/a}\right), \quad \gamma = \min\{c/\max(b,1), 1\} .$$

This is easily obtained from Theorem 3.1 by using (2.10) for the operators characterized by (2.9).

4 Numerical Experiment

The numerical experiment in this section deals with the the problem of determining the temperature, $f(t)$, at the left end, $x = 0$, of the rod, $0 \leq x \leq 1$, from temperature readings, $u_k = u(x_0, t_k)$, $k = 1, 2, ..., 50$, at the point $x_0 = 0.1$, assuming that the initial temperature and the temperature at the right end $x = 1$, are both zero. More specifically, the temperature $u(x, t)$ satisfies:

$$\frac{\partial u}{\partial t} = \frac{\partial^2 u}{\partial x^2}, \ 0 < x < 1, \ t > 0$$
$$u(0, t) = f(t), \quad u(1, t) = 0, \ t \geq 0$$
$$u(x, 0) = 0, \ 0 \leq x \leq 1$$

and the data consists of approximate values

$$v_k \approx u(x_0, t_k), \ t_k = 0.02k, \ k = 1, 2, ..., 50.$$

This is equivalent to the deconvolution problem (cf. [4, 9, 11, 18, 19]):

$$f = K * u(x_0, \cdot)$$

where the Fourier transform

$$|\hat{K}(s)| \approx \exp(-x_0 \sqrt{|s|/2}).$$

Then Corollary 3.1 is directly applicable. Thus each value of p, called the degree of regularization, in Corollary 3.1, will produce an approximate solution. In this experiment, $f(t) = \sin(2\pi t)$, for $0 \leq t \leq 1$, and is zero otherwise. Thus $f \in H_q$ if and only if $q < 3/2$. The data was generated with an accuracy of 0.001, and white noise with the same standard deviation was added. Regularized solutions, and their errors, were obtained for various degrees of regularization (i.e. values of the parameter p in Corollary 3.1), using the value $q = 1$.

The graph of these errors against the corresponding values of p is given on the next page.

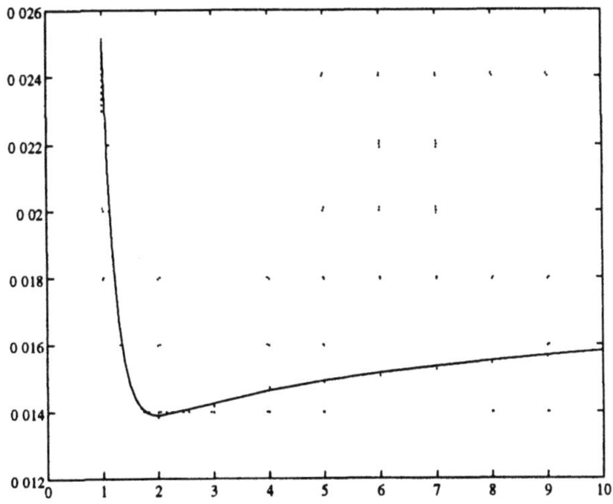

Figure 3.1: Error vs. Degree of Regularization

These, and other similar numerical tests, confirm the expectation that over–regularization improves the numerical error in solving even severely ill–posed inverse problems.

References

[1] D. ANG, J. LUND, and F. STENGER, "Complex variable and regularization methods of inversion of the Laplace transform," *Math. Comp.*,v. 53, 1989, pp. 589–608.

[2] J. BAUMEISTER, "Stable Solution of Inverse Problems," *Friedr. Vieweg & Sohn*, Braunsweig/Wiesbaden, 1987.

[3] J.V. BECK, B. BLACKWELL, and C.R. ST. CLAIR Jr., "Inverse Heat Conduction: Ill-posed Problems," Wiley, 1985.

[4] A. CARASSO, "Determining surface temperatures from interior observations," *SIAM J. Appl. Math.*, v. 42, 1982, pp. 558–574.

[5] R. J. CARROLL, A. C. M. VAN ROOIJ, and F. H. RUYMGAART, "Theoretical aspects of ill-posed problems in Statistics," *Acta Appl. Math.* v. 24, 1991, pp. 113–140.

[6] A. EL JAI, and A.J. PRITCHARD, "Sensors and Controls in the Analysis of Distributed Systems," Wiley, 1988.

[7] H. W. ENGL and H. GFRERER, "A posteriori parameter choice for general regularization methods for solving linear ill–posed problems," *Applied Numer. Math.*, v. 4, 1988, pp. 395–417.

[8] D. S. GILLIAM, Z. LI, and C. F. MARTIN, "Discrete observability of the heat equation on bounded domains," *Int. J. Control*, v. 48, 1988, pp. 755–780.

[9] D. S. GILLIAM, J. R. LUND, B. A. MAIR, and C. F. MARTIN, "A regularization method for inverse heat conduction problems," in *Computation and Control II, K. Bowers& J. Lund (eds.)* , Birkhäuser, 1991, pp. 135–139.

[10] D. S. GILLIAM, B. A. MAIR, and C. F. MARTIN, "Determination of initial states of parabolic systems from discrete data," *Inverse Problems*, v. 6, 1990, pp. 737–767.

[11] D. S. GILLIAM, B. A. MAIR, and C. F. MARTIN, , "An inverse convolution method for regular parabolic equations," *SIAM J. Control and Opt.*, v. 29, 1991, pp. 71–88.

[12] M. M. LAVRENTE'V, V. G. ROMANOV and S. P. SHISHAT·SKIĬ, " Ill-Posed Problems of Mathematical Physics and Analysis," *Translations of Mathematical Monographs*, v. 64, AMS, 1986.

[13] J. LOCKER and P. PRENTER, "Regularization with differential operators, I: General theory," *Math. Anal. and Appl.*, v. 74, 1980, pp. 504–529.

[14] B. A. MAIR, "Tikhonov regularization for finitely and infinitely smoothing operators," *preprint*.

[15] K. MILLER, "Least squares methods for ill-posed problems with a prescribed bound," *SIAM J. Math Anal.*, v. 1, 1970, pp. 52–74.

[16] F. NATTERER, "Error bounds for Tikhonov regularization in Hilbert scales," *Applicable Analysis*, v.18, 1984, pp. 29–37.

[17] A. NEUBAUER, "An a posterior parameter choice for Tikhonov regularization in Hilbert scales leading to optimal convergence rates," *SIAM J. Numer. Anal.*, v. 25, 1988, pp. 1313–1326.

[18] G. TALENTI and S. VESSELLA, "A note on an ill-posed problem for the heat equation," *J. Austral. Math. Soc. (Series A)*, v. 32, 1982, pp. 358–368.

[19] A. N. TIKHONOV and V. Y. ARSENIN, "Solutions of Ill-Posed Problems," Wiley, 1977.

A MODEL FOR THE OPTIMAL CONTROL OF A MEASLES EPIDEMIC

C. Martin, L. Allen, M. Stamp*, M. Jones, R. Carpio[†]
Department of Mathematics
Texas Tech University
Lubbock, TX 79409

1 Introduction

The number of cases of measles in the United States has risen dramatically in the last few years. In 1983, the number of measles cases was at an all time low, 1497 reported cases, but in 1988, there were 3411 cases and in 1989 there were approximately 17,850 cases (Centers for Disease Control 1990). The state of Texas reported the second highest number of cases in 1989 (3201 cases) and one of the worst outbreaks in the United States occurred in the Houston area (1802 cases reported from late 1988 to September 1989) (Centers for Disease Control 1990; Canfield 1989). A significant number of cases of the measles have occurred on university and college campuses and many of the epidemics have centered around public schools. The adult population in the United States is largely immune to the measles because of infection prior to the beginning of vaccination programs. The purpose of this paper is to present some data analysis and to develop a model suitable for the analysis of an optimal strategy for the distribution of vaccine during an epidemic.

There are basically two strategies employed by public health officials to control an epidemic of measles on a university campus. The first is vaccination. If a person has not been exposed to the measles and is vaccinated the probability of becoming infected after exposure is very low (less than .06) and even if a susceptible person has been exposed if that person is vaccinated in the first two or three days after exposure there is significant probability that person will not become infected. Thus, the immediate distribution of vaccine is a critical need for the control of an epidemic. The second strategy is to minimize the probability of exposure. During recent epidemics that have occurred on university campuses there has been a tendency to minimize the opportunities for student to student contact by such means has canceling attendance at sporting events such as basketball that are held in enclosed buildings. However, most universities have built in contact through large classes, dormitories and social events. We have

*Current address: Department of Mathematics, Worcester Polytechnic Institute, Worcester, MA 01609

[†]Supported in part by a Summer Research Grant from the Graduate School, Texas Tech University

found that at Texas Tech University a critical point of contact is in the university dining halls. It seems that the students when they are becoming infectious but not acutely symptomatic do not attend classes but they do go to the dining room for meals. The dining rooms in the cafeterias are large seating several hundred students concentrate in a period of about two hours. This provides an almost laboratory setting for the transmission of airborne viruses. Thus, while it may be effective on some types of campuses to limit student contact, on campuses such as Texas Tech University the only feasible mode of attack is the distribution of vaccine. Of course, the optimal strategy would be to enforce vaccination prior to coming to the university but this has not been implemented at most universities in the United States.

Measles epidemics are very expensive both in the actual outlay of dollars and in the loss of productive time by the infected. There are a significant number of deaths and debilitating side effects associated with infection by the measles virus. The goal of this project at Texas Tech University is to develop vaccination strategies that will minimize the number of cases of measles in Texas and consequently to minimize both the financial and personal loss caused by the measles.

When an epidemic occurs, typically the first vaccination is given 25 days after the first infection occurs. This is because there is 10 to 14 day latency period after the first person is infected. Often the first case is not recognized immediately and the second wave of the epidemic is in progress before it is recognized that there is an epidemic. The problem we consider is how should the vaccinations be distributed on the campus in order to minimize the total number of cases. This problem was considered in a previous paper (Allen, et al, 1992). In that paper the problems was studied by using extensive computer simulations that were based on a simulation model that was reported in (Allen, et al, 1990). In this paper those results will be summarized. However, since then it was realized that it is possible to formulate the model as a stochastic process and that the process is markov under certain reasonable conditions. This leads to the possibility of formulating the problem as a bilinear control problem and hence to the possibility of determining an optimal vaccination strategy.

Some theoretical studies have confirmed the need for a higher rate of immunity from measles than that achieved by a single vaccination given during the first two years of life (Hethcote 1983, 1988). These investigations are based on the analysis of epidemiological models of *SEIR* type (S–Susceptible, E–Exposed, I–Infectious, R–Removed). Analyses of these models have provided a theoretical foundation for studying many questions related to the spread of an epidemic (Anderson and May 1979; Bailey 1975; Greenhalgh 1990; Hethcote 1976, 1983, 1988; Hethcote and Yorke 1984; Hoppensteadt 1975; May 1986; McLean and Anderson 1988; Waltman 1974). There has also been epidemiological studies that have confirmed that if the first vaccination is given too early that there is reduced immunity, (Allen, et al, 1993).

In this paper, *SEIR* models (stochastic and deterministic) are applied to a specific epidemic in a university setting. Data from a 1989 measles epidemic that occurred on the campus of Texas Tech University in Lubbock is thoroughly investigated. Calculations from the data and simulations of the models provide an estimate of the level of immunity that is necessary in this university population to prevent a measles epidemic and provide a basis for the formulation of a model suitable for optimization.

2 Description of Data

An outbreak of measles (rubeola) occurred on the Texas Tech University campus and in the city of Lubbock in January, February, and March of 1989. The first case of measles was reported on January 13, 1989. Evidently, the student with the first reported case of measles infected a small group of students within his dormitory (dormitory 5) during spring semester enrollment. The epidemic continued until about the middle of March. After this point no more cases were reported at the university. Vaccination began on the university campus on February 1 (20 days after the first reported case). A total of 303 confirmed cases occurred in Lubbock; 198 of these were students at Texas Tech University which included 159 cases among students residing on campus (campus population size = 6104) and 39 cases among students residing off campus (off-campus population size = 17,396).

The data collected included onset date (first appearance of rash), age, race, sex, college of enrollment and living quarters, and were entered into one of the computer systems available at Texas Tech University. The dormitories at Texas Tech University were divided into ten complexes on the basis of common dining hall; students generally eat in the dining hall associated with their complex.

The data in Table 2.1 were used to estimate the level of immunity at the end of the epidemic. Two simplifying assumptions were made to estimate this level of immunity: (1) the susceptible proportion of the population equaled .06 before the epidemic began and (2) the individuals receiving vaccinations represented a random sample of the entire population. The infection rate was highest in dormitory 5 (where the epidemic began); it reached 5.8%. Therefore, an estimate of .06 for the susceptible proportion seems reasonable. (Estimates of .06 and .07 were used in the deterministic simulation; however, a susceptible proportion equal to.06 provided a closer fit to the data.) The high proportion of vaccine coverage (Table 2.1) is a good indication that individuals received vaccinations regardless of their vaccination history (ignorance of vaccination history or the possibility of a previous vaccine failure may have prompted many students to obtain the vaccine). Thus, based on our assumptions only 6% of the individuals receiving vaccinations were susceptible. We used the proportion vaccinated just during the first week of February as the proportion vaccinated during the epidemic, since they represented more than 90% of the total proportion vaccinated (Table 2.1) and the individuals vaccinated late in the epidemic probably did not affect the course of the epidemic. Thus, the level of immunity reached by the end of the epidemic is estimated as follows. The proportion infected (f_I) plus .06 of those individuals receiving vaccinations during the first week of February (f_V) equals the proportion of the susceptible class that became immune during the course of the epidemic. Adding this proportion to .94 (the proportion immune at the beginning of the epidemic) gives an estimate of the proportion immune or removed (f_R) at the end of the epidemic; i.e.,

$$f_I + .06 f_V + .94 = f_R$$

(Table 2.1). The average level of immunity is $\bar{f}_R=.981$. A 95% confidence interval was calculated for the average level of immunity, $\bar{f}_R$, at the end of the epidemic:

$$(.977, .985).$$

This estimate indicates (under the above assumptions) that a high level of immunity ($\approx 98\%$) must be achieved in this population for a measles epidemic to conclude. It also gives an indication of the level of immunity necessary to prevent an epidemic ($\approx 98\%$) in this population; the epidemic concluded due to the reduced proportion of the susceptible in the population.

The data was analyzed to determine the significance of the various statistics. It was found that the only statistic that was relevant was the dormitory complex in which the student lived. Again we feel that this reflects the fact that the students will go to the cafeteria to eat even if they are sick while they do not go to class or to social events with the same frequency. This is probably a clear indication that the strategy of isolation is not feasible for universities with large on campus populations. Therefore, in the following model, the population was subdivided according to place of residence; no other demographic variables were included. A brief summary of the model and the simulation results are presented in the next sections; more complete descriptions can be found in the references (Allen, Jones, and Martin 1991; Allen, Lewis, Martin, and Stamp 1990).

Dorm	Popul. Size	% Infected f_I	% Vacc. f_V (2/1-2/7)	Total % Vacc.	% Removed f_R
1	628	.022	.443	.467	.989
2	420	.019	.488	.507	.988
3	354	.025	.381	.395	.988
4	477	.010	.413	.470	.975
5	1355	.058	.409	.422	——
6	670	.010	.481	.500	.984
7	633	.027	.395	.425	.991
8	867	.008	.249	.273	.963
9	144	.007	.514	.542	.978
10	55	.018	.354	.379	.979

Table 2.1: Estimate Of The Proportion Immune At The End Of The Epidemic (Dormitory 5 is not included since most of the cases in dormitory 5 occurred prior to the commencement of the vaccination program.)

3 Model Formulation

The model is based on a discrete-time *SEIR* model formulated by Rvachev and Longini (1985) for the global spread of influenza. The model consists of an *SEIR* submodel for each dormitory complex and an additional one for those students who do not reside in a dormitory (off-campus students). Each of the states S, E, I, and R is defined below:

$$
\begin{aligned}
S_i(t) \;&=\; \text{number of susceptible individuals on day } t, \\
E_i(\tau,t) \;&=\; \text{number of individuals on day } t \text{ who were exposed on} \\
&\quad\; \text{day } t-\tau,\ \tau = 0,\ldots,\tau_1, \\
I_i(\tau,t) \;&=\; \text{number of individuals that are infectious on day } t \text{ who} \\
&\quad\; \text{were exposed on day } t-\tau,\ \tau = 0,\ldots,\tau_2, \\
R_i(t) \;&=\; \text{number of individuals that are removed from the in-} \\
&\quad\; \text{fection process by immunity or isolation on day } t,
\end{aligned}
$$

where $i = 1, 2, \ldots, 11$, representing the 10 dorms plus the off-campus students, τ_1 is the maximum length of the latent period (assumed to be 11 days) and τ_2 is the maximum length of the latent plus infectious period (assumed to be 14 days). The population size in each dormitory is assumed to be time invariant (no births or deaths), i.e.,

$$
S_i(t) + \sum_{\tau=0}^{\tau_1} E_i(\tau,t) + \sum_{\tau=0}^{\tau_2} I_i(\tau,t) + R_i(t) = P_i, \tag{3.1}
$$

where P_i is the population size within each dormitory, $i = 1, \ldots, 10$ (Table 2.1) and $P_{11} = 17,396$ is the off-campus population size.

Rvachev and Longini (1985) used an $SEIR$ submodel for each major city and linked the cities via a transportation operator. We do not use a transportation operator; in our model there are three sources of daily contact, within a dormitory, between dormitories, and within the entire student population and it is through these contacts that the epidemic is spread. The basic time unit of the model is one day (reported number of cases are known on a daily basis).

A basic assumption in most $SEIR$ models is that the population mixes homogeneously. In a homogeneous-mixing model, it is assumed that the number of newly exposed individuals, $E(t)$, on day t is proportional to the product of the number of susceptible individuals, $S(t)$, and the number of infectious individuals, $I(t)$, on day t, i.e., $\alpha S(t)I(t)$. The constant of proportionality α is the contact rate, i.e., the average number of individuals with whom an infectious individual will make sufficient contact (to pass infection) in one day (Rvachev and Longini 1985).

Initially it was assumed that the population mixed homogeneously; however, agreement between observed and simulated results was poor. Therefore, it was assumed that α decreased with time. This is a reasonable assumption for many epidemics, since as the epidemic becomes more severe, individuals become more careful of their social contacts and avoid those individuals known to be infectious and those places where contagious individuals reside. In addition, this measles epidemic began at the start of spring semester, at a time when there is probably more socializing and a greater number of contacts than during midsemester when the epidemic concluded. We chose $\alpha = \bar{\alpha}S(t)$, where $\bar{\alpha}$ is a constant, α decreases with time since the susceptible group S decreases with time (Allen, Jones, and Martin 1991; Allen, Lewis, Martin, and Stamp 1990). The choice of α is specific to this population. At the beginning of the semester there was most likely a very high rate of contact (large $S(t)$) and as the semester progressed the rate of contact probably decreased to a constant level ($\bar{\alpha}S(t) \approx \alpha$). There are other reasonable forms that may be proposed for a decreasing contact rate (e.g., a contact rate dependent on $N - I(t)$) and further research is needed to test these various forms.

There are three parameters, $\bar{\alpha}_1$, $\bar{\alpha}_2$, and $\bar{\alpha}_3$ associated with the three sources of contact: within a dormitory ($\bar{\alpha}_1$), within the campus population, dorms 1-10, ($\bar{\alpha}_2$), within the entire student population, dorms 1-11, ($\bar{\alpha}_3$). Contact rates for measles (under the assumption of a homogeneous-mixing population) have been estimated for the United States, North America and Great Britain; they range from 12.5 to 15 (Hethcote 1983). However, depending on the structure of the population, the rates can be significantly different. Therefore, values for the parameters $\bar{\alpha}_i$, $i = 1, 2, 3$ were selected within a reasonable range of values to provide the best fit to data (minimize the absolute differences). After estimation of $\bar{\alpha}_i$ an average contact rate (corresponding to a homogeneous-mixing model) was calculated for the deterministic model, i.e., the average of $\sum_{i=1}^{3} \alpha_i$ over the course of the epidemic (60 days), and it was found to be ≈ 10, a value which agrees with previous estimates.

Measles is contagious from one to two days before the onset of symptoms (e.g., fever) and from three to five days before the appearance of a rash (Committee on Infectious Diseases 1988). It was assumed that an exposed individual remains in the exposed state from 9 to 11 days and that an infectious individual remains infectious for approximately three days. If the measles vaccine is given within 72 hours after exposure, it can still provide protection against infection (Committee on Infectious Diseases 1988). Therefore, it was assumed that if a susceptible individual is successfully vaccinated (95% efficacy (Immunization Practices Advisory Committee 1989)), or if an individual in the exposed class is successfully vaccinated within three (or four) days after exposure, that individual becomes immune, but if the vaccination is given four (or five) or more days after exposure, the vaccination has no effect. In the deterministic model four days were used and in the stochastic model five days were used.

The state equations for the deterministic model are defined below (similar equations hold for the stochastic model). The removed state, $R_i(t)$, can be determined from equation (3.1) and is omitted from the state equations.

$$S_i(t+1) = [1 - v_i(t)][S_i(t) - E_i(0, t+1)],$$

$$E_i(\tau+1, t+1) = \begin{cases} [1 - v_i(t)]E_i(\tau, t), & \tau = 0, \ldots, \beta_0, \\ k(\tau)E_i(\tau, t), & \tau = \beta_0 + 1, \ldots, \tau_1 - 1, \end{cases}$$

$$I_i(\tau+1, t+1) = \begin{cases} [1 - k(\tau)]E_i(\tau, t) + r(\tau)I_i(\tau, t), & \tau = 0, \ldots, \tau_1, \\ r(\tau)I_i(\tau, t), & \tau = \tau_1 + 1, \ldots, \tau_2 - 1, \end{cases}$$

where β_0 (3 or 4) is the number of days during the latent period in which the vaccine is effective and $v_i(t)$ is the fraction of susceptible students in dorm i who become immune due to vaccinations on day t (calculated from the data). The functions $k(\tau)$ and $r(\tau)$ are transition probabilities; $k(\tau)$ is the probability that an individual is latent on day $\tau + 1$ given that the individual was latent on day τ, and $r(\tau)$ is the probability that an individual is infectious on day $\tau + 1$ given that the individual was infectious on day τ. Specification of initial and boundary conditions completes the model formulation. The boundary conditions for $E_i(0, t+1)$, $t = 0, \ldots,$ the number of newly exposed individuals on day $t+1$, are specified via the contact rates (discussed previously). The boundary conditions for I_i are $I_i(0, t) = 0$,

$t = 0, \ldots$, because it takes more than one day to become infectious. The initial conditions used in the simulations, $S_i(0)$, $E_i(\tau, 0)$, $I_i(\tau, 0)$, and $R_i(0)$, $\tau = 0, \ldots, \tau_2$ are specified in the next section. The model formulation and the functions v_i, k, and r are discussed in more detail in the references (Allen, Jones, and Martin 1991; Allen, Lewis, Martin, and Stamp 1990; Jones 1990; Longini 1986, 1988; Rvachev and Longini 1985).

4 Computer Simulations

For purposes of simulation it was assumed that 6% of the population was susceptible in each dormitory (5.8% infection rate was reported in dormitory 5). In dormitory 8 the fewest number of vaccinations were given (Table 2.1); therefore, we should expect a relatively high number of cases, but the reverse occurred. Therefore, the proportion susceptible was reduced to .03 in dormitory 8. (There is most likely some variation from the .06 level in all of the dormitories, but in dormitory 8 this variation was most evident.) Since the epidemic began in dormitory 5, the simulation was initiated (initial conditions) by putting four infected students in dormitory 5. Vaccination began on the twenty-fifth day of the simulation (twentieth day since the first reported cases of measles). The three parameters $\bar{\alpha}_1$, $\bar{\alpha}_2$, and $\bar{\alpha}_3$ were estimated by trial and error until the simulated results were in close agreement with the observed results. In the deterministic model the estimates were chosen to minimize the absolute difference between the simulated and the reported results at 80 days (Allen, Jones, and Martin 1991).

The simulations were used to estimate the level of immunity necessary to prevent an epidemic (herd immunity). Vaccinations were eliminated and the rate of immunity was varied from .95 to .99 (all other parameters were kept fixed). The results for each model, number of cases at each rate of immunity (95% to 99%), are given in Table 4.1 (only one simulation output at each rate of immunity in the stochastic model). An immunization rate above 98% does not generate any new cases in the stochastic model (four students were initially infected); in the deterministic model two cases accumulate over a period of 80 days. This slight discrepancy in the two models is due to the fact that the stochastic output generates integer values, whereas, the deterministic output generates real values. The epidemic continues in the deterministic model with only a small infected population (< 1), but this is impossible in the stochastic model since values < 1 are zero (signifying the end of the epidemic). The simulation results indicate a high level of immunity ($> 98\%$) is required to end the epidemic. This estimate agrees with the estimate obtained from the raw data.

Percent Immunity	Stochastic		Deterministic	
	Total	On Campus	Total	On Campus
95	252	223	276	234
96	160	135	168	151
97	55	53	73	69
98	19	17	23	22
99	4	4	6	6

Table 4.1 Number Of Cases After 80 Days For Each of The Models With Various Immunity Rates And No Vaccinations

5 Vaccination

The optimal strategy would be to vaccinate all of the students on the first hint of an epidemic. Unfortunately this is not possible because of the expense of the vaccination and because of the difficulty of getting the student to be vaccinated. There are also real limitations as to the number of personnel available to give the vaccine, the number of doses available on any given day and the total number of doses available in the state at any given time. In this project theses limitations are strictly considered.

For this paper the following aspects will be considered:

I. The distribution of the vaccines. In order to analyze this aspect we will keep constant the day when the vaccination begins. It will be the same as in the real epidemic, that is on the 25th day of the simulation. The number of vaccines to be given per dormitory complex will be changed. The following strategies will be tested, considering the aspect above.

a)No vaccination will be given in the dormitory complex where the epidemic started (host dorm) since it spread rapidly through this dormitory complex. Concentrate the vaccines in other dormitories, following the schedule that was actually used.

b)No vaccination will be given in the host dorm. Concentrate the vaccines in the dormitory complexes with fewer cases of infected people.

c)No vaccination will be given in the host dorm. Concentrate the vaccines in the dormitory complexes that have a larger percentage of infected people.

d)Some vaccines will be given in the host dorm and for the other dormitory complexes, strategies a,b, or c will be considered, depending on which one gives the fewest number of total cases.

II. The time when to vaccinate. In order to analyze this aspect, the same strategies a,b,c, and d will be considered as in I, but the vaccination will begin on different days. The following strategies will be tested, considering this second aspect:

a)Begin the vaccination on the 15th day of the simulation, that is ten days before the real case. This is after the first wave of cases of infections. Notice that this strategy will be considered four times (i.e., with each one of the strategies in I).

b)Begin the vaccination after the first cases of infection appear, that is, on the 4th day of the simulation. This strategy will also be considered with each one of the strategies in I.

c)Begin the vaccination after the first cases of infection appear, that is, on the 4th day of the simulation. Increase the number of vaccines for dormitory complex number 2 with respect to strategy d in I.

III. The number of vaccines available. The following strategies will be tested considering this aspect:

a)Consider the 1st day of the vaccination to be as in the real case, the 25th day of the simulation. For this day use the vaccines available for the actual first three days, shift the others up.

b)Consider the strategy that gives the least number of total cases in II. Use the vaccines available for the first three days and shift the others up.

The idea is to measure the effect of each vaccination strategy by considering the total number of cases obtained after the vaccination.

In order to test each vaccination strategy, fifty runs of the simulation program were made for each. An average of the output was taken and used to compare the vaccination strategies.

An interpretation of the results obtained for each one of the vaccination strategies tested by using the simulation program is summarized here.

For Strategy Ia through Id we want to test in which dormitory complexes to concentrate the vaccines. The day the vaccination program starts will be constant and the number of vaccines to be given per dormitory was changed according to the vaccination strategy. It is assumed that slightly more than 94% of the population was immune at the beginning of the epidemic. Thus 6% of the population was susceptible to the measles.

In Strategies Ia, Ib, and Ic, no vaccination was given to dormitory complex 2. This is because the 6% susceptible implies that there were approximately 81 susceptible students in dormitory complex 2. ¿From these 81 susceptible students approximately 67 students were already infected when the vaccination program started.

Strategy Ia. The vaccination program starts on the 25th day of the simulation, that is 25 days after the epidemic began. By this time the epidemic has already spread. In the dormitory complex where the epidemic started (dorm 2), it spread rapidly. Thus, no vaccines are given to dormitory 2 and instead they are used for other dormitories but following the schedule that was actually used.

Using all the vaccines available (5895), a total of 356 students became immune due to the vaccination, slightly more than 25% of the susceptible students. And there was a total of 189 cases, which is 13.4% of the susceptible students.

In order to reduce the cost of the epidemic, the number of cases must decrease. The worst case had 217 cases and the best had 156 cases.

Strategy Ib. As stated before, the starting day for the vaccination program is the same, the 25th day of the simulation. No vaccines are given to dormitory complex 2. The vaccines are concentrated in the dormitories with fewer numbers of cases.

Using all the vaccines available (5895 vaccines) with this vaccination strategy, 303 students became immune due to this vaccination, that is 21.49% of the susceptible students. But more important for the purposes of this study is the number of cases that needs to be reduced. Using this strategy, there were 187 cases, which corresponds to a 13.26% of the susceptible students.

Using this strategy we obtain slightly better results than the real case, even though the worst case of the fifty runs of the simulation (i.e., the one with more number of cases) falls above the real case, with 218 cases.

Strategy Ic. The vaccination program also starts on the 25th day of the simulation. No vaccines are given to dormitory complex 2. The vaccines are distributed proportionally to the percentage of cases in each dormitory complex, except for dormitory complex 2.

Using the 5895 vaccines with this vaccination strategy, a total of 302 students became immune due to the vaccination, that is 21.42% of the susceptible students, and a total of 173 cases, which is 12.27% of the susceptible students.

The best case with 143 cases and the worst with 207 cases.

Comparing Strategies Ia, Ib, and Ic, Strategy Ic is the one that gives the fewest number of cases.

Strategy Id. For this vaccination strategy the same criterion will be used for the distribution of the vaccines as in Strategy Ic, since it gives the fewest number of cases. The only difference will be that this strategy will consider the distribution of some vaccines to dormitory complex 2 (where the epidemic started).

It will be interesting to see if it is beneficial to give some vaccines in the dormitory where the epidemic started, especially considering that by the time the vaccination program started, approximately 83% of the susceptible students (i.e, 67 out of 81 students) were already infected.

The day when the vaccination program starts stays constant. Some vaccines are given to the host dorm and the distribution given in Strategy Ic will be considered for the other dormitory complexes (i.e., more vaccines to the dormitories with higher percentage of infected people).

278 vaccines were given in dormitory complex 2 and only 2 students out of the 14 susceptible became immune due to that vaccination. There was a total of 75 cases in dormitory complex 2, only 2 less than the number of cases considering no vaccination for that dormitory.

In order to give some vaccines in dormitory complex 2, the number of vaccines in other dormitories needed to be reduced. This fact made the number of cases increase in some dormitory complexes.

Considering the totals, there were 175 cases, that is 12.41% of the susceptible students. Of the fifty runs of the simulation, the best case was the one with 134 cases and the worst with 205 cases.

It seems that giving some vaccines to dormitory complex 2 after the epidemic is already spread, may reduce the number of cases in that dormitory (in a very small percentage) but increases the number of cases in other dormitories. In general, it increases the total number of cases.

If the number of vaccines to be given in dormitory complex 2 increases, then the number of vaccines available for the other dormitories decreases, increasing in this way the total number of cases.

We see that if the number of vaccines given in dormitory complex 2 increases (from 278 to 408, that is 47%) the number of cases in that dormitory stays at 75. But the number of cases in the dormitories where the vaccines were reduced increases; thus, the total number of cases increases. This is what actually happened on the Texas Tech University campus. More vaccines were given in dormitory complex 2 and so the total number of cases was 198.

Strategies IIa, IIb, IIc, IId. The same vaccine distribution used in Strategies Ia, Ib, Ic and Id will be used for Strategies IIa, IIb, IIc, and IId, respectively. The only difference will be the day when the vaccination program begins. Instead of beginning on the 25th day of the simulation program, the vaccination program will begin on the 15th day of the simulation. This is after the first wave of cases.

Comparing the results for these strategies with the results for Strategies Ia, Ib, Ic and Id it can be seen very clearly that the number of cases has been reduced. Notice that Strategy IIa considers no vaccination in dormitory complex 2 and for the other dormitories uses the same distribution used for the real epidemic. Strategy IIb considers no vaccination in dormitory complex 2 and concentrates the vaccines in the dormitory complexes with fewer cases. Strategy IIc considers no vaccination in dormitory complex 2 and concentrates the vaccines in the dormitory complexes with more percentage of infected people. Finally, Strategy IId considers some vaccines in dormitory complex 2 and for the other dormitories uses the same vaccine distribution as in Strategy IIc. For all these four strategies the day when the vaccination program starts is on the 15th day of the simulation.

Using Strategy IId, the total number of cases is reduced by a 48% (i.e., from 198 cases to 102 cases.) Notice that if the vaccination program starts earlier, then it is beneficial to give some vaccines in dormitory complex 2.

Using Strategy IId, 278 vaccines were given in dormitory complex 2 (the same number of vaccines were given when using Strategy Id) and the number of cases decreased by 12% with respect to the number of cases obtained between Strategy Id and IId is the day when the vaccination program begins.

Strategies IIIa, IIIb, IIIc, IIId. The same vaccine distribution used in Strategies IIa, IIb, IIc, and IId will be used for Strategies IIIa, IIIb, IIIc, and IIId, respectively. The only difference is the day when the vaccination program starts. This time, the vaccination program begins on the fourth day of the simulation, that is, after the first cases of infection appear.

If the day when the vaccination program begins is early enough, like for these strategies on the fourth day of the simulation of the epidemic (21 days before than in the real case) then the effect of the vaccines given in dormitory complex 2 is better. Also, it is better for the other dormitories.

Using Strategy IIId, (the one that considers giving some vaccines in dormitory complex 2 and for the other dormitory complexes it concentrates the vaccines in the dormitories with a higher percentage of infected people), the total number of cases decreases from 198 cases to 71 cases, that is 64%.

Strategy IV. From the results obtained by using Strategy IIId, it seems that if the number of vaccines to be given in dormitory complex 2 increases, then the number of cases will decrease even more.

Strategy IV considers the beginning of the vaccination program on the fourth day of the simulation. The number of vaccines to be given in dormitory complex 2 will be increased from 278 (considered in Strategy IIId) to 408.

Using this strategy, the total number of cases is reduced to 60 cases, with respect to the real epidemic, that is 70%. The effect of the vaccine is better, the sooner the vaccination program begins.

Strategy V. This strategy considers the first day of the vaccination to be on the 25th day of the simulation. For this day it uses the vaccines available for the actual first three days, that is, in stated of 1399 vaccines, 3526 vaccines will be distributed among the students on the day the vaccination program begins. On the second day of the vaccination, the vaccines that were available on the fourth day of the vaccination for the real case will be used. In other words, the other vaccines will be shifted up.

The same distribution as in the real epidemic is used. With this strategy, the possibility of decreasing the total number of cases by increasing the number of vaccines to be used on the first day of the vaccination will be tested.

These results were analyzed using SAS for significance and it was concluded that the only statistically significant differences occurred when the vaccinations were begun early. In fact, every day that the vaccination is moved forward makes a very significant difference in the total number of cases in the epidemic. It must be emphasized that the results reported here are experimental in nature. A variety of strategies were simulated and within the class of strategies considered the only significant improvement was to vaccinate at an earlier date. In reality this not possible and the very real possibility is that the vaccination program used by the public health officials is as good as any that can be formulated. However, there is always the intriguing possibility that there is a strategy that will reduce the total number of cases in an epidemic on a university campus. In the next section we will formulate a model that has the potential to answer the question of what is the optimal strategy.

6 A Stochastic Process

We begin by noting that an epidemic is not a deterministic process. Individuals react differently to exposure to viruses and the constants associated with various infectious diseases are averages that have been observed and have either become part of the medical literature or part of the medical folklore. We recall from Section 2 that an individual remains *in the exposed state for 9 to 11 days*, remains *infectious for approximately 3 days*, is *infectious from 1 to 2 days before the onset of symptoms* and is *infectious from 3 to 5 days before the rash appears*. In a complete simulation each of these parameters should be randomized but this requires knowledge of the distribution of these random variables. Information for estimating the distributions is not readily available. In most deterministic models the above constants are fixed and indeed good estimates of the spread of an epidemic can be obtained from good deterministic models, (Allen, Jones and Martin, 1991).

Lets consider the standard $SEIR$ model from a stochastic viewpoint. The populations is initially divided between the susceptible S state and the removed R state. The number in the S state is a random variable with mean $.96P$ where P is the total population. At time zero an individual is

moved from the S state into the exposed E state and the individual remains
in that state for time period that is a random variable with mean 10 days.
The distribution of this random variable should have the form of $xe^{-\lambda x}$
with the maximum occurring at the 10th day and it should be very close
to zero before the 9th day and after the 11th day and it should be 0 before
the 0th day. At the end of this random time period the individual moves
from the E state to the infectious state and remains there for a time period
which is a random variable with mean 4. If vaccine is being distributed the
individual may transition from state S to state R or from state E to state
R. With this formulation it is virtually impossible to analyze the stochastic
process because the individuals in the states E and I transition depending
on when they entered the process. This shortcoming can be overcome at
the expense of dimensionality. The state E is partitioned into infinitely
many states E_n where this will mean *exposed for a period of n days* and the
state I will be likewise be partitioned into an infinite sequence of states I_n.
An individual in E_n will transition to one of three states with probability
1; E_{n+1}, I_1 or R. For example, the probability of transitioning from E_{12}
to I_1 is large compared to the probability of transitioning from E_{12} to R or
E_{13}. Let $P(Y, X)$ denote the probability of transitioning from state Y to
state X. The process is known when the probabilities $P(S, R)$, $P(S, E_1)$,
$P(E_n, R)$, $P(E_n, I_1)$ and $P(I_n, R)$ are known for each n. Formulated in this
manner the state of the system at time $N + 1$ depends only on the state
of the system at time N and the transition probabilities. Thus, there is no
delay in the system and the process is Markov.

The problem becomes a problem with control through the introduction
of vaccination. The probability $P(S, R)$ is zero if there is no vaccination
but is positive if vaccination is introduced. Certain of the probabilities
are independent of the vaccination program, for example $P(I_n, R)$ is not
influenced by vaccination. When a vaccination strategy is introduced the
probabilities become dependent on time and should be denoted by $P_k(X, Y)$
to denote the probability of transitioning from state X to state Y on the
kth day of the epidemic. The cost of the epidemic should be the

$$\sum_{k=0}^{\infty} (b^2 I_1^k + a^2 V_k)$$

where a and b are constants, I_1^k is the number of individuals in the state
I_1 on day k and V_k is the number of vaccinations given on the kth day.
The basic model is a linear infinite dimensional system in discrete time and
the controls enter through the matrix of transition probabilities. A major
difficulty is that there are hard constraints on the number of vaccinations
that can be given in any fixed time period. The optimization problem in
this setting can probably be done. However, the problem as formulated

in Section 3 is much more difficult in that here we have been considering a very homogeneous population and in the case of the epidemic at Texas Tech University the populations is known to be nonhomogeneous. The population is divided into ten groups depending on the dormitory of residence. Thus, the above model must be subdivided into ten groups and the attendant probabilities are much more complicated to estimate.

This model has the advantage that it is a better reflection of reality than either the deterministic model or the simulation models considered in the earlier section. However, it has the extreme disadvantage that it is infinite dimensional. Most of the states are empty with probability very close to one so it is possible to approximate the model with a finite dimension model. However, even the finite dimensional model will have very high dimension because of the number of dormitory complexes. In the future this model will be considered in an idealized case when there are only two subgroups in the population.

There is also the problem of determining the sensitivity of the model to assumptions on the distributions of the various random variables. This is a very difficult problem in general but there is a certain amount of literature devoted to just this issue. Again in a low dimensional approximation it should be possible to determine the sensitivity.

In this section we have given a partial formulation of a model that has the potential to answer the question of whether or not the optimal strategy is better than the strategy used by the public health officials in combating an epidemic. Based on the simulation results reported in this paper it is not probable that the optimal strategy is statistically better than the random vaccination used during the epidemic at Texas Tech University, but it would be very significant if the question could be answered one way or the other.

7 Summary

It appears that the present level of immunity obtained with one vaccination is insufficient to prevent measles epidemics. Hopefully, the recommendation by the ACIP (Immunization Practices Advisory Committee 1989) of two measles vaccines before entry into school will increase the percentage immune to a sufficiently high level to prevent measles outbreaks. An estimate of herd immunity in various population settings will be useful in determining whether the present strategy (if achieved) will eliminate measles epidemics in these settings.

A number of simplifying assumptions were made in this measles model. Although many of the assumptions appear reasonable, they require further validation and verification from other investigations. Our analysis raises some questions. Is the assumption of nonhomogeneous-mixing a valid one for a measles epidemic in a university setting? Is behavior modification (or other factors) an important consideration in modeling social contacts? How does the structure of the population affect the spread of the epidemic? These and other important questions which have been addressed in other model simulations (Ackerman, Elveback, and Fox 1984) require further investigation to justify the model assumptions and the conclusions derived from them.

References

[1]ACKERMAN, E., ELVEBACK, L. R., and FOX, J. P. (1984), *Simulation of Infectious Disease Epidemics*, Springfield, Illinois: Charles C. Thomas Publisher.

[2]ALLEN, L. J. S., JONES, M. A., and MARTIN, C. F. (1991), "A Discrete-Time Model with Vaccination For a Measles Epidemic," *Mathematical Biosciences*, 105, 111-131.

[3]ALLEN, L. J. S., KELLEY, M., MARTIN, C. F., WAY, A., POTI-TADKUL, K., and HIGGINS, H. (1993), *Center for Applied Systems Analysis, Texas Tech University* **Report # 203**.

[4]ALLEN, L. J. S., LEWIS, T., CARPIO, R., JONES, M. A., MARTIN, C. F., STAMP, M., MUNDEL, G., and WAY, A. (1992), "Stochastic Analysis of Vaccination Strategies," Stochastic Theory and Adaptive Control: Proceedings of a Workshop in Lawrence, KS. Sept. 26-28, 1991, *Lecture Notes in Control and Information Sciences* **184**, Eds. T.E. Duncan and B. Pasick-Duncan, Springer-Verlag, pp 1-12.

[5]ALLEN, L., LEWIS, T., MARTIN, C. F., and STAMP, M. (1990), "A Mathematical Analysis and Simulation of a Localized Measles Epidemic," *Applied Mathematics and Computation*, 39, 61-77.

[6]ANDERSON, R. M., and MAY, R. M. (1979), "Population Biology of Infectious Diseases: Part I," *Nature*, 280, 361-367.

[7]BAILEY, N. J. T. (1975), *The Mathematical Theory of Infectious Diseases*, (2nd ed.) New York: Macmillan.

[8]CANFIELD, M. (1989), "Measles Epidemic in Houston/Harris County, 1988-89: A Perspective From the Harris County Health Department," *Texas Preventable Disease News*, October 21, 1989, 49, 1-4.

[9]CARPIO, R. "Intervention strategies for local epidemics", Masters Thesis, Texas Tech University, 1991.

[10]Centers for Disease Control. (1990), "Measles–United States, 1989 and First 20 Weeks 1990," *Morbidity and Mortality Weekly Report*, 39, 353-363.

[11]Committee on Infectious Diseases. (1988), *Report of the Committee on Infectious Diseases*, 21st ed., American Academy of Pediatrics.

[12]GREENHALGH, D. (1990), "Vaccination Campaigns for Common Childhood Diseases," *Mathematical Biosciences*, 100, 201-240.

[13]GROSS, A. J., and CLARK, V. A. (1975), *Survival Distributions: Reliability Applications in the Biomedical Sciences*, New York: John Wiley & Sons.

[14]HETHCOTE, H. W. (1976), "Qualitative Analyses for Communicable Disease Models," *Mathematical Biosciences*, 28, 335-356.

[15]HETHCOTE, H. W. (1983), "Measles and Rubella in the United States," *American Journal of Epidemiology*, 117, 2-13.

[16]HETHCOTE, H. W. (1988), "Optimal Ages of Vaccination for Measles," *Mathematical Biosciences*, 89, 29-52.

[17]HETHCOTE, H. W., and Yorke, J. A. (1984), *Gonorrhea Transmission Dynamics and Control*, Lecture Notes in Biomathematics, Berlin: Springer-Verlag.

[18]HOPPENSTEADT, F. (1975), *Mathematical Theories of Populations: Demographics, Genetics and Epidemics*, Regional Conference Series in Applied Mathematics, Philadelphia, Penn.: SIAM.

[19]Immunization Practices Advisory Committee. (1989), "Measles Prevention: Recommendations of the Immunization Practices Advisory Committee," *Morbidity and Mortality Weekly Report*, 38(no.S-9).

[20]JONES, M. A. (1990), "A Deterministic Mathematical Model of a Measles Epidemic, " Master's Report, Department of Mathematics, Texas Tech University.

[21]LANDWEHR, J. M., PREGIBON, D., and SHOEMAKER, A. C. (1984), "Graphical methods for assessing logistic regression models," *Journal of American Statistical Association*, 79, 61-74.

[22]LO, C. K. (1989), "The Statistical Data Analysis of the 1989 Measles Outbreak on the Texas Tech Campus," Master's Report, Department of Mathematics, Texas Tech University.

[23]LONGINI, Jr., I. M. (1986), "The Generalized Discrete-Time Epidemic Model with Immunity: a Synthesis," *Mathematical Biosciences*, 82, 19-41.

[24]LONGINI, Jr., I. M. (1988), "A Mathematical Model for Predicting the Geographic Spread of New Infectious Agents," *Mathematical Biosciences*, 90, 367-383.

[25]MAY, R. M. (1986), "Population Biology of Microparasitic Infections",

In: *Biomathematics*, Vol. 17, T. G. Hallam and S. A. Levin (Eds.) pp. 405-442. Berlin: Springer-Verlag.

[26]MCLEAN, A. R., and ANDERSON, R. M. (1988), "Measles in Developing Countries. Part II. The Predicted Impact of Mass Vaccination," *Epidemiol. Infect.* , 100, 419-442.

[27]RVACHEV, L. A., and LONGINI, Jr, I. M. (1985) "A Mathematical Model for the Global Spread of Influenza," *Mathematical Biosciences*, 75, 3-22.

[28]WALTMAN, P. (1974), *Deterministic Threshold Models in the Theory of Epidemics*, Lecture Notes in Biomathematics, Berlin: Springer-Verlag.

CONDITION NUMBERS FOR THE SINC MATRICES ASSOCIATED WITH DISCRETIZING THE SECOND-ORDER DIFFERENTIAL OPERATOR

Kelly M. McArthur*

Department of Mathematics
Colorado State University
Fort Collins, CO 80523

1 Introduction

The Sinc-Galerkin discretizations of the model equations

$$\mathcal{L}u(x) \;\equiv\; -u''(x) = f(x) \;\;, 0 < x < 1$$

$$u(0) \;=\; u(1) \;= 0 \tag{1.1}$$

and

$$\mathcal{L}u(x,y) \;\equiv\; -\left(\frac{\partial^2}{\partial x^2} + \frac{\partial^2}{\partial y^2}\right) u(x,y) = f(x,y) \;\;, (x,y) \in (0,1)^2 \tag{1.2}$$

$$u|_{\partial(0,1)^2} \;=\; 0$$

are briefly outlined below in Sections 2 and 3, respectively. Detailed discussions of the discretization of more general, linear, second-order spatial operators are found in [9, 10, 5, 8]. The emphasis here is presenting results regarding condition numbers for the linear systems affiliated with (1.1) and (1.2). These condition numbers vary widely depending on how the linear systems are posed. Adroit changes of variables can lead to well-conditioned linear systems while poor choices can magnify existing problems. The importance of these choices is especially evident when developing a Sinc-Galerkin ADI scheme [7]. It should be pointed out that conditioning problems are not unique to this spectral scheme. Canuto [2] and Deville and Mund [3] have implemented finite element preconditioners for their spectral methods applied to (1.2). The changes of variables represented below may be thought of as preconditioners.

Besides examining condition numbers based on changes of variables, condition numbers based on three parameters are also detailed. The three parameters include one based on the behavior of the right hand side on the boundary, one linked to the choice of weight function for the Galerkin

---

*Supported in part by NSF grant DMS-8907895

inner product, and the last parameter which may be viewed as either the size of the discretized system or its stepsize. The data gives notice of the care needed to solve the linear systems when exact relationships between parameters are unknown.

While it is true that the condition numbers for the linear systems pared with, say, (1.1) do not carry over exactly to those for

$$-u''(x) + p(x)u' + q(x)u(x) = r(x),$$

if p and q are of reasonable magnitude then condition numbers for (1.1) should provide guide lines. This principle must be used with care. For instance, the advection-diffusion equation

$$\frac{\partial u}{\partial t} + c\frac{\partial u}{\partial x} = \nu\frac{\partial^2 u}{\partial x^2} \quad, c, \nu > 0$$

is theoretically parabolic but when $|\nu/c| << 1$ it exhibits a strong hyperbolic flavor. Researchers in finite difference methods have long known that the choice of discretization must be adjusted for such problems. It is only reasonable to expect that spectral methods must also undergo fine tuning.

2   The Sinc-Galerkin Discretization of (1.1)

The necessary framework to carry out the Sinc-Galerkin discretization of (1.1) includes defining basis functions, a trial function, a quadrature rule, and an orthogonalization procedure. The condition numbers of the derived linear system and transformed versions of it are a major topic of this paper. Hence to begin, on $(0, 1)$ let $S_j \equiv S(j, h) \circ \phi$ be a typical sinc basis function where ϕ is the conformal map

$$\phi(z) = \log\left(\frac{z}{1 - z}\right),$$

$$S(j, h)(x) \equiv \text{sinc}\left(\frac{x - jh}{h}\right) \quad h > 0, \ x \in \Re.$$

and

$$\text{sinc}(x) \equiv \frac{\sin(\pi x)}{\pi x}.$$

Given the basis, the trial function is defined as

$$u_m(x) = \sum_{j=-M}^{N} u_j S_j(x) \ , \ m = M + N + 1$$

where the coefficients are denoted u_j to call attention to the identity $u_m(\phi^{-1}(jh)) = u_j$. Since the points $\{x_j \equiv \phi^{-1}(jh), \ j = M : N\}$ play a role similar to the

Gauss points used in other spectral methods, they are called sinc points here. Besides appearing in the definition of the sinc points and the basis $\{S_j\}$, the map ϕ is incorporated into the weighted inner product

$$(u,t) \equiv \int_0^1 t(x)u(x)v(x)dx$$

via the weight $v(x) = (1/\phi'(x))^\gamma$, $0 < \gamma \le 1$. The standard choice for the paramter γ is $\gamma = 1$ [9] while the choice $\gamma = 1/2$ is developed in [5]. Shortly we consider how γ affects matrix structure and conditon number.

Referring to the definitions above, a system of unknowns in $\{u_j\}_{j=-M}^{N}$ is specified by orthogonalizing the residual $\mathcal{L}u_m - f$ with respect to $\{S_j\}_{j=-M}^{N}$. Two integrations by parts are performed so that the differentiation of u is entirely transferred to $S_j v$. Mild restrictions guarantee that the boundary terms vanish. Finally, the integrals are evaluated using the sinc quadrature rules presented in [10]. This yields the linear system

$$B^{(2)}\mathbf{u} = h^2 D(v/\phi')\mathbf{f} \tag{2.1}$$

where the coefficient matrix $B^{(2)}$ is given by

$$
\begin{aligned}
B^{(2)} \;=\; & \left\{ -I_m^{(2)} - (1 - 2\gamma)h I_m^{(1)} D(2x - 1) \right. \\[2mm]
& \left. + \gamma h^2 D\left(2x(1 - x) + (1 - \gamma)(2x - 1)^2\right)\right\} D(v\phi'),
\end{aligned}
\tag{2.2}
$$

the elements of $I_m^{(2)}$ are

$$\delta_{jk}^{(2)} = \begin{cases} -\frac{1}{3}\pi^2 & j = k \\ -2(-1)^{k-j}/(k - j)^2 & j \ne k, \end{cases}$$

those of $I_m^{(1)}$ are

$$\delta_{jk}^{(1)} = \begin{cases} 0 & j = k \\ (-1)^{k-j}/(k - j) & j \ne k, \end{cases}$$

and $D(g)$ in general refers to a diagonal matrix whose diagonal entries are $g_k = g(x_k)$. The elements of the vector $\mathbf{u}$ are u_j, those of $\mathbf{f}$ are $f(x_j)$, and in all cases the indices j, k run from $-M$ to N. The choice of stepsize h for this problem involves balancing the truncation error related to the trial function and the quadrature error related orthogonalizing the residual. Assuming that the true solution $u(x)$ satisfies

$$|u(x)| \le C\left(x(1 - x)\right)^\alpha, \quad 0 < x < 1$$

for some $\alpha > 0$ [10] suggests balancing error by using

$$h = \pi/\sqrt{\alpha M} \tag{2.3}$$

and $N = M$. Lund [5] proposes another choice of h but for our discussion (2.3) is selected. At some future time a similar analysis to that carried out here should be performed using the Lund stepsize.

Having at last presented all parameters needed to fully define $B^{(2)}$ the notation $B^{(2)}(\gamma, M, \alpha)$ is used henceforth. The form of $B^{(2)}(\gamma, M, \alpha)$ in (2.2) suggests a simple change of variables for the system (2.1) that reduces the operation count. Namely, if

$$\begin{aligned}
\hat{\mathbf{u}} &= D(v\phi')\mathbf{u} &= D\left((\phi')^{1-\gamma}\right)\mathbf{u} \\
\mathbf{g} &= D(v/\phi')\mathbf{f} &= D\left((\phi')^{-1-\gamma}\right)\mathbf{f}
\end{aligned}$$

and

$$\begin{aligned}
A^{(2)}(\gamma, M, \alpha) &= -I_m^{(2)} - (1 - 2\gamma)h I_m^{(1)} D(2x - 1) \\
&\quad + \gamma h^2 D(2x(1 - x) + (1 - \gamma)(2x - 1)^2) \\
&= B^{(2)}(\gamma, M, \alpha) D\left((\phi')^{\gamma-1}\right)
\end{aligned} \tag{2.4}$$

then (2.1) becomes

$$A^{(2)}(\gamma, M, \alpha)\hat{\mathbf{u}} = h^2\mathbf{g}. \tag{2.5}$$

Notice that when $\gamma = 1$,

$$A^{(2)}(1, M, \alpha) = B^{(2)}(1, M, \alpha)$$

and that when $\gamma = 1/2$

$$A^{(2)^T}(1/2, M, \alpha) = A^{(2)}(1/2, M, \alpha) = -I_m^{(2)} + \frac{h^2}{4}I_m$$

where I_m is the $m \times m$ identity matrix. Analysis of the method's convergence rate when $\gamma = 1/2$ is found in [5] and when $\gamma = 1$ is found in [9, 10].

While systems (2.1) and (2.5) are theoretically equivalent, their condition numbers exhibit radically different behavior and hence the computed solution ensuing from the two approaches may vary significantly. The 2-norm or spectral condition number of a matrix A is denoted $\kappa(A) = ||A|| \, ||A^{-1}||$, $|| \cdot || = || \cdot ||_2$. All calculations reported here were performed in MATLAB using the default double precision arithmetic. Figure 4.1 constitutes four plots where the dimension of the matrix $A^{(2)}(\gamma, M, \alpha)$ appears on the horizontal axis and its condition number $\kappa(A^{(2)}(\gamma, m, \alpha))$ appears on the vertical axis. The parameter α is held fixed to 2 in all plots while γ is held fixed in a single plot but then varies from plot to plot. Numerical trials indicate that the qualitative outcome is not affected by the choice of α. The graphs in each plot strongly indicate that $\kappa(A^{(2)}(\gamma, M, \alpha))$ increases linearly in m when γ and α are held fixed. Figure 4.2 documents four plots similar to those in Figure 4.1 but the 2-norm condition number of $B^{(2)}(\gamma, M, \alpha)$, $\kappa(B^{(2)}(\gamma, m, \alpha))$, is now on the vertical axis. The graphs corresponding to the smaller values of γ certainly do not exhibit linear growth. In fact, given that the plots in Figure 4.3 correspond to those in Figure 4.2 except that $\left(\log \kappa(B^{2)}(\gamma, M, \alpha))\right)^2$ now appears on the vertical axis, the evidence suggests $\kappa(B^{(2)}(\gamma, m, \alpha))$ increases exponentially in $\sqrt{M}$ for γ away from 1.

Analysis supports the above conclusion regarding $\kappa(B^{(2)}(\gamma, m, \alpha))$. Given the definiton of the sinc points x_k, $\phi'(x_k) = e^{kh} + 2 + e^{-kh}$. This implies that

$$\max_{k=M:M} \phi'(x_k) \;=\; e^{Mh} + 2 + e^{-Mh}$$

$$=\; e^{\pi\sqrt{M/\alpha}} + 2 + e^{-\pi\sqrt{M/\alpha}}$$

assuming that $h = \pi/\sqrt{\alpha M}$. Hence the 2-norm of $D\left((\phi')^{1-\gamma}\right)$ behaves like $e^{(1-\gamma)\pi\sqrt{M/\alpha}}$ as M ($m = 2M + 1$) increases while $\left\|D\left((\phi')^{\gamma-1}\right)\right\| = 1/4$. Therefore away from $\gamma = 1$, $\kappa(D\left((\phi')^{1-\gamma}\right))$ grows like $e^{(1-\gamma)\pi\sqrt{M/\alpha}}$ in M and since

$$\kappa(B^{(2)}(\gamma, M, \alpha)) \leq \kappa(A^{(2)}(\gamma, m, \alpha))\,\kappa(D\left((\phi')^{1-\gamma}\right))$$

the behavior in Figures 4.2 and 4.3 in not unexpected. Further results showing changes in $\kappa(B^{(2)}(\gamma, M, \alpha))$ with respect to γ and α are presented below. For a given problem the dependence of condition number on the fineness of the discretization is probably of most interest.

3   The Sinc-Galerkin Discretization of (1.2)

We next turn our attention to Poisson's equation (1.2) and its discretization. Although the conformal maps here are the same in both variables, for bookkeeping purposes the conformal map in the y variable is referred to as ψ while ϕ remains the map in x. The basis functions are now tensor products

$$S_{jk}(x, y) = S_j^x(x)S_k^y(y)$$

where $S_j^x(x) = S(j, h_x) \circ \phi$ and $S_k^y = S(k, h_y) \circ \psi$ and the affiliated trial function is

$$u_{mn} = \sum_{jk} u_{jk} S_{jk} \quad j = -M_x : N_x \ , \quad k = -M_y : N_y$$

with $m = M_x + N_x + 1$ and $n = M_y + N_y + 1$. The inner product is

$$(u, t) \equiv \int \int_{(0,1)^2} t(x, y) u(x, y) v(x) w(y) dA$$

where v is as before and $w = (1/\psi')^\delta$, $0 < \delta \le 1$. The residual $\mathcal{L}_{mn} - f$ is again orthogonalized with respect to the basis which yields the linear system

$$B_x^{(2)}(\gamma, M, \alpha) U D_y(w/\psi') + D_x(v/\phi') U B_y^{(2)^T}(\delta, M_y, \beta)$$

$$(3.1)$$

$$= h_x^2 h_y^2 D_x(v/\phi') F D_y(w/\psi').$$

$U = [u_{jk}]$, $j = -M_x : N_x$, $k = -M_y : N_y$ and the subscripts x and y are used to indicate the dimension of a particular matrix and the spatial variable to which it is coupled. That is, $B_x^{(2)}(\gamma, M, \alpha)$ and $D_x(v/\phi')$ are $m \times m$, $B_y^{(2)}(\delta, M_y, \beta)$ and $D_y(w/\psi')$ are $n \times n$. $D_x(v/\phi')$ is evaluated at x_j and $D_y(w/\psi')$ at y_k. Letting

$$\begin{aligned} \hat{U} &= D_x(v) U D_y(w) \\ G &= D_x(v) F D_y(w) \\ \mathcal{A}_x^{(2)}(\gamma, M, \alpha) &= D_x(\phi') A_x^{(2)}(\gamma, m, \alpha) D_x(\phi') \end{aligned}$$

and

$$\mathcal{A}_y^{(2)}(\delta, M_y, \beta) = D_y(\psi') A_y^{(2)}(\delta, M_y, \beta) D_y(\psi')$$

equation (3.1) becomes

$$\mathcal{A}_x^{(2)}(\gamma, M, \alpha)\hat{U} + \hat{U}\mathcal{A}_y^{(2)}(\delta, n, \beta) = h_x^2 h_y^2 G. \qquad (3.2)$$

The change of variables here has a different flavor than in the single variable case to accomodate a discretized version of separation of variables.

Various solution techniques have been applied to (3.2) for the case $\gamma = \delta = 1/2$ [1, 6]. The author believes these were successful because the principle step in those algorithms involved diagonalizing $\mathcal{A}_x^{(2)}(1/2, M, \alpha)$ and/or $\mathcal{A}_y^{(2)}(1/2, n, \beta)$. Since $\mathcal{A}_x^{(2)}(1/2, M, \alpha)$ and $\mathcal{A}_y^{(2)}(1/2, n, \beta)$ are symmetric they are diagonalizable via orthogonal transformations and this is a well-conditioned operation (see [4], Chapter 7). Recent work the author has done applying ADI methods to (3.2) indicates that performing LU decompositions or inverting $\mathcal{A}_x^{(2)}(\gamma, M, \alpha)$ (and $\mathcal{A}_y^{(2)}(\delta, n, \beta)$) is not well-conditioned.

Like the previous figures, Figures 4.4 and 4.5 each include four plots with matrix dimension on the horizontal axis. All plots in Figure 4.4 show $\kappa(\mathcal{A}^{(2)}(\gamma, m, \alpha))$ on the vertical axis while those in Figure 4.5 show $\left(\log \kappa(\mathcal{A}^{(2)}(\gamma, M, 2))\right)^2$. Unlike $\kappa(B^{(2)}(\gamma, M, 2))$ the exponential growth in $\sqrt{M}$ of $\kappa(\mathcal{A}^{(2)}(\gamma, M, 2))$ is consistent for all values of γ. This is not surprising given that $\kappa(D(\phi')) \approx e^{\pi\sqrt{M/\alpha}}$ and

$$
\begin{aligned}
\kappa(\mathcal{A}^{(2)}(\gamma, M, \alpha)) &\leq \kappa(A^{(2)}(\gamma, m, \alpha))\left(\kappa(D(\phi'))\right)^2 \\
&\approx e^{2\pi\sqrt{M/\alpha}}\kappa(\mathcal{A}^{(2)}(\gamma, M, \alpha)).
\end{aligned}
$$

For $\gamma = 1/4$ and $1/2$ it may not at first appear that $\kappa(B^{(2)}(\gamma, M, 2))$ and $\kappa(\mathcal{A}^{(2)}(\gamma, M, 2))$ are that different. Their qualitative behavior as m changes is not different but the raw numbers mean that double precision must be used for calculations involving $\mathcal{A}^{(2)}(\gamma, M, 2)$ while single precision may suffice for $B^{(2)}(\gamma, M, 2)$.

The values $\kappa(A^{(2)}(\gamma, M, \alpha))$, $\kappa(B^{(2)}(\gamma, m, \alpha))$, and $\kappa(\mathcal{A}^{(2)}(\gamma, M, \alpha))$ also depend on the parameters γ and α. Results for the parameter α are discussed next. Figure 4.6 presents four plots where $\kappa(A^{(2)}(\gamma, M, \alpha))$ is graphed against $\alpha = .1 : .1 : 2$. The parameter γ is held fixed in each plot but varies from plot to plot. For $\gamma = 1/4$ and $1/2$ the graphs appear to grow linearly in α while at $\gamma = 3/4$ and 1 the growth is less than linear. When compared with Figure 4.1, $\kappa(A^{(2)}(\gamma, M, \alpha))$ is smaller for small values of α (more singular solutions) than for large values of m (finer discretizations); that is, $\kappa(A^{(2)}(\gamma, M, \alpha))$ is less sensitive to α than to m.

Figure 4.7 corresponds to Figure 4.6 with $\kappa(A^{(2)}(\gamma, M, \alpha))$ replaced by $\kappa(B^{(2)}(\gamma, M, \alpha))$. Obviously, $\kappa(B^{(2)}(\gamma, M, \alpha))$ is sensitive to changes is α, the more so when γ is away from 1. Since $\kappa(D((\phi')^{1-\gamma}))$ behaves like $e^{(1-\gamma)\pi\sqrt{M/\alpha}}$, the author expected that graphs corresponding to those in Figure 4.7 except with $1/\left(\log\kappa(B^{(2)}(\gamma, M, \alpha))\right)^2$ on the vertical axis would show linear growth. In fact, these graphs were very flat but showed some decay in α rather than growth.

The condition number $\kappa(\mathcal{A}^{(2)}(\gamma, M, \alpha))$ did exhibit this $1/(\log)^2$ linear growth in α. Figure 4.8 includes the graphs for $\kappa(\mathcal{A}^{(2)}(\gamma, m, \alpha))$ that correspond to Figures 4.6 and 4.7. The graphs in Figure 4.9 are those in Figure 4.8 but using a $1/(\log)^2$ scale. When Figures 4.7 and 4.8 are compared it is clear that $\kappa(\mathcal{A}^{(2)}(\gamma, M, \alpha))$ is more sensitive to small values of α than is $\kappa(B^{(2)}(\gamma, M, \alpha))$. Indeed, as α approaches zero the condition number of $\kappa(\mathcal{A}^{(2)}(\gamma, M, \alpha))$ exceeds double precision.

The last parameter considered is γ where $\gamma = .05 : .05 : 1$. No output is shown for $\kappa(\mathcal{A}^{(2)}(\gamma, M, \alpha))$ because for smaller values of γ MATLAB issued warning errors indicating division by zero. In both Figures 4.10 and 4.11 $\alpha = 2$ is held fixed. Figures 4.10(a) and 4.11(a) show the graphs of $\kappa(A^{(2)}(\gamma, 16, 2))$ and $\kappa(B^{(2)}(\gamma, 16, 2))$, respectively. Figures 4.10(b) and 4.11(b) replace 16 with 32. When compared, $\kappa(B^{(2)}(\gamma, 16, 2))$ is not significantly greater than $\kappa(A^{(2)}(\gamma, 16, 2))$ in magnitude. Figures 4.10(c) and 4.11(c) display graphs of $\log \kappa(A^{(2)}(\gamma, 16, 2))$ and $\log \kappa(B^{(2)}(\gamma, 16, 2))$, respectively, while Figures 4.10(d) and 4.11(d) replace 16 with 32. Although these graphs are not as markedly linear as some in other figures, the vertical scale does provide evidence of exponential decay in $\kappa(A^{(2)}(\gamma, M, \alpha))$ and $\kappa(B^{(2)}(\gamma, M, \alpha))$ with respect to γ. From the form of $A^{(2)}(\gamma, M, \alpha)$ it is not obvious why this should be so; however, $\kappa(D((\phi')^{1-\gamma})) \approx e^{(1-\gamma)\pi\sqrt{M/\alpha}}$ does provide reason to expect such behavior in $\kappa(B^{(2)}(\gamma, M, \alpha))$.

4 Conclusions

When discretizing a ordinary differential or partial differential equation with the Sinc-Galerkin method care must be exercised when posing the final linear system that is numerically solved. Limiting the condition number of the coefficient matrix with respect to inversion may be a more important objective than seeking a structure such as symmetry. Further, the values of the parameters discussed in detail above should also be considered when selecting a form for the linear system.

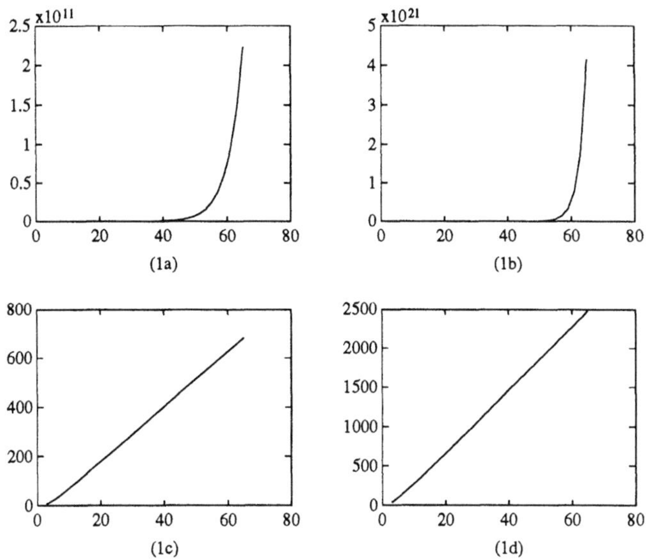

Figure 4.1: $\kappa(A^{(2)}(\gamma, M, 2))$ for $M = 1 : 32$.

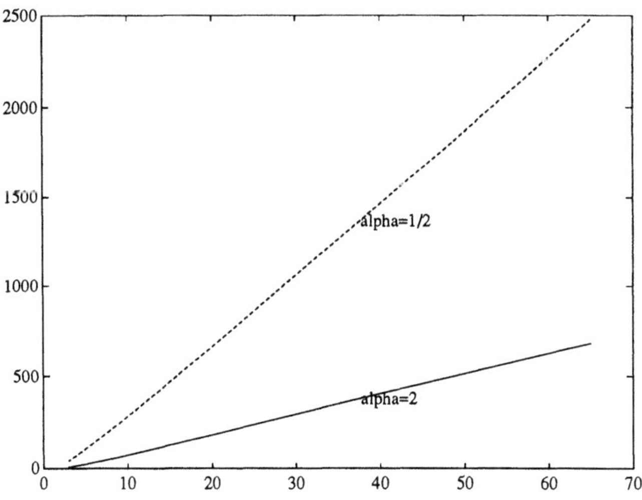

Figure 4.2: $\kappa(B^{(2)}(\gamma, M, 2))$ for $M = 1 : 32$.

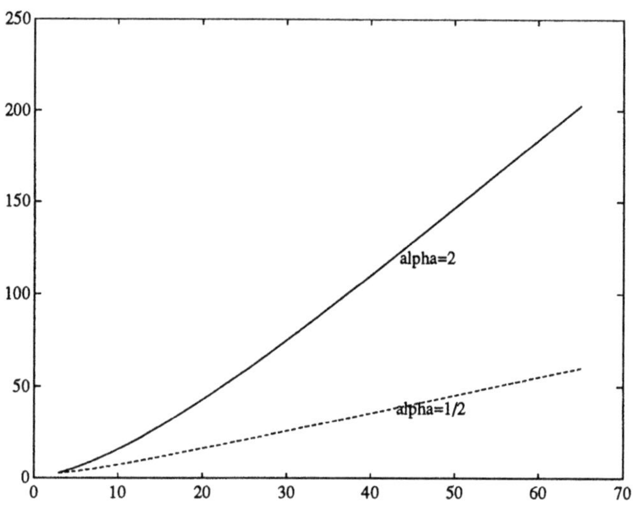

Figure 4.3: $(\log \kappa(B^{(2)}(\gamma, M, 2)))^2$ for $M = 1 : 32$.

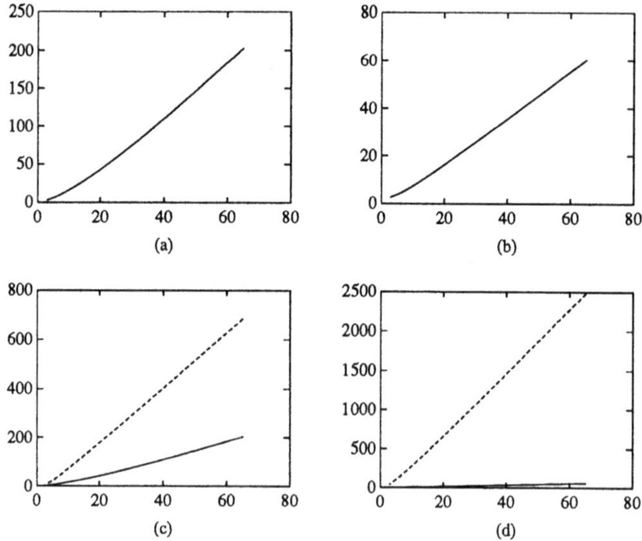

Figure 4.4: $\kappa(\mathcal{A}^{(2)}(\gamma, M, 2))$ for $M = 1 : 32$.

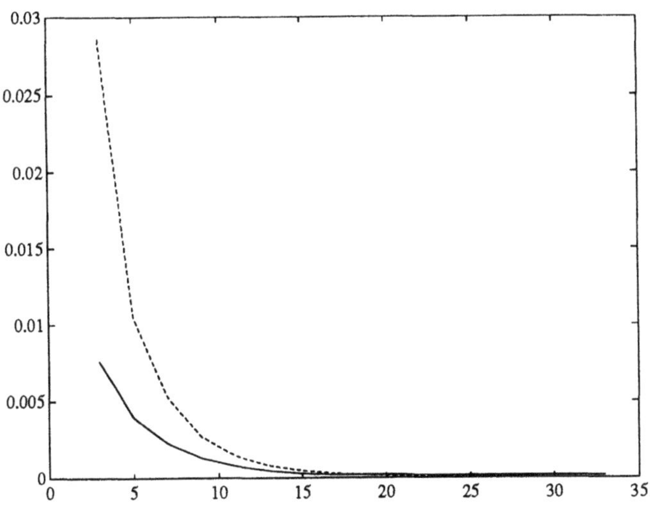

Figure 4.5: $(\log \kappa(\mathcal{A}^{(2)}(\gamma, M, 2)))^2$ for $M = 1 : 32$.

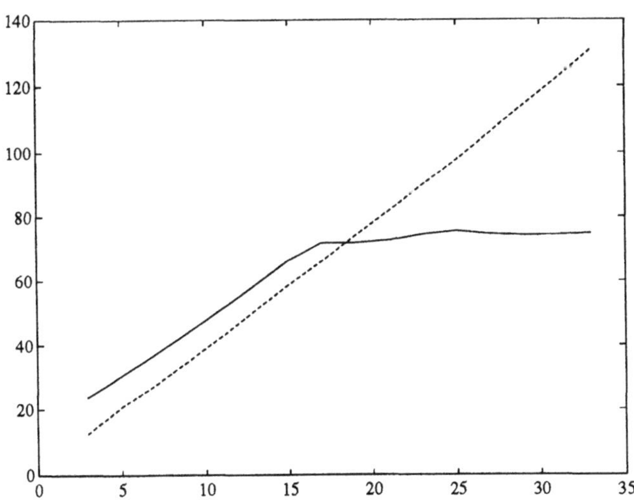

Figure 4.6: $\kappa(A^{(2)}(\gamma, 16, \alpha))$ for $\alpha = .1 : .1 : 2$.

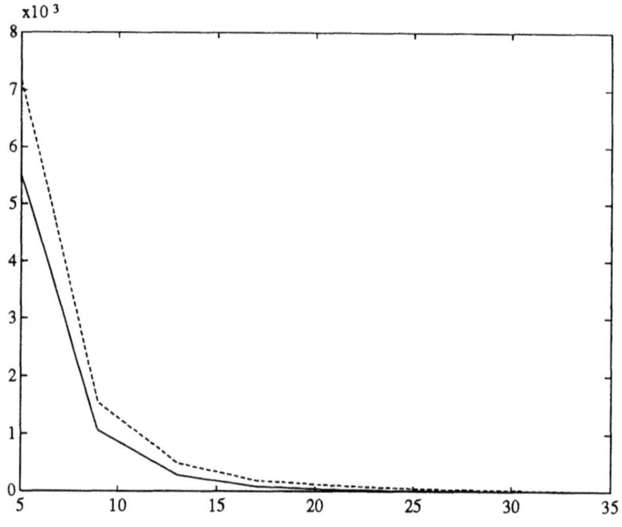

Figure 4.7: $\kappa(B^{(2)}(\gamma, 16, \alpha))$ for $\alpha = .1 : .1 : 2$.

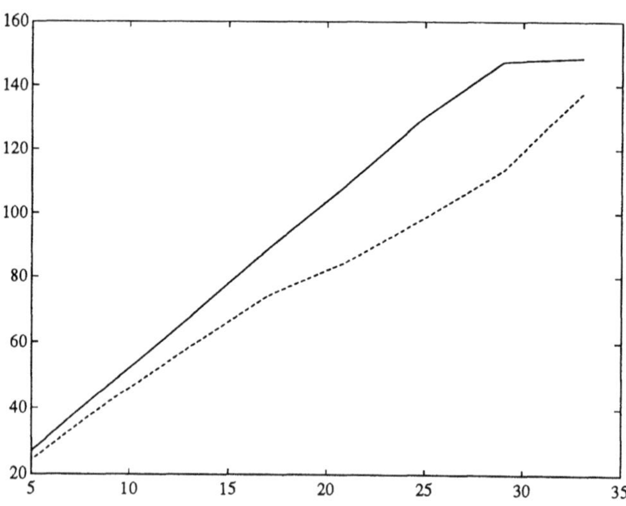

Figure 4.8: $\kappa(\mathcal{A}^{(2)}(\gamma, 16, \alpha))$ for $\alpha = .1 : .1 : 2$.

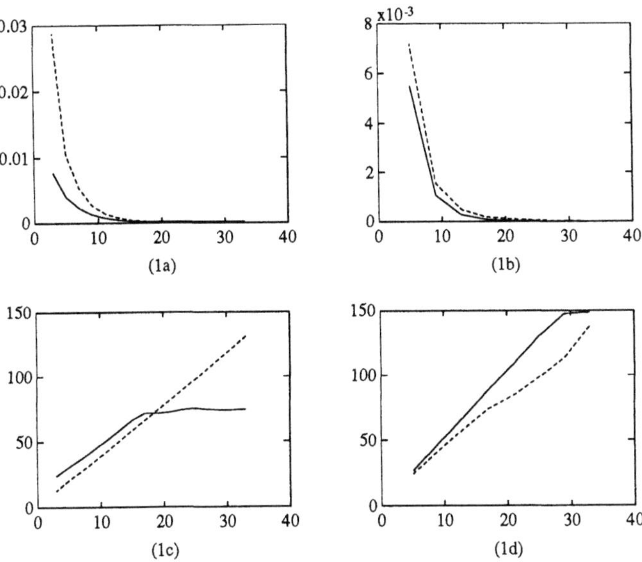

Figure 4.9: $1/(\log \kappa(\mathcal{A}^{(2)}(\gamma, 16, \alpha)))^2$ for $\alpha = .3 : .1 : 2$.

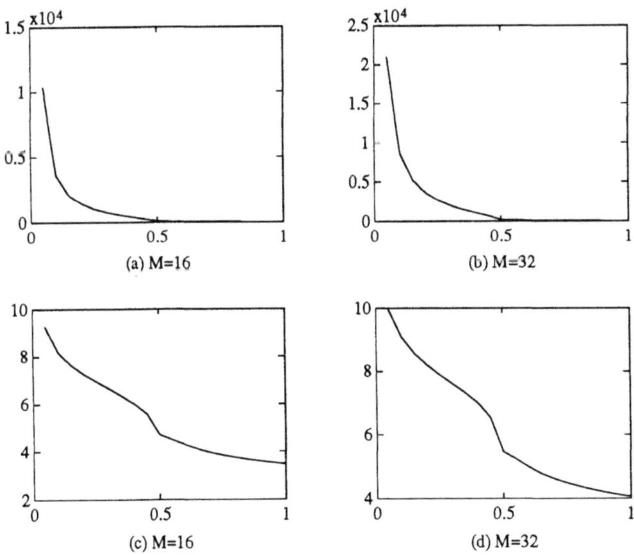

Figure 4.10: For $\gamma = .05 : .05 : 1$: (a) $\kappa(A^{(2)}(\gamma, 16, 2))$, (b) $\kappa(A^{(2)}(\gamma, 32, 2))$, (c) $\log \kappa(A^{(2)}(\gamma, 16, 2))$, (d) $\log \kappa(A^{(2)}(\gamma, 32, 2))$

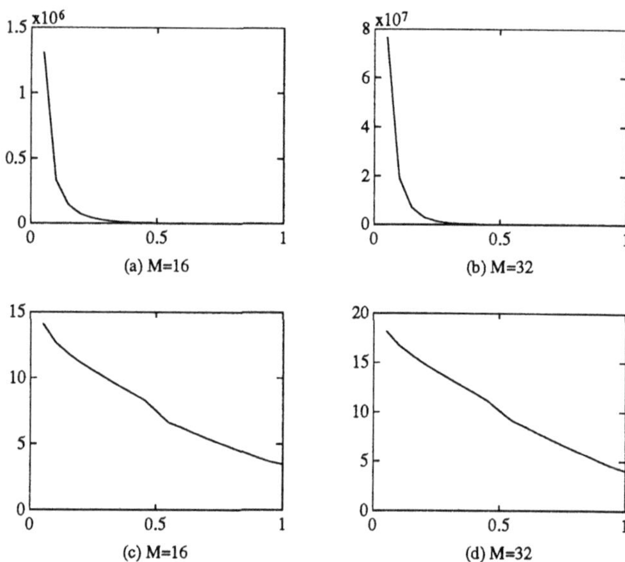

Figure 4.11: For $\gamma = .05 : .05 : 1$: (a) $\kappa(B^{(2)}(\gamma, 16, 2))$, (b) $\kappa(B^{(2)}(\gamma, 32, 2))$, (c) $\log \kappa(B^{(2)}(\gamma, 16, 2))$, (d) $\log \kappa(B^{(2)}(\gamma, 32, 2))$

References

[1] K. L. Bowers and J. Lund. Numerical solution of singular Poisson problems via the Sinc-Galerkin method. *SIAM J. Numer. Anal.*, 24(1):36–51, 1987.

[2] C. Canuto and P. Pietra. Boundary and interface conditons within a finite element preconditioner for spectral methods. *Journal of Comp. Phys.*, 91:310–343, 1990.

[3] M.O. Deville and E.H. Mund. Finite-element preconditioning for pseudospectral solutions of elliptic problems. *SIAM J. Sci. Stat. Comput.*, 11(2):311–342, 1990.

[4] G.H. Golub and C.F. Van Loan. *Matrix Computations*. The Johns Hopkins University Press, Baltimore, 2nd edition, 1989.

[5] J. Lund. Symmetrization of the Sinc-Galerkin method for boundary value problems. *Math. Comp.*, 47:571–588, 1986.

[6] K. M. McArthur, K. L. Bowers, and J. Lund. The sinc method in multiple space dimensions: Model problems. *Numer. Math.*, 56:789–816, 1990.

[7] K.M. McArthur. The Sinc-Galerkin ADI method for elliptic equations. preprint.

[8] K.M. McArthur, R.C. Smith, K.L. Bowers, and J. Lund. The Sinc-Galerkin method for parameter-dependent self-adjoint problems. *Appl. Math. and Comp.*, 50(2 and 3):175–202, 1992.

[9] F. Stenger. A Sinc-Galerkin method of solution of boundary value problems. *Math. Comp.*, 33:85–109, 1979.

[10] F. Stenger. Numerical methods based on Whittaker cardinal, or sinc functions. *SIAM Review*, 23(2):165–224, 1981.

COMPUTATIONAL MODELS FOR LATTICE STRUCTURES

Robert E. Miller*

*Department of Mathematical Sciences
University of Arkansas
Fayetteville, Arkansas 72701*

1 Introduction

Lattice structures such as trusses are used frequently in engineering applications, so an important mathematical problem is finding approximate models which are suitable for computations. This paper considers the use of asymptotic methods for deriving such models and focuses on the problem of computing the eigenvalues for a truss. One way to describe a truss is to model each element as a beam, but such an approach clearly leads to intractable numerical difficulties. Instead, we consider the truss as a three-dimensional object with two dimensions small compared to the third. Thus, there is a naturally occurring small parameter in the problem. By letting this parameter tend to zero, we obtain, in place of the original three-dimensional problem, two "beam-like" equations describing the transverse motions of the truss and a second order equation for the axial motions. These equations have periodic coefficients, so homogenization techniques can be used to simplify the problem further.

In [3] we obtained an approximation to the eigenvalue problem using rigorous variational techniques. As a means of testing our model, we computed several eigenvalues and compared them to the frequencies observed in experiments with a truss conducted at NASA Langley Research Center. When we found that the predicted frequencies were higher than those observed in the experiments, our original hope was that the agreement could be improved by including a damping mechanism in the model. Further study has revealed, however, that the discrepancy is greater than can be accounted for by the effects of damping. Hence it appears that either the linear model is inadequate due to the complicated structure of the truss (we remark that the maximum deflection at the tip of the truss was very small compared to the length) or that too much information was lost in passing to the limit. In an effort to determine whether the latter is the case, we consider in this paper an alternative derivation of the approximate model. We assume an asymptotic expansion in the small parameter

*This research was carried out while the author was with the Center for Applied Mathematical Sciences at the University of Southern California and was supported in part by the Air Force Office of Scientific Research under grant AFOSR-90-0091 and in part by the National Aeronautics and Space Administration under NASA grant NAG-1-1116.

e of the eigenvalues and eigenfunctions and derive equations satisfied by the leading terms. This method has been used extensively, especially in studying plates; see the references in [1]. Although this approach is formal, we obtain the same model derived rigorously in [3]. Thus, we propose next to extend this derivation to obtain higher order "corrector" terms to the eigenvalues.

2   The Three-Dimensional Problem

Let ω^e be a region in $\mathbb{R}^2$, and set $Q^e = \omega^e \times (0, L)$. We consider structures $\Omega^e_{\delta\epsilon}$ contained in Q^e. Observe that we actually have three small parameters: e is related to the size of ω^e (e.g., we may take $\omega^e = (-e, e) \times (-e, e)$), ϵ is the period in the x_3^e direction (the truss is composed of $N = L/\epsilon$ identical "cells"), and δ indicates the amount of material (δ may be the diameter of the cross-section of one of the elements making up the truss, for example). For $0 \leq x_3^e \leq L$, define

$$S^e_{\delta\epsilon}(x_3^e) = \left\{ (x_1^e, x_2^e) \mid (x_1^e, x_2^e, x_3^e) \in \overline{\Omega}^e_{\delta\epsilon} \right\},$$

and define

$$\Gamma^e_\delta = S^e_{\delta\epsilon}(0) = S^e_{\delta\epsilon}(\epsilon).$$

Figure 2.1 depicts the set $S^e_{\delta\epsilon}$ for the case in which the representative period

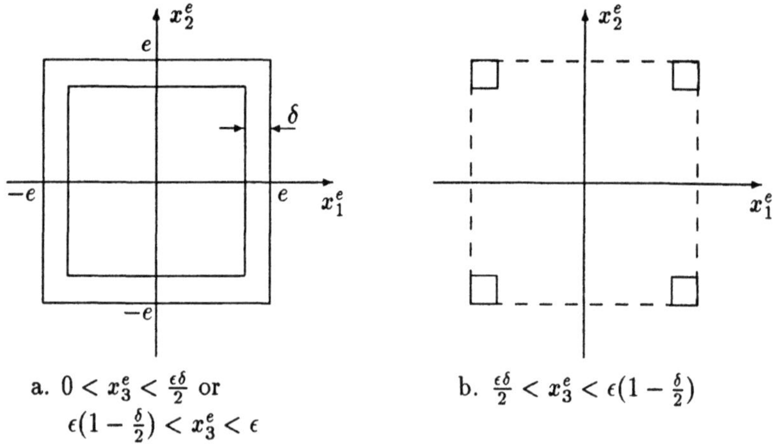

a. $0 < x_3^e < \frac{\epsilon\delta}{2}$ or
$\epsilon(1 - \frac{\delta}{2}) < x_3^e < \epsilon$

b. $\frac{\epsilon\delta}{2} < x_3^e < \epsilon(1 - \frac{\delta}{2})$

Figure 2.1: The set $S^e_{\delta\epsilon}(x_3^e)$

of the truss is a cube with material distributed only along the edges. Observe that $S^e_{\delta\epsilon}$ is periodic with period ϵ. We make the following geometric assumptions on the structure:

$$(2.1) \qquad \int_{S^e_{\delta\epsilon}(x^e_3)} x^e_1 \, dx^e_1 dx^e_2 = \int_{S^e_{\delta\epsilon}(x^e_3)} x^e_2 \, dx^e_1 dx^e_2 = 0$$

and

$$(2.2) \qquad \int_{S^e_{\delta\epsilon}(x^e_3)} x^e_1 x^e_2 \, dx^e_1 dx^e_2 = 0$$

for all $0 \leq x^e_3 \leq L$. The first assumption just says that the centroid of each cross section lies on the same straight line; the second assumption will allow the one-dimensional equations we derive to decouple. In this paper we will also assume that the structure has the same configuration when rotated 90 degrees about the x^e_3-axis. Other representative periods which also satisfy the above assumptions are shown in Figure 2.2. Later on we

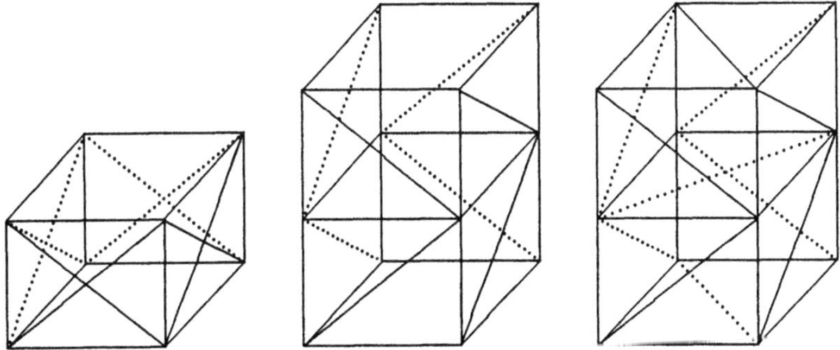

Figure 2.2: Representative periods of other trusses

will scale the variables so that for each e the problem is defined on the same fixed domain. The superscript "e" will indicate the original unscaled variables; variables without the superscript are scaled. Since we are mainly interested in the dependence on e, we will drop the ϵ and δ for simplicity of notation.

We denote by $\gamma^e_{kh}(v^e)$ the linearized strain tensor:

$$\gamma^e_{kh}(v^e) = \frac{1}{2}(\partial^e_h v^e_k + \partial^e_k v^e_h)$$

where $\partial^e_h v^e = \frac{\partial v^e}{\partial x^e_h}$, and we assume the following linear relation betweeen the stress tensor σ^e and the displacement u^e (here and throughout the paper we adopt the summation convention on repeated indices with the added convention that Latin indices take on the values 1, 2 and 3 and Greek the values 1 and 2):

$$\sigma^e{}_{ij} = a_{ijkh}\gamma^e_{kh}(u^e).$$

We make the following assumptions on the elasticity coefficients:

(i) $a_{ijkh} = a_{jikh} = a_{khij}$ for all i, j, k, h.

(ii) There exists $\alpha_0 > 0$ such that $\alpha_0 v_{ij} v_{ij} \leq a_{ijkh} v_{kh} v_{ij}$ for all $v = (v_{ij})$ with $v_{ij} = v_{ji}$.

Set $\Gamma^e_0 = \Gamma^e \times \{0\}$ and $\Gamma^e_1 = \partial\Omega^e \setminus \Gamma^e_0$. Then the problem we wish to consider is the following (see [3]). Find real numbers Λ^e and functions $u^e : \overline{\Omega}^e \to \mathbb{R}^3$ that solve

$$
\begin{aligned}
-\partial^e_j \sigma^e_{ij}(u^e) &= \rho^e \Lambda^e u^e_i && \text{in } \Omega^e, \\
u^e &= 0 && \text{on } \Gamma^e_0, \\
\sigma^e_{ij}(u^e) n_j &= 0 && \text{on } \Gamma^e_1,
\end{aligned}
$$

(2.3)

where ρ^e denotes the mass of the material and $n = (n_j)$ denotes the outward unit normal. If we define the space V^e by

$$V^e = \{\, v = (v_i) \in H^1(\Omega^e) \mid v^e = 0 \text{ on } \Gamma^e_0 \,\},$$

then we can write (2.3) in weak form as

$$(2.4)\quad \int_{\Omega^e} a_{ijkh}\gamma^e_{kh}(u^e)\gamma^e_{ij}(v^e)\,dx^e = \Lambda^e \int_{\Omega^e} \rho^e u^e_i v^e_i\,dx^e \text{ for all } v^e \in V^e.$$

Each eigenvalue is positive and of finite multiplicity, and the corresponding eigenfunctions are in V^e.

3   Derivation of the One-Dimensional Problems

In order to simplify the derivation we will assume that the truss is made of an isotropic, homogeneous elastic material so that the elasticity coefficients are given by

$$a_{ijkh} = \lambda \delta_{ij}\delta_{kh} + \mu(\delta_{ik}\delta_{jh} + \delta_{ih}\delta_{jk})$$

where λ and μ are the Lamé constants. In order to obtain the one-dimensional problems we take the limit as $e \to 0$, but as e changes, the domain changes. Thus we begin by transforming the problem to the fixed domain $\Omega \subset Q = Q^1$ as follows. To each point $x \in \overline{Q}$ we associate the point $\pi^e x^e \in \overline{Q}^e$ by

$$\pi^e : x = (x_1, x_2, x_3) \mapsto x^e = (ex_1, ex_2, x_3).$$

COMPUTATIONAL MODELS FOR LATTICE STRUCTURES

We define the scalings $\Lambda^e = \Lambda(e)$, $\rho^e = e^2\rho$ and

$$\left.\begin{array}{rcl} u_\alpha^e(\boldsymbol{x}^e) & = & eu_\alpha(e)(\boldsymbol{x}) \\[2mm] u_3^e(\boldsymbol{x}^e) & = & e^2 u_3(e)(\boldsymbol{x}) \end{array}\right\} \text{ for all } \boldsymbol{x}^e = \boldsymbol{\pi}^e\boldsymbol{x} \in \overline{\Omega}^e.$$

By substituting these scalings into the weak form of the problem (equation (2.4)) and cancelling e^6, we obtain the scaled problem:

$$(3.1) \qquad \int_\Omega \sigma_{ij}(e, \boldsymbol{u}(e))\partial_j v_i\, dx = \Lambda(e)\rho \int_\Omega [u_\alpha(e)v_\alpha + e^2 u_3(e)v_3]\, dx$$

for all $\boldsymbol{v} = (v_i) \in V$, where $V = \{\,\boldsymbol{v} \in H^1(\Omega) \mid \boldsymbol{v} = 0 \text{ on } \Gamma_0\,\}$, and

$$(3.2) \qquad \begin{aligned} \sigma_{\alpha\beta}(e, \boldsymbol{u}(e)) &= e^{-4}[\lambda\gamma_{\eta\eta}(\boldsymbol{u}(e))\delta_{\alpha\beta} + 2\mu\gamma_{\alpha\beta}(\boldsymbol{u}(e))] \\ &\qquad + e^{-2}\lambda\gamma_{33}(\boldsymbol{u}(e))\delta_{\alpha\beta}, \\ \sigma_{\alpha 3}(e, \boldsymbol{u}(e)) &= e^{-2}\,2\mu\gamma_{\alpha 3}(\boldsymbol{u}(e)) = \sigma_{3\alpha}(e, \boldsymbol{u}(e)), \\ \sigma_{33}(e, \boldsymbol{u}(e)) &= e^{-2}\lambda\gamma_{\eta\eta}(\boldsymbol{u}(e)) + (\lambda + 2\mu)\gamma_{33}(\boldsymbol{u}(e)). \end{aligned}$$

Next, we assume asymptotic expansions of the form

$$(3.3) \qquad \begin{aligned} \boldsymbol{u}(e) &= \boldsymbol{u}^0 + e\boldsymbol{u}^1 + e^2\boldsymbol{u}^2 + \cdots, \text{ with each } \boldsymbol{u}^i \in V, \\ \Lambda(e) &= \Lambda^0 + e\Lambda^1 + \cdots. \end{aligned}$$

The expansion of $\boldsymbol{u}(e)$ induces a formal expansion of the stress tensor:

$$\Sigma(e, \boldsymbol{u}(e)) = e^{-4}\Sigma^{-4} + e^{-3}\Sigma^{-3} + \cdots.$$

Compontentwise we have (from equations (3.2))

$$(3.4) \qquad \begin{aligned} \sigma_{\alpha\beta}^{-4} &= \lambda\gamma_{\eta\eta}(\boldsymbol{u}^0)\delta_{\alpha\beta} + 2\mu\gamma_{\alpha\beta}(\boldsymbol{u}^0), \\ \sigma_{\alpha\beta}^{-3} &= \lambda\gamma_{\eta\eta}(\boldsymbol{u}^1)\delta_{\alpha\beta} + 2\mu\gamma_{\alpha\beta}(\boldsymbol{u}^1), \\ \sigma_{\alpha\beta}^{-2} &= \lambda[\gamma_{33}(\boldsymbol{u}^0) + \gamma_{\eta\eta}(\boldsymbol{u}^2)]\delta_{\alpha\beta} + 2\mu\gamma_{\alpha\beta}(\boldsymbol{u}^2), \\ \sigma_{\alpha 3}^{-2} &= 2\mu\gamma_{\alpha 3}(\boldsymbol{u}^0), \\ \sigma_{33}^{-2} &= \lambda\gamma_{\eta\eta}(\boldsymbol{u}^0), \end{aligned}$$

and so forth. Observe that $\sigma_{ij}^{-4} = \sigma_{ij}^{-3} = 0$ when at least one of $i,\,j$ is 3.

Now, in order to derive the equations satisfied by the leading terms, we simply match powers of e in equation (3.1). First for e^{-4}, we get

$$\int_\Omega \sigma_{\alpha\beta}^{-4}\partial_\beta v_\alpha\, dx = 0 \text{ for all } \boldsymbol{v} \in V.$$

Taking $v = u^0$ and using the symmetry, we get

$$0 \leq \int_\Omega \left[\lambda(\gamma_{\eta\eta}(u^0))^2 + 2\mu\gamma_{\alpha\beta}(u^0)\gamma_{\alpha\beta}(u^0)\right]\, dx = 0.$$

Thus, we must have

$$\gamma_{\alpha\beta}(u^0) = 0 \quad \text{and} \quad \sigma_{\alpha\beta}^{-4} = 0.$$

Similarly, for e^{-3}, we get

$$\gamma_{\alpha\beta}(u^1) = 0 \quad \text{and} \quad \sigma_{\alpha\beta}^{-3} = 0.$$

For e^{-2}, we have

$$\int_\Omega \sigma_{ij}^{-2}\partial_j v_i\, dx = \int_\Omega \left[\sigma_{\alpha\beta}^{-2}\partial_\beta v_\alpha + \sigma_{\alpha3}^{-2}\partial_3 v_\alpha + \sigma_{3\beta}^{-2}\partial_\beta v_3 + \sigma_{33}^{-2}\partial_3 v_3\right]\, dx$$

$$= \int_\Omega \left[\sigma_{\alpha\beta}^{-2}\gamma_{\alpha\beta}(v) + 2\sigma_{\alpha3}^{-2}\gamma_{\alpha3}(v) + \sigma_{33}^{-2}\gamma_{33}(v)\right]\, dx = 0$$

for all $v \in V$. We already know that $\gamma_{\alpha\beta}(u^0) = 0$, so $\sigma_{33}^{-2} = \lambda\gamma_{\eta\eta}(u^0) = 0$. Thus, taking $v = u^0$, we have

$$0 = \int_\Omega 2\sigma_{\alpha3}^{-2}\gamma_{\alpha3}(u^0)\, dx = 4\mu \int_\Omega \gamma_{\alpha3}(u^0)\gamma_{\alpha3}(u^0)\, dx.$$

Hence, $\gamma_{\alpha3}(u^0) = 0$, so we get finally,

$$\int_\Omega \sigma_{\alpha\beta}^{-2}\gamma_{\alpha\beta}(v)\, dx = 0 \text{ for all } v \in V.$$

We wish to show that $\sigma_{\alpha\beta}^{-2} = 0$ for all α, β. Let $\varphi \in H_L^1(0, L)$ where

$$H_L^1(0, L) = \{\, \varphi \in H^1(0, L) \mid \varphi(0) = 0 \,\}.$$

We will take test functions $v \in V$ of the form

$$(3.5) \qquad\qquad v = \begin{pmatrix} a_1 x_1^{n_1} x_2^{m_1} \varphi(x_3) \\ a_2 x_1^{n_2} x_2^{m_2} \varphi(x_3) \\ v_3 \end{pmatrix}$$

where $a_\alpha \in \mathbb{R}$, and $n_\alpha, m_\alpha \in \mathbb{N}$. First, for $n_1 = 0$, $m_1 = n+1$, $a_1 = 1/(n+1)$ and $a_2 = 0$, we have $\gamma_{11}(v) = \gamma_{22}(v) = 0$ and $\gamma_{12}(v) = \gamma_{21}(v) = \frac{1}{2}x_2^n\varphi(x_3)$. Thus, $\sigma_{\alpha\beta}^{-2}\gamma_{\alpha\beta}(v) = 2\mu x_2^n\varphi(x_3)\gamma_{12}(u^2)$. Interchanging the subscripts, we get $\sigma_{\alpha\beta}^{-2}\gamma_{\alpha\beta}(v) = 2\mu x_1^n\varphi(x_3)\gamma_{12}(u^2)$. Thus,

$$\int_0^L \varphi(x_3) \left(\int_{S(x_3)} x_\alpha^n\gamma_{12}(u^2)\, dx_1dx_2\right) dx_3 = 0 \text{ for all } \varphi \in H_L^1(0, L)$$

which implies that

$$(3.6) \qquad \int_{S(x_3)} x_\alpha^n \gamma_{12}(u^2)\, dx_1 dx_2 = 0 \text{ a.e. } x_3 \in (0, L) \text{ and for all } n.$$

Now, by the assumption that the structure is the same when rotated 90 degrees, the roles of the variables x_1 and x_2 are interchangeable, so we must have that $\gamma_{11}(u^2) = \gamma_{22}(u^2)$. Thus, taking first $n_1 = n + 1$, $m_1 = 0$, $a_1 = 1/(n+1)$ and $a_2 = 0$ and then $m_2 = n + 1$, $n_2 = 0$, $a_2 = 1/(n+1)$ and $a_1 = 0$ in (3.5), we get

$$\int_\Omega x_\alpha^n \varphi(x_3) \left[\lambda \gamma_{33}(u^0) + 2(\lambda + \mu)\gamma_{11}(u^2)\right]\, dx = 0 \text{ for all } n.$$

Hence, just as above, we get

$$(3.7) \qquad \int_{S(x_3)} x_\alpha^n \left[\lambda \gamma_{33}(u^0) + 2(\lambda + \mu)\gamma_{11}(u^2)\right]\, dx_1 dx_2 = 0 \text{ for all } n.$$

Now suppose that

$$(3.8) \qquad \int_{S(x_3)} x_1^n x_2^m \gamma_{12}(u^2)\, dx_1 dx_2 = 0$$

and

$$(3.9) \qquad \int_{S(x_3)} x_1^n x_2^m \left[\lambda \gamma_{33}(u^0) + 2(\lambda + \mu)\gamma_{11}(u^2)\right]\, dx_1 dx_2 = 0$$

for all $n, m \in \mathbb{N}$ with $m \leq M$. By equations (3.6) and (3.7) our assumption is true for $M = 0$. Now if we take $n_1 = n + 1$, $m_1 = M + 1$, $a_1 = 1/(n+1)$ and $a_2 = 0$ in (3.5), we get

$$\sigma_{\alpha\beta}^{-2}\gamma_{\alpha\beta}(v) = \left[\lambda \gamma_{33}(u^0) + 2(\lambda + \mu)\gamma_{11}(u^2)\right] x_1^n x_2^{M+1}\varphi(x_3)$$
$$+ \frac{2(M+1)}{n+1}\mu\gamma_{12}(u^2)x_1^{n+1} x_2^M \varphi(x_3).$$

By equation (3.8), $\int_\Omega \gamma_{12}(u^2)x_1^{n+1} x_2^M \varphi(x_3)\, dx = 0$, so (3.9) holds for $m = M + 1$. Next, taking $a_1 = 0$, $m_2 = M + 1$, $n_2 = n + 1$ and $a_2 = 1/(n+1)$ and using (3.9), we see that (3.8) also holds for $m = M + 1$. Thus, by induction, equations (3.8) and (3.9) hold for all $n, m \in \mathbb{N}$. Hence, by the Weierstrass Approximation Theorem, $\gamma_{12}(u^2) = 0$ and

$$(3.10) \qquad \gamma_{11}(u^2) = \gamma_{22}(u^2) = -\frac{\lambda}{2(\lambda + \mu)}\gamma_{33}(u^0).$$

It follows immediately that $\sigma_{\alpha\beta}^{-2} = 0$. For e^{-1} we get the same results as above but with the subscript 0 replaced by 1 and the 2 replaced by 3.

Finally, matching the e^0 terms, we get

$$(3.11) \quad \int_\Omega [\sigma_{\alpha\beta}^0 \partial_\beta v_\alpha + \sigma_{\alpha 3}^0 \partial_3 v_\alpha + \sigma_{3\beta}^0 \partial_\beta v_3 + \sigma_{33}^0 \partial_3 v_3]\, dx$$

$$= \Lambda^0 \rho \int_\Omega u_\alpha^0 v_\alpha\, dx \text{ for all } v \in V.$$

From equations (3.2), we see that $\sigma_{\alpha\beta}^0$ is a function of u^2 and u^4 and $\sigma_{\alpha 3}^0$ is a function of u^2, and from (3.2) and (3.10)

$$(3.12) \quad \begin{aligned} \sigma_{33}^0 &= \lambda \gamma_{\eta\eta}(u^2) + (\lambda + 2\mu)\gamma_{33}(u^0) \\ &= \left[-\frac{\lambda^2}{\lambda + \mu} + (\lambda + 2\mu)\right]\gamma_{33}(u^0) = E\gamma_{33}(u^0). \end{aligned}$$

We have seen that $\gamma_{\alpha j}(u^0) = 0$. If we restrict v in equation (3.11) so that $\gamma_{\alpha j}(v) = 0$, then we get

$$E \int_\Omega \gamma_{33}(u^0)\gamma_{33}(v)\, dx = \Lambda^0 \rho \int_\Omega u_\alpha^0 v_\alpha\, dx.$$

We need the following lemma.

Lemma 3.1 *If $v \in V$ satisfies $\gamma_{\alpha j}(v) = 0$, then there exist $V_\alpha(x_3) \in H_L^2(0, L)$ where*

$$H_L^2(0, L) = \{\, \varphi \in H^2(0, L) \mid \varphi(0) = \varphi'(0) = 0 \,\},$$

and $V_3(x_3) \in H_L^1(0, L)$ such that

$$v_\alpha(x) = V_\alpha(x_3),$$

and

$$v_3(x) = -x_\alpha V_\alpha'(x_3) + V_3(x_3).$$

Proof: This follows from taking specific choices for α and j. For example, from $\alpha = j = 1$ we get $\partial_1 v_1 = 0$, so v_1 is independent of x_1. For remaining details, see [2] and [3].

Thus,

$$\gamma_{33}(u^0)\gamma_{33}(v)$$
$$= x_1^2 U_1''(x_3)V_1''(x_3) + x_1 x_2[U_1''(x_3)V_2''(x_3) + U_2''(x_3)V_1''(x_3)]$$
$$+ x_2^2 U_2''(x_3)V_2''(x_3) - (x_1 + x_2)[U_3'(x_3) + V_3'(x_3)] + U_3'(x_3)V_3'(x_3).$$

Integrating over Ω and using (2.1) and (2.2), we get

$$E \int_0^L \int_{S(x_3)} \left(x_1^2 U_1'' V_1'' + x_2^2 U_2'' V_2'' + U_3' V_3' \right) dx_1 dx_2 dx_3$$

$$= \Lambda^0 \rho \int_0^L \int_{S(x_3)} U_\alpha V_\alpha \, dx_1 dx_2 dx_3$$

for all $V_\alpha \in H_L^2(0, L)$ and all $V_3 \in H_L^1(0, L)$. Taking $V_1 = V_2 \equiv 0$, we get $U_3(x_3) \equiv 0$. Then taking $V_2 \equiv 0$ and then $V_1 \equiv 0$, we see that (U_1, Λ^0) and (U_2, Λ^0) satisfy:

(3.13)

$$E \int_0^L \left(\int_{S(x_3)} x_1^2 \, dx_1 dx_2 \right) U'' V'' \, dx_3 = \Lambda^0 \rho \int_0^L |S(x_3)| \, U V \, dx_3$$

for all $V \in H_L^2(0, L)$. Thus we have proved the following theorem.

Theorem 3.2 *If $(\Lambda_\ell(e), u_\ell(e))$ is an eigenvalue-eigenfunction pair of equation (3.1) having asymptotic expansions in e as given by equations (3.3), then*

$$\Sigma(e, u(e)) = \Sigma^0 + e \Sigma^1 + \cdots$$

(i.e., $\Sigma^{-4} = \cdots = \Sigma^{-1} = 0$), and the leading terms $(\Lambda_\ell^0, u_\ell^0)$ satisfy:

(i) $\gamma_{\alpha j}(u_\ell^0) = 0$, *so*

$$u_\ell^0 = (U_{1,\ell}(x_3), U_{2,\ell}(x_3), -x_1 U_{1,\ell}'(x_3) - x_2 U_{2,\ell}'(x_3))^T$$

where $U_{1,\ell}, U_{2,\ell} \in H_L^2(0, L)$.

(ii) $(\Lambda_\ell^0, U_{\alpha,\ell})$ *satisfy equation (3.13).*

In [3] we showed that if $(\Lambda_\ell(e), u_\ell(e))$ is an eigenvalue-eigenfunction pair of equation (3.1), then there is a subsequence of $\{ (\Lambda_\ell(e), u_\ell(e)) \}$ such that

$$\Lambda_\ell(e) \rightarrow \Lambda_\ell$$

and

$$u_\ell(e) \rightharpoonup u_\ell^* \text{ weakly in } V$$

where $\gamma_{\alpha j}(u_\ell^*) = 0$, and (Λ_ℓ, u_ℓ^*) satisfy (i) and (ii) of Theorem 3.2. Hence, the results of the theorem can be established rigorously without the assumption of the asymptotic expansions (3.3).

4 The Homogenized Equation and Numerical Results

Observe that if $S(x_3) = \omega$ for all x_3, then equation (3.13) is just an Euler-Bernoulli equation. As it is, the coefficients are periodic, so we restore the ϵ and δ we dropped earlier and homogenize. Set

$$a_\delta^\epsilon(x_3) = \int_{S(x_3)} x_1^2 \, dx_1 dx_2.$$

If $(\Lambda_\ell^{\epsilon\delta}, U_\ell^{\epsilon\delta})$ is an eigenvalue-eigenfunction pair for equation (3.13), then there is a subsequence of ϵ such that

$$\Lambda_\ell^{\epsilon\delta} \to \Lambda_\ell^\delta$$

and

$$U_\ell^{\epsilon\delta} \rightharpoonup U_\ell^\delta \text{ weakly in } H_L^2(0, L)$$

where

(4.1) $$E q_1^\delta (U_\ell^\delta)'''' = \rho \Lambda_\ell^\delta q_2^\delta U_\ell^\delta$$

with

$$q_1^\delta = \frac{1}{\frac{1}{\epsilon} \int_0^\epsilon \frac{1}{a_\delta^\epsilon(x)} \, dx},$$

$$q_2^\delta = \frac{1}{\epsilon} \int_0^\epsilon |S_{\epsilon\delta}(x)| \, dx.$$

See [2] and [3] for details of the homogenization process.

As we remarked above, the frequencies predicted by the homogenized equation do not agree with those observed in the experiment. Since there are two limit processes involved in obtaining equation (4.1), we computed three eigenvalues of each of equations (3.13) and (4.1) to determine, if possible, which limit contributes most to the error. We considered a steel truss of length 10 ft. having the period depicted on the right in Figure 2.2. We considered the case where the truss has five periods ($\epsilon = 1/5$) and the case where the truss has ten periods ($\epsilon = 1/10$). Table 4.1 shows the first three frequencies predicted by the first limit equation compared with those of the homogenized equation for $\epsilon = 1/5$. A typical truss may have δ in the range of $1/40$ to $1/50$, so these results indicate that homogenization gives a good approximation to equation (3.13). As seen in Table 4.2, we obtained similar results for the truss with ten periods.

δ	Equation (3.13)	Equation (4.1)	Relative Error
1/40	34.26	33.45	2.43%
	214.23	209.64	2.19%
	598.86	587.06	2.01%
1/50	34.04	33.39	1.95%
	212.78	209.24	1.69%
	594.70	585.93	1.50%
1/100	33.54	33.26	0.85%
	209.78	208.43	0.65%
	586.02	583.67	0.40%

Table 4.1: First three frequencies for a truss with five periods

δ	Equation (3.13)	Equation (4.1)	Relative Error
1/50	17.02	16.69	1.98%
	106.68	104.62	1.97%
	298.58	292.97	1.91%

Table 4.2: First three frequencies for a truss with ten periods

References

[1] P. G. CIARLET, *Plates and Junctions in Elastic Multi-Structures,* RMA 14, Paris: Masson, 1990.

[2] D. CIORANESCU and J. SAINT JEAN PAULIN, "Elasticity Problems for Towers," preprint.

[3] R. E. MILLER, "The Eigenvalue Problem for a Class of Long, Thin Elastic Structures with Periodic Geometry," *Q. Appl. Math.* to appear.

[4] R. E. MILLER, "Modeling of Long, Thin Elastic Structures with Periodic Geometry," Proceedings of the Thirty-first IEEE Conference on Decision and Control, to appear.

THE PARTIAL DIFFERENTIAL EQUATIONS OF CONTROLLED INVARIANCE

Viswanath Ramakrishna

Frick Laboratories
Princeton University
Princeton, NJ 08544

1 Introduction

In recent years there has been a veritable rash of applications of partial differential equations to the study of *finite dimensional nonlinear control systems*. The most well known of these partial differential equations is, perhaps, the Hamilton-Jacobi Bellman equation for the value function in optimal control, as also is its variant in the theory of differential games and the so-called state approach to the H^∞ optimization problem -the Hamilton-Jacobi-Bellman-Isaacs equation. Both equations are, of course, very much inspired(at least in their geometric aspects) by the Hamilton-Jacobi equation of classical mechanics and the eikonal equation in (geometrical) optics. Then there is the coupled system of nonlinear PDEs and analytic(as opposed to differential)equations ([9]) arising in the analysis of nonlinear regulators, dubbed the Francis-Byrnes-Isidori or FBI equations by ([24]). These are equations of the "centre-manifold" type- in fact, they are equations for the graph of a mapping which is to be an invariant manifold for the closed loop system. Then there is a Riccatti-type equation arising in the study of factorization of nonlinear systems (See ([3])). The study of feedback linearization and observer error linearization also lead to PDEs which are overdetermined (See ([2, 20, 29]) and the numerous references cited therein). These equations can be solved, under favourable circumstances, algebraically, *so long as one is not interested in solutions with special features*. Finally, there are the overdetermined systems of PDEs of the controlled invariance problem- and their study will be the subject of this paper.

The controlled invariance problem, which seeks to find feedback laws under the influence of which the closed loop system will leave an entire foliation, not just a manifold, invariant is of central importance in control theory arising as it does in several very important synthesis problems such as disturbance -decoupling, non-interacting control and even stabilization. Furthermore, the genesis of the notion of nonlinear zeroes lies in the controlled invariance problem (see ([23]). For these reasons, if not any other, the controlled invariance problem merits study and has been , in fact, been intensely studied (See ([17, 21, 31, 12, 13, 11, 32, 43, 7, 8, 10, 14, 38, 35]) -and countless other publications).

The controlled invariance problem, as mentioned before, leads naturally to partial differential equations. However, there has been in the literature, quite understandably, a leaning towards seeking a purely algebraic approach to its solution. This has lead to the so-called controlled invariant subdistribution algorithm (See ([2, 20, 29]). This algorithm breaks down in the presence of singularities. There is yet another reason for attempting to solve these partial differential equations directly and not resorting to purely algebraic methods, namely the latter often fail to detect all solutions, especially those with certain desired properties (See ([38, 36])). Since, under very mild hypothesis, all solutions to the problem must necessarily solve these PDEs one can expect, by analysing the symmetry aspects of these PDEs, to uncover solutions with special properties.

The theory of point and generalised symmetries of differential equations is a venerable subject which has enjoyed, in the recent past, a belated resurgence (See ([33, 5]). In particular the possibility of detecting the so-called group invariant solutions via group theoretic methods is a very important reason for undertaking such a study. Not only are group invariant solutions easier to come by, but they also have, in engineering problems, tremendous practical value. Furthermore, if one is forced to deal with singularities, the presence of symmetry is often of use. If the symmetry and singularity interact in a suitable manner one can find solutions to a system of differential equations in spite of the singularity. In most cases the the presence of any symmetry in the problem gives rise to, at the very least, a microscoping effect inasmuch as the singularity becomes "lower dimensional" and thereby more benign. We will see an instance of this in the sequel.

The rest of the paper is organised as follows. We first set up the controlled invariance problem by motivating it via the so-called disturbance decoupling problem. The aforementioned PDEs are also introduced in the same section. In the next section we examine the symmetry aspects of these PDEs -in particular, the existence of group invariant solutions. Finally, we discuss the solvability of the controlled invariance problem in the presence of singularities. Most of these results have appeared or will appear elsewhere and hence we will not, for reasons of space, present the proofs in this paper. The only assertions which will be proved here are those which, are perhaps well known, but to the best of our knowledge have not been explicitly demonstrated anywhere else. The bibliography does not follow any particular order. The author would like to apologise to the reader for any inconvenience this may give rise to.

2 Notations and Convention

Unless explicitly stated to the contrary: 1) The word distribution will be used in the differential geometric sense of that word, i.e. to mean a vector field system (see ([1, 6])); 2) the distribution which one seeks to render controlled invariant will always be involutive, but will in general, have singularities; 3) The distribution of 2) above will be denoted Δ. It is assumed to be locally finitely generated, i.e. is the span (over either the smooth functions or analytic functions defined on a small open set about the reference point) of a set of d vector fields $X_1, \ldots, X_d$; 4) The only smoothness categories we will be concerned with are i) $C^\omega(U)$, or ii) $C^\infty(U)$; 5) All existence results are local, i.e. are valid only on some small open set containing the reference point, whose size, we will arbitrarily shrink to our convenience.

3 The Controlled Invariance Problem and Its Applications

Consider the following affine nonlinear system:

$$\dot{x} = f(x) + \sum_{i=1}^{p} g_i(x)u_i \tag{3.1}$$

Here, for simplicity of exposition we assume that the state x lies in some open set $U \subset R^n$ containing the reference point which we assume is the origin. The controls u_i are functions, explicitly or implicitly, of time. The vector field representing the drift is $f(x)$ and the $g_i(x)$ denote the control couplings. All the vector fields are assumed to be C^r, where r is either ∞ or ω. The controlled invariance problem consists of finding a feedback control law of the form

$$u(x) = \alpha(x) + \beta(x)v_i \tag{3.2}$$

which will cause the closed loop to leave $\Delta = \mathrm{span}\{X_1, \ldots, X_d\}$, a given distribution. The span of the vector fields $X_i, i = 1, \ldots, d$ is over either $C^\omega(U)$ or $C^\infty(U)$. Here $\alpha(x)$ is a vector of functions of dimension p, β is a $p \times p$ matrix of functions and v represents the new inputs to be used for future control action. One would prefer that the feedback functions , namely the components of α and the matrix β be of the same degree of smoothness as the vector fields $f(x)$ and $g_i(x)$. In addition, one desires that $\beta(x)$ be nonsingular in some neighborhood of the reference point, so that one has as much control as one started with.

To motivate the need for solving the controlled invariance problem, let us discuss its archtypal application, the disturbance decoupling problem. We will follow the excellent treatment in ([20]). The disturbance decoupling problem consists of the following data:

$$\dot{x} = f(x) + \sum_{i=1}^{p} g_i(x)u_i + \sum_{k=1} qp_i(x)w_i, \, y = h(x) \tag{3.3}$$

In (3.3), the f, g_i and u_i are as in (3.1). The new ingredients , w_k consist of a set of deleterious inputs ¿from which we wish to insulate the outputs $h(x)$ of the system, and $p_k(x)$ are also vector fields describing how the deterministic disturbances w_k influence the evolution of the system. The control problem, of course, is to find a feedback law of the form (3.2) which will cause all possible $w_k(t)$ (so long as they are admissible, i.e.the differential equation with these choices of the w's has unique trajectories for every initial condition in U) to have absolutely no effect on the outputs $y = h(x)$ *for all times.*

The solution consists of finding a distribution Δ contained in the distribution *ker dh* and which also contains the p_k's and then solve the controlled invariance problem for this distribution.

Why does this work? First of all notice that, by virtue of the Jacobi identity for the Lie bracket of vector fields, we may as well assume that the distribution Δ is involutive. Indeed, if Δ is going to be left invariant by some vector field, then so is its involutive closure(See ([20])). Now if, in addition, this distribution is either analytic or nonsingular then global versions of the Frobenius theorem say that Δ will give rise to a foliation(See ([15, 27, 41, 42])). It is well known that if some vector field leaves Δ invariant, then this vector field's flow locally preserves the foliation(See ([20])) . In symbols, if x_1 and x_2 belong to the same leaf then so do $\theta(s, x_i), i = 1, 2$ for all sufficiently small s (all s, if the given vector field is complete).

We now assume that $[f, \Delta] \subseteq \Delta$ and $[g_i, \Delta] \subseteq \Delta$ for $1 = 1, \ldots, p$. We will now present an argument, inspired by that in (([14])), which will show why this leads to a resolution of the disturbance decoupling problem. We need to show that the responses of the systems (3.1) and (3.3) lie, for each t, in the same leaf of Δ for every admissible choice for the pair u_i and w_k. This works because the leaves of Δ are contained in the level sets of the output. Hence, if either all vector fields of the control system , with or without disturbances, are complete or if the output is analytic the output will be impervious to the disturbances. Since admissible controls can be approximated by a concatenation of constant inputs we will demonstrate the above property only for such inputs. The result will, for these inputs, follow almost immediately from the following lemma (see ([16])):

Lemma 3.1 *The solution to the differential equation, with IC x_0, :*

$$\dot{x} = X_1(x) + X_2(x) \tag{3.4}$$

can be written as $\phi^1(t, \phi^2(t, x_0))$ *, where* ϕ^1 *is* X_1*'s flow, if and only if* $\phi^2(., x_0)$ *solves the initial value problem :*

$$\dot{x} = \sum_{n=0} \frac{(-t)^n}{n!}(ad^n X_1, X_2), x(0) = x_0$$

To finish the "proof" we note that for a given pair of constant inputs, u and w, the vector field $f + \sum_{i=1}^{p} u_i g_i$, which is to be the X_1 of 3.4, brackets back into Δ, whereas $\sum_{k=p}^{q} p_k w_k$, which is going to be the X_2 of 3.4, takes values in Δ. Thus $\phi^2(.,.)$ is always tangent to the leaves of Δ, and the flow $\phi^1(.,.)$ locally preserves the leaves. This yields the desired result.

We now turn to the controlled invariance problem per se. In presenting the formulation of the PDEs which need to be resolved and the necessary and sufficient conditions for finding a solution to the same, we will modify *the lucid presentation of the topic in ([20])* to accommodate singularities. The first hypothesis we impose on the data is referred to as weak(f, g)-invariance. It is also alluded to as local controlled invariance. In symbols it consists of the following two equations:

$$[f, X_i] = V_i + \sum_{j=1}^{p} c_j^i(x)g_j, i = 1, \ldots, d \tag{3.5}$$

$$[g_j, X_i] = W_{ij} + \sum_{k=1}^{p} b_{jk}^i(x)g_j, i = 1, \ldots, d, j = 1, \ldots, p \tag{3.6}$$

where $V_i, W_{ij} \in \Delta.$, and the $c_j^i(x), b_{jk}^i(x)$ are functions defined on all of U and are of the same degree of smoothness as the control system. Under certain regularity hypotheses weak (f, g)-invariance can be cast in a coordinate free fashion. However, since even in that case, one needs the functions c_j^i and b_{jk}^i to formulate the PDEs, we will stick to terra firma and content ourselves with the above description.

The next hypothesis is that G, the span of the vector fields g_k's has maximal rank p on a dense subset of U, and that the distributions G and Δ intersect in the zero section alone on a dense subset of U. In the presence of regularity assumptions these two conditions can be replaced by the weaker condition that $(G + \Delta)/\Delta$ is nonsingular on all of U. We remark that of the two hypotheses we have made so far the former is almost necessary, in that if a nonsingular feedback $\beta(x)$ is required then weak(f, g)-invariance becomes necessary. The latter hypothesis is not necessary. See ([14, 10]) for a solution to the controlled invariance problem without this stipulation. However the solutions presented in these references require nonsingularity of Δ.

Under these assumptions the requisite feedback may be found by solving the following equations:

$$X_i(\beta_{jk}) = \sum_{r=1}^{p} b_{rk}^i \beta_{jr}, j, k = 1, \ldots, p, i = 1, \ldots, d \qquad (3.7)$$

If (3.7) can be solved the vector fields g_k will bracket the distribution Δ into Δ. This is alluded to as input-insensitivization. The controlled invariance problem was first studied, under the assumption that Δ is input-insensitive, in ([17]). The terminology "input-insensitive" is due to ([28]). Note that the equations above are an overdetermined system of first order PDEs . Observe, also, that (3.7) consists of p sets of equations which are separate, i.e. do not interact with each other - one for each row of β. Once (3.7) has been solved the feedback α can be found by the following sets of overdetermined, first order, linear PDEs:

$$X_i \alpha_j = c^i - j(x), i = 1, \ldots, d, j = 1, \ldots, p \qquad (3.8)$$

We note that if $p = 1$ the equations (3.7) can be cast into the same form as (3.8) via logarithmic derivatives. The necessary integrability conditions for both systems of PDEs , (3.8) and (3.7), follow from the hypotheses we made above. The argument (which is an elementary application of the Jacobi identity for the Lie bracket) in ([20]) can be adapted to show that the integrability conditions hold on a dense subset, and hence, everywhere on U.

Now, if in addition Δ is nonsingular on U, a solution to either system can, in fact, be found. We will outline the argument for the system (3.8) (See ([20])). Define an extended distribution, $\hat{\Delta}$, on the space of dependent and variables by defining this distribution to be the span of $\hat{X}_i, i = 1, \ldots, d$, where:

$$\hat{X}_i = X_i + c^i(x)\frac{\partial}{\partial \alpha}$$

(the subscript on the c's have been omitted, since 3.8 consists of p *separate* equations).

The involutivity of $\hat{\Delta}$ is a consequence of the involutivity of the distribution Δ and the integrability conditions. The nonsingularity of $\hat{\Delta}$ is a consequence of that of Δ. Consequently,by the theorem of Frobenius, there passes, through each point of this space, a unique integral manifold of $\hat{\Delta}$. This integral manifold may even be written as the common level set of some functions The implicit function theorem is then invoked to write this level set as the graph of a function- which is, then, the solution of (3.8). Conversely, any solution $\hat{\alpha}(x)$ to (3.8) determines a first integral of the distribution $\hat{\Delta}$. In fact, the first integral is nothing other than the function:

$$\alpha - \hat{\alpha}(x) \tag{3.9}$$

Whilst, the converse of the prescription above works for distributions with singularities as well, the forward and the constructive direction need not. It is true that there is a version of the global formulation of Frobenius' theorem(see, for instance, ([15, 27, 42, 41])) under the hypothesis on Δ that we have already made. However, the remaining steps of this construction need not go through (especially the final step).

4 Symmetry Aspects of the Controlled Invariance Problem

Let us suppose that there is a smooth action of a connected Lie group H on U. We will denote the diffeomorphism of U corresponding to $h \in H$ by $\phi_h(x)$. Furthermore, let $W_i, i = 1, \ldots, m$, $m = dim\ H$ be the infinitesimal generators of this group action, and Δ_H, the distribution spanned by these infinitesimal generators. We refer the reader to ([33]) for these definitions. One would like to know if this information can be used to any end in the task of solving the equations (3.7) and (3.8). We will follow ([33]) for all appropriate definitions and notions. The notion most appropriate for the description of the interaction of the Lie group action and the task of resolving a system of differential equations is that of a symmetry group.

Definition 4.1 *H is a symmetry group for (3.8)(resp (3.7)) if, whenever $u = u(x)$ solves 3.8(resp 3.7, so does $u^h(x)$, for all the $h \in H$ for which it is defined. Here, $u^h(x)$ is defined to be $u(\phi_h(x))$.*

As an immediate application of this notion we will prove a simple result which gives a simple criterion for H to be a symmetry group for a system of PDEs whose solutions are the first integrals of a given foliation. When one bears in mind the discussion of the previous section, especially the equation 3.9, the relevance of this result becomes apparent.

Proposition 4.2 *Let $\hat{X}_i, i = 1, \ldots, d$ be a basis for a smooth distribution $\hat{\Delta}$. If H leaves this distribution invariant then H is a symmetry group of the system of equations :*

$$\hat{X}_i \alpha = 0, i = 1, \ldots, d$$

Proof: Let $\alpha(x)$ be a solution of the system of equations in the statement of the proposition. Consider the functions $W_k \alpha(x)$, for $k = 1, \ldots, m$. The connectedness of H implies that $[W_k, X_i] \in \hat{\Delta}$, for all $i = 1, \ldots, d$, $k = 1, \ldots, m$. Hence $[W_k, X_i]\alpha(x) = 0$. Hence $X_i(W_k\alpha(x)) = 0$ for all i and k. Thus for each k, $W_k\alpha$ also solves the above system of equations. The same argument shows that all higher order, iterated Lie derivatives of the solution $\alpha(x)$ along all the infinitesimal generators also solve the above system of equations. This proves that $\alpha^h(x)$, by virtue of its Lie series expansion in terms of the $W_k \alpha(x)$ (see ([33])), is also a solution of this system of equations. This yields the requisite result.

Remark 4.1 *Observe that we assumed neither the involutivity nor the non-singularity of $\hat{\Delta}$ in the proposition above. This is because the definition of a symmetry group is not concerned with the existence of a solution.*

Before proceeding with our interest in this notion it must be emphasised that we have restricted ourselves to group actions only on the space of independent variables. One can, in general, let H act on the space of dependent variables and even its derivatives and gain much out of that (see ([33, 5])).

If we take the circle of ideas in the previous section culminating in (3.9), and Proposition (4.2) in conjunction we arrive at the following criteria for H to be a symmetry group for (3.8) (a similar result can be obtained for (3.7)).

Lemma 4.3 *Assume Δ is H-invariant. Then H is a symmetry group for (3.8) if the following holds:*

$$W_k c_j^i = \sum_{n=1}^{d} t_{ijkn}(x) W_k c_j^n$$

Here the $t_{ijkn}(x)$ are determined by the equation

$$[W_k, X_i] = \sum_{n=1}^{d} t_{ijkn}(x) X_n$$

which expresses the H -invariance of Δ.

A sufficient condition for this to happen is given by:

Theorem 4.4 *The equations (3.8) admit H as a symmetry group if: i) Δ is an H-invariant distribution which is also input-insensitive, ii) $f(x)$ is H-invariant mod Δ, which just means that $[f, W_k] \in \Delta$ for all k, and iii) g_j 's for $j = 1, \ldots, p$ are also H-invariant mod Δ.*

This result was proved in ([38]) via a prolongation formula approach (see ([33])). The advantage of a prolongation approach is that it is also gives a necessary condition under local solvability assumptions(see ([33])). The same result for (3.7) holds under the same hypotheses in the theorem above, except that one does not need ii).

We now address a slightly different problem. In view of the theorem above one would like to know if it is possible to *simultaneously* induce Δ to be input-insensitive and the control vector fields to be group invariant mod Δ. One can view a part of this problem as *symmetrization by feedback*. The symmetry here , of course, is the one relevant to the controlled invariance problem. An affirmative answer to this is contained in(see ([38])):

Theorem 4.5 *Let us assume that the distributions Δ and $\Delta + G$ are both H-invariant. Suppose Δ and Δ_H res nonsingular and so is the distribution $\Delta \cap \Delta_H$. Under these hypotheses it is possible to render (via nonsingular feedback) Δ input-insensitive and the control fields H-invariant mod Δ.*

We remark that the proof of this last result also involves solving overdetermined systems of PDEs. The philosophical import of the "mod Δ" invariance assumptions in the penultimate theorem is that in the controlled invariance problem one is really impervious to what happens within Δ.

Group Invariant Solutions: A group invariant solution to a system of differential equations is a solution which is constants on the orbits of the action of the group. In order for a system of differential equations to possess H-invariant solutions, it is not necessary that the differential equations have H as a symmetry group. However, things are best when that does happen. Typically H-invariant solutions are obtained by projecting the entire system onto the quotient manifolds U/H, provided this quotient exists. A sufficient condition for the existence of a quotient manifold is that H act freely and properly on H (see ([6, 1]). There are several results on the existence of a quotient in the literature. However, one has to bear the purpose for seeking a quotient in mind, before one eagerly invokes these results. Our primary purpose is that we wish to do analysis on the quotient, specifically, *solve differential equations on the quotient*. To the best of the author's knowledge the aforementioned condition is the only one which suffices. It must be pointed out that there are some results on solving differential equations on spaces with corners(whatever that may mean), but we are not aware of any result which will give the quotient such a structure. Even, if this be the case one has to make sense of projecting the system of differential equations onto the quotient.

We will now state our main result on H-invariant solutions to the equations (3.8). A similar result holds for (3.7). We will not state this result in its fullest generality because the statement of the theorem then appears quite verbose. *The important thing to bear in mind is that H-invariant solutions are going to be found by projecting all data onto the quotient manifold U/H, and the hypotheses stated below are only one of many situations which ensure that all data project and that the quotient system of differential equations possesses a non-trivial solution. For example, nonsingularity of Δ is not always required.* In the next section we will present results on the resolution of (3.8) and (3.7) in the presence of singularities. All of this, of course, can be applied to the quotient system. Furthermore, even if Δ is nonsingular, there is no guarantee that the quotient system will be nonsingular as well. Once again this does not, in itself, preclude the existence of a group invariant feedback. . We will leave the burden of putting the results of this section with those of the next to the reader.

Theorem 4.6 *Suppose: a)* Δ *is* H-*invariant, nonsingular and possesses an* H-*invariant basis, denoted* $X_i, 1 = 1, \ldots, d$, *b) The* $g_j, j = 1, \ldots, p$ *are* H-*invariant mod* Δ *and that* Δ *is input-insensitive, c) The* $[f, X_i], i = 1, \ldots, d$ *are* H-*invariant mod* Δ, *d)* $\Delta \cap \Delta_H$ *is nonsingular and* f *is* H-*conditioned invariant, i.e.* $[f, \Delta \cap \Delta_H] \in \Delta$, *and e)* U/H *is a smooth manifold and the canonical projection of the manifold* U *onto* U/H *is a smooth submersion. Then the system of partial differential equations (3.8) has an* H-*invariant solution.*

A proof of this result may be found in ([38]). With regard to hypotheses a) of Theorem (4.6) one may wonder if the existence of an H-invariant basis for Δ is always a coordinate-free notion. To that end, note that the condition H-invariance of Δ implies, because of its involutivity, that the distribution $\Delta + \Delta_H$ is also involutive. This is easy to verify. Under regularity hypotheses a coordinate system and a basis change was produced in ([39]), in which Δ and Δ_H had basis which commuted with each other. Actually the result in ([39]) is not concerned with group actions, but has more to do with obtaining a simultaneous Frobenius chart for many involutive distributions, satisfying certain inclusion and regularity hypotheses. We do not want to use this result to obtain an H-invariant basis for an H-invariant Δ, since we wish to preserve the structure of the infinitesimal generators. Recall that the infinitesimal generators are not only the basis of Δ_H but also constitute a *finite- dimensional* Lie algebra. There is a way to circumvent this problem. In ([36]) it is shown that *if* H'*s action is free and has a slice then there is indeed an* H-*invariant basis for* Δ *on some open subset of* U. This result makes no use of either *the nonsingularity or the involutivity of* Δ. All that is required of Δ is that it be smooth and H-invariant. There exist certain non-free actions which give rise to H-invariant basis for Δ too(see ([35])).

The terminology H-conditioned invariance has been adapted from the better known use of that word in the case when $\Delta_H = ker(dh)$, where $h(x)$ represents the output or the measurements made on the control system (3.1) (. Finally note that condition c) in Theorem (4.6) is much weaker than demanding that f also be H-invariant mod Δ. In fact this weaker condition turned out to be quite essential in a rigid body example analysed in ([38]), taken ¿from ([30]), where this example was first presented, but without any reference to the theory of symmetry groups of partial differential equations.

Uses of H-invariant Solutions: Group invariant solutions are extremely important. For a very readable account of their importance we can do no better than place the reader in the hands of the eminently well written manuscript ([33]). Not only are they easier to come by (in view of the reduction of the number of variables required for their search) but, as mentioned before, they have great practical relevance. As an example conditions for the existence of a feedback which achieves controlled invariance but only uses the measurements can be obtained from this angle. This works even for distribution with singularities (see ([36]). Similar, and more extensive, results (under some regularity hypotheses) were obtained in ([31]) for general nonlinear systems. Along the same lines one can seek conditions for finding feedback for the controlled invariance problem which does not depend on the constant but unknown parameters in the system (see ([36]). Group invariant systems are also handy in obtaining global (i.e. on all of U) solutions. Indeed suppose that U is simply connected, then so is U/H. This is because H is connected (and the long exact sequence for a Serre fibration, if one wishes to be fancy!). Therefore, if in addition, $\Delta \cap \Delta_H$ is 0 and $dim\ U\ =\ dim\ (\Delta)\ +\ dim\ H$ then the quotient system becomes an overdetermined system which seeks the integral of a closed one -form, which is exact because of the simple connectivity of U/H, and such a system has, of course, a global solution. In the more general case the only result that we are aware of is a construction suggested to us by Dr.Dayawansa which is based on a symmetry analysis of his excellent paper ([12]). A very obvious, but nevertheless important, observation is that if f and the g_j's are already H-invariant than they will continue to be so upon the imposition of group invariant feedback, which is usually of great value for future synthesis and design problems. Also note that in seeking H-invariant feedback we do not have to avail of Frobenius' theorem (for distributions which are nonsingular) right from the very start. In fact, the correct coordinates (which is the Frobenius for Δ_H) , typically suggests itself for the more well known group actions. This last point brings out yet another important matter. Most of the interesting actions do have fixed points. Therefore group invariant solutions are typically sought for away ¿from these fixed points (Δ_H has maximal rank on an open and dense subset). If this solution is to be extended across the fixed points then one is confronted with two problems. First, coordinates found on the quotient, away from the fixed points, may themselves not extend across the fixed point. for example all scaling actions(see ([33, 5]) belong to this category. However, the rotation group and in fact all compact. linear Lie group actions have invariants which can only be smooth functions of a polynomial . So the only question is whether this function itself extends across the fixed points. We will illustrate this via an example shortly. Finally, note that, as observed before, the quotient system may have singularities because the original system, before reduction was performed, did. The quotient system though singular has fewer variables to be dealt with. Thus group reduction has a microscoping effect on the singular behavior. In fact one can write examples where all this happens even for non-free actions. Let us now look at a simple-minded example where the presence of symmetry yields a solution despite the singularity in the data.

Example Let the plant be given by:

$$\dot{x} = f(x) + ug(x)$$

where $f(x)$ is any vector field which vanishes at the origin and $g(x)$ is the vector field which generates the action of the rotation group on the plane. In coordinates it is given by

$$g(x) \stackrel{def}{=} R = x_2 \frac{\partial}{\partial x_1} - x_1 \frac{\partial}{\partial x_2}$$

Let Δ be the distribution spanned by the vector field which generates the scaling action on the plane which we will denote by S. In coordinates Δ is the span of:

$$S(x) = x_1 \frac{\partial}{\partial x_1} + x_2 \frac{\partial}{\partial x_2}$$

Δ is involutive and becomes singular at the origin. Since the R and S commute, it is also input-insensitive. Furthermore, since R and S span the distribution on the plane which is two dimensional everywhere except at the origin where it is zero dimensional, it follows that $f(x)$ admits a representation of the form $f(x) = a(x)S + B(x)R$.

$$[f, S] = -(L_S a(x))S - (L_S b(x))R$$

Let us examine if the vector field $[f, S]$ is $SO(2)$ invariant mod Δ. Since R is the infinitesimal generator of $SO(2)$'s action on the plane, we compute:

$$[[f, S], R] = (L_R L_S a(x))S + (L_R L_S b(x))R$$

By virtue of the fact that R and S are generically linearly independent it is clear that if the above vector field is going to lie in Δ then the second summand on the right hand side of the above equation must vanish, i.e.

$$L_R L_S b(x) = 0$$

Thus the function $L_S b(x)$ is itself rotation invariant. Note that this function is exactly $c^1(x)$. To find the requisite feedback we have to solve the equation $L_S a = c^1(x)$. To do this we switch to polar coordinates. Denote, by ρ^2, the function $x_1^2 + x_2^2$. Since c^1, as we saw above, is also $SO(2)$ invariant it must be of the form $c(\rho^2)$, where c is some smooth(resp analytic) function of its argument. The vector field S has the coordinate representation $\rho\frac{\partial}{\partial\rho}$, and hence the partial differential equation$S\alpha = c^1(x)$ is transformed into an ordinary differential equation with regular singularities, to wit:

$$\rho\frac{d\alpha}{d\rho} = c(\rho^2)$$

which can be easily solved. Note that in this example the Lie group H $(SO(2))$ has a fixed point at the origin, and this does not cause any problem even at the origin. If, on the other hand, H had been $R^+ - \{0\}$ acting via scaling transformations we would not have been able to find a solution which works at the origin.

Remark 4.2 *Reduction is not the only mechanism via which group invariant solutions exist. One can show, for instance, that under suitable assumptions on the invariance properties of the data the Hamilton-Jacobi Bellman equation has H-invariant solutions for Lie groups H acting linearly (see ([37]). As a matter of fact, it is precisely the fixed point which produces the H-invariant solution.*

5   Controlled Invariance in the Presence of Singularities

We have already encountered examples where one could successfully extend the solutions to 3.8 across the singularities of the given distribution Δ. In what follows we will describe some situations where there are concrete theoretical results on the existence (and computation) of solutions to the controlled invariance problem for *locally finitely generated* distributions with varying rank. Of course, one can significantly enhance these results by combining them with results on symmetry groups, the blowing-up construction ([7]), and perhaps results on extension of analytic functions past sets which are " thin" in some sense of the word. *Whilst we do not have any concrete results, just a few examples, on the last mentioned approach -it should be stressed that existence of a solution is really a question of being able to extend solutions, in as smooth a manner as possible, past the singularities.* Our first result is about a canonical basis for a locally finitely generated distribution with varying rank. The proof of this theorem, which appears in ([35]). can itself be construed as rendering a *nonsingular distribution, but with singularities in the control,* controlled invariant, as was suggested to us by Dr.W.P. Dayawansa (see [10, 14, 34]). We remark that this is not the only possible choice for a basis, but for the problem at hand, is the one which is most expedient. The theorem asserts that a distribution which is generated, locally, by d vector fields (d can conceivably be bigger than n) and which has rank d_0 (d_0 is no bigger than either d or n) at the reference point, has a local basis, in a particular coordinate system, consisting of d_0 *coordinate vector fields* and $d - d_0$ other vector fields tangent to R^{n-d_0}, all of which vanish at the origin of R^{n-d_0}. The proof of this theorem is "quite" constructive.

Let us call the equations in (3.8) (resp (3.7)), corresponding to the coordinate vector fields in the above basis the coordinate system, and the remaining equations the degenerate system. The next set of results assert that if we are able to obtain a solution to the degenerate system for a particular value (appropriately chosen) of the first d_0 coordinates, of (3.8) (resp. (3.7)), one can then solve the entire system (3.8) (resp. (3.7)) (see [34]). Hence the problem boils down to one's ability to solve the degenerate system.

We will first state our results for the solvability (amongst classical solutions) of a degenerate system of the type (3.8) and (3.7) and then put forth sufficient conditions, in control theoretic terms, for the hypotheses postulated in these results to hold good. We will assume, for simplicity, that $d_0 = 0$. However all the results below go through even if $d_0 \neq 0$, with appropriately modified hypotheses. As a matter of fact, they go through under situations which are, control-theoretically speaking, less demanding.

Notation: Let $Y_i, i = 1, \ldots, d$ be the linear parts of the vector fields $X_i, i = 1, \ldots, d$. We denote, by $\hat{\Delta}$, the Lie algebra spanned by the Y_i's. Note that this is a finite-dimensional Lie algebra. If the $g_i, i = 1, \ldots, p$ vanish at the reference point, we denote, by $\hat{G}$, the span over the real numbers of the linear parts of the control vector fields. If $f(x)$ vanishes at the origin we denote , by $\hat{f}(x)$, its linearization. Note that most of these objects depend on the choice of coordinates. However, the properties of these constructs that we shall invoke in the sequel, do not.

Proposition 5.1 *Consider a degenerate system of the kind (3.8), where all data are assumed to be smooth. Assume: a) There is one vector field in Δ, say X_1, whose equilibrium point at the origin is either stable or anti-stable and whose linearization , denoted Y_1, lies in the center of the Lie algebra $\hat{\Delta}$, b) The functions $c_j^i(x)$ vanish at the origin for $i = 1, \ldots, d$ and $j = 1, \ldots, p$, c) the dimension of $\hat{\Delta}$, as a vector subspace of the tangent space to a point in U, is d on a dense subset of points in U.*

Then there is a smooth solution, in some neighbourhood of the reference point, to the system of partial differential equations (3.8).

It must be noted that the result above does not guarantee that there exists an analytic solution even if everything in the system of equations is analytic. For analyticity one needs the eigenvalues of X_1 to satisfy certain Poincare-Siegel type estimates(see ([4]).

Proposition 5.2 *Consider a degenerate system of the type 3.7 with all data analytic. Assume that condition c) of 5.1 holds. Assume also that condition a) of the preceding proposition holds with the extra requirement that the eigenvalues of Y_1 satisfy a Poincare-Siegel estimate. In addition, assume: b') $C^i_{jk}(0) = 0$, for all $1 = 1, \ldots, d$ and $j, k = 1, \ldots, p$. Then one can find an analytic solution, in some neighbourhood of the reference point, to the system of partial differential equations (3.7).*

We believe that the above result can be extended to both when the eigenvalues do not satisfy a Poincare-Siegel estimate (to yield a smooth solution) and when all data are only smooth. A result of this kind is being explored currently.

These results were proved in ([35, 34]), and they draw upon some work of ([4, 18]). Once again, we have not presented the results in their full generality. For more details on this and the connection of our work to the nonlinear regulation problem ([9, 19, 24]) see ([34]). We now turn to the control theoretic characterization of the hypotheses used in the penultimate two propositions. If, as we have assumed here, $d_0 = 0$ an elementary calculation shows that the presence of even one hyperbolic vector field in the distribution and the requirement that this distribution be input-insensitive forces $G(0) = 0$. Hence, the need for for the object $\hat{G}$. A sufficient condition, now, for condition b) of Proposition (5.1) is that i) $f(0) = 0$, ii) $\hat{\Delta} \cap \hat{G} = 0$ on a dense subset, iii) $\hat{\Delta}$ and $\hat{G}$ have maximal rank on a dense subset, and that iv) $[\hat{f}, \hat{\Delta}] \in \hat{\Delta}$. As for the condition b') of 5.2, a sufficient condition for it to be true, is that i) $G(0) = 0$, ii) conditions ii) and iii) of the preceding paragraph hold, and iii) the linearizations of the control vector fields leave $\hat{\Delta}$ invariant under the operation of Lie bracketing.

The proofs of these sufficient conditions may be found in ([35, 34]). We also refer the reader to the papers ([25, 40]) for ideas which are related to those above. Note that even if Δ is not input-insensitive the sufficient conditions above require $G(0)$ to be 0. On the other hand if it is input-insensitive then (almost) necessarily $G(0) = 0$, as observed above. Thus, this may be viewed as the devil's alternative - the resolution of the controlled invariance problem for singular distributions requires singularities in the control vector fields as well. Of course, if $d_0 \neq 0$ singularities in the control vector fields are no longer necessary. Note that the condition $C^i_{jk}(0) = 0$ could well be necessary both for the existence of a solution to even a determined (as opposed to overdetermined system) system of the type (3.7). , and if existence is guaranteed, for being able to prescribe the initial (i.e. at the origin) value of the solution. In ([34]) examples are presented, of vector fields X, which have the property that the only solution to $X\beta = C\beta$, where β is a *scalar* unknown and C is a non-zero scalar, is the trivial solution. The second phenomenon ought to be self-explanatory.

Finally, we note that having nontrivial first integrals is an impediment, not only to the existence of a solution to an overdetermined system, but is also an obstruction to the existence of even generalised (though not multi-valued) solution to a single equation of the type (3.8). Loosely speaking, the transpose of a differential operator, such as the vector field which generates the rotation group's action on the plane, is again an operator of the same type. Since this operator is not injective, there is no hope that the former would be surjective or even have closed range (see ([34]). Thus solutions exist only in specific situations.

6 Conclusion

First order, overdetermined systems of partial differential equations are of importance in control theory, as, we hope, this article demonstrates. There are several problems of interest in the topic.One such problem, which to the best of the author's knowledge, has received no attention is seeking necessary and sufficient conditions for the solvability of the controlled invariance problem for codistributions. As is well known, the controlled invariance problem for distributions , in applications, arises from actually dualising a related problem for codistributions. The main difficulty lies in obtaining an analog of weak (f, g)-invariance. Some very preliminary work shows that , under restrictive hypotheses, a suitable necessary condition is that a certain one-form be closed. However, the benefits of being able to arrive at such a theory (the ability to analyse robustness of solutions, singularities in solutions as compared to singularities in the data - to name but a few) warrant such an undertaking.

References

[1] R. ABRAHAM, J. MARSDEN and T. RATIU, *Manifolds, Tensor Analysis and Applications*, Springer-Verlag, New York, 1983.

[2] S.P. BANKS, *Mathematical Theories of Nonlinear Systems*, Prentice Hall, 1989.

[3] A. BEN ARTZI and J. HELTON, preprint.

[4] G. BENGEL and R. GERARD, "Formal and Convergent Solutions to Singular Partial Differential Equations," *Manuscripta Mathematica*, v. 38, 1982, pp. 343-373.

[5] BLUMAN and KUMEI, *Symmetry & Differential Equations*, Springer-Verlag, New York, 1989.

[6] W. BOOTHBY, *An Introduction to Differentiable Manifolds and Riemannian Geometry*, Academic Press, 1975.

[7] C. BYRNES, "Towards a global theory of (f,g)-invariant distributions with singularities," *Proc MTNS*, Beer-Sheva, Israel, Springer, 1983.

[8] C. BYRNES, "Feedback Decoupling of Rotational Disturbances for Spherically Constrained Systems," *Proc 23rd IEEE CDC*, Las Vegas, 1984, pp. 421-426.

[9] C. BYRNES and A. ISIDORI, "Output Regulation of Nonlinear Systems," *IEEE Trans Aut Control*, v. 35, 1990, pp. 131-140.

[10] D. CHENG and T. TARN, "New Result on (f,g)-invariance," *System-Control Letters*, Dec 1989, pp. 319-326.

[11] W.P. DAYAWANSA, "On the Existence of a Homogeneous Solution to the Disturbance Decoupling Problem for Homogeneous Systems," *Proc ECC*, Grenoble, July 1991.

[12] W.P. DAYAWANSA, D. CHENG, W. BOOTHBY and T.TARN, "Global (f,g)-invariance of Nonlinear Systems," *SIAM Journal of Control and Optimization*, v. 26, No. 5, 1988, pp. 1119-1132.

[13] W.P. DAYAWANSA and C.F. MARTIN, "Disturbance Decoupling, (f,g)-invariance and Controllability Subspaces for a Class of Homogeneous Polynomial Systems," *SIAM Journal of Control and Optimization*, v. 1, 1989, pp. 108-119.

[14] K. GRASSE, "Controlled Invariance for Fully Nonlinear Systems," to appear in *Int Journal of Control*.

[15] R. HERMANN, "The Differential Geometry of Foliations," *J. Math and Mech*, v. 11, 1966, pp. 302-316.

[16] H. HERMES, "Nilpotent and Higher Order Approximations of Vector Field Systems," *SIAM Review*, v. 33, 1991, pp. 238-264.

[17] R.M. HIRSCHORN, "(A, B) Invariant Distributions and the Disturbance Decoupling for Non-linear Systems", *SIAM Journal of Control*, v. 19, 1981, pp. 1-19.

[18] L. HORMANDER, *Analysis of Linear Partial Differential Operators III*, Springer-Verlag, New York, 1985.

[19] J. HUANG and W. RUGH, "Stabilization on Zero-Error Manifolds and the Nonlinear Servomechanism Problem," *Proc 29th IEEE CDC*, Honolulu, 1990, pp. 1262-1267.

[20] A. ISIDORI, *Nonlinear Control Systems*, Springer-Verlag, New York, 1989.

[21] A. ISIDORI, A. KRENER, C. GORI-GEORGI and S. MONACO, "Nonlinear Decoupling via Feedback- a Differential Geometric Approach," *IEEE Trans Aut Control*, v. 26, 1981, pp. 331-345.

[22] A. ISIDORI, A. KRENER, C. GORI-GIORGI and S. MONACO, "Locally (f, g)-invariant Distributions," *System-Control Letters*, v. 1, 1981, pp. 331-345.

[23] A. ISIDORI and C. MOOG, "On the Equivalents of the Notion of Transmission Zeroes, in Modelling and Adaptive Control," Byrnes and Kurzhanskii eds, *Lecture Notes in Control and Information Sciences*, No 105, Springer-Verlag, New York.

[24] A. KRENER, "The Construction of Optimal Linear and Nonlinear Regulators," *Systems, Models and Feedback- Theory and Applications*, Birkhauser, Boston, 1992.

[25] E. LIVINGSTON and D. ELLIOTT, "Linearization of Families of Vector Fields," *Journal of Differential Equations*, v. 3, 1984, pp. 289-299.

[26] D. LUKES, "Optimal Regulation of Nonlinear Systems", *SIAM J of Control and Optimization*, v. 7, 1969, pp. 75-100.

[27] T. NAGANO, "Linear Differential Systems with Singularities and an Application to Transitive Lie Algebras," *J. Math Soc Japan*, v. 18, 1966, pp. 398-404.

[28] H. NIJMEIER, "Controlled Invariance for Affine Nonlinear Control Systems," *Int J Of Control*, v. 24, 1981, pp. 825-833.

[29] H. NIJMEIER and A. VAN DER SCHAFT, *Nonlinear Dynamical Control Systems*, Springer-Verlag, New York, 1991.

[30] H. NIJMEIER and A. VAN DER SCHAFT, "Controlled Invariance for Nonlinear Systems - Two Worked Examples," *IEEE Trans Aut Control*, v. 29, 1984, pp. 361-364.

[31] H. NIJMEIER and A. VAN DER SCHAFT, "Controlled Invariance by Static Output Feedback for Nonlinear Systems," *System-Control Letters*, v. 1, 1982, pp. 39-48.

[32] H. NIJMEIER and A. VAN DER SCHAFT, "Controlled Invariance for Nonlinear Systems," *IEEE Trans Aut Control*, v. 27, 1982, pp. 904-914.

[33] P. OLVER, *Applications of Lie Groups to Differential Equations*, GTM 107, Springer-Verlag, New York, 1986.

[34] V. RAMAKRISHNA, "On the Solvability of Overdetermined, Degenerate Systems of Partial Differential Equations - A Control Perspective," submitted for publication.

[35] V. RAMAKRISHNA, "Controlled Invariance for Singular Distributions," to appear in *SIAM Journal of Control*.

[36] V. RAMAKRISHNA, "Robustness in the Controlled Invariance Problem," submitted.

[37] V. RAMAKRISIINA, "Group Invariant Solutions to Equations of the Hamilton-Jacobi Type in the Presence of Fixed Points," under preparation.

[38] V. RAMAKRISHNA and H. SCHAETTLER, "Controlled Invariant and Group Invariant Distributions," *Journal of Mathematical Systems, Estimation and Control*, v. 1, No. 2, 1991, pp. 201-240.

[39] W. RESPONDEK, "On Decomposition of Nonlinear Systems," *System-Control Letters*, v. 1, 1982, pp. 301-309.

[40] J.L. SEDWICK and D. ELLIOTT, "Linearization of Analytic Vector Fields in the Transitive Case," *Journal of Differential Equations*, v. 25, No. 3, 1977, pp. 377-390.

[41] P. STEFAN, "Accessible Sets, Orbits and Foliations with Singularities," *Proc London Math Soc*, v. 29, 1974, pp. 699-713.

[42] H. SUSSMAN, "Orbits of Families of Vector Fields and the Integrability of Distributions," *Transactions AMS*, v. 180, 1973, pp. 171-190.

[43] W. M. WONHAM, *Linear Multivariable Control- A Geometric Approach*, Springer-Verlag, New York, 1985.

WHAT IS THE DISTANCE BETWEEN TWO AUTOREGRESSIVE SYSTEMS?

Joachim Rosenthal[*]
Department of Mathematics
University of Notre Dame
Notre Dame, IN 46556

Xiaochang Wang[†]
Department of Mathematics
Texas Tech University
Lubbock, TX 79409

1 Introduction

In the recent control literature there has been a great interest in controller design techniques which are robust with respect to plant perturbations. Crucial for all robustness studies is of course the availability of a "good metric" defined on the set of all plants. A metric is considered good if it can be easily computed and if it gives a measure of numerical robustness.

On the set of proper transfer functions several metrics have been proposed. Most prominently we want to mention the gap metric introduced by Zames and El-Sakkary [15], the graph metric introduced by Vidyasager [13] and the pointwise gap metric introduced recently by Qiu and Davison [7].

All those metrics have in common that they induce the same topology on the set of proper transfer functions with a fixed McMillan degree. For recent contributions to the subject of gap metric and graph metric we refer to [3, 7, 11] and in particular to the survey article of Glüsing-Lüerßen [4], where a comparative study between those metrics is provided.

A natural generalization of the set of proper transfer functions is the set of autoregressive systems. The main contribution of this short paper is a new metric on the space of autoregressive systems. Because those systems arc not so widely known and for the convenience of the reader we summarize in the following section the relevant notions. For simplicity we will develop the theory over the real numbers $\mathbb{R}$.

2 The set of autoregressive systems

Consider the space $(\mathbb{R}^q)^{\mathbb{R}}$ of all vector valued functions with domain the real line and range the vector space $\mathbb{R}^q$. $(\mathbb{R}^q)^{\mathbb{R}}$ has in a natural way the structure of a real vector space. Let $\mathcal{H} \subset (\mathbb{R}^q)^{\mathbb{R}}$ be a linear subspace. If the elements of $\mathcal{H}$ are differentiable and if $\mathcal{H}$ is invariant under the linear transformation $\frac{d}{dt}$ we will call $\mathcal{H}$ a *signal space*. (Compare with [2, page 76].) Spaces which are signal spaces are e.g. the set of polynomial functions, the set of smooth functions $C^{\infty}(\mathbb{R}, \mathbb{R}^q)$ or the set of vector valued distributions.

[*]Supported in part by NSF grant DMS-9201263.

[†]Supported in part by the Research Enhancement Fund from College of Art and Sciences, Texas Tech University.

In linear dynamical systems theory we are interested in particular subspaces of $\mathcal{H}$ which Willems [6, 14] calls the behavior of a system.

In order to explain this notion consider a $p \times q$ matrix $R(s)$ with entries in the polynomial ring $\mathbb{R}[s]$. The polynomial matrix $R(s)$ induces a system of autoregressive equations in the sense of [6, 14] given through:

$$R(\frac{d}{dt})w(t) = 0 \tag{2.1}$$

Note that (2.1) is a linear transformation on the signal space. Using the language of Willems [14] we call the kernel $\mathcal{B}$ of this transformation the *behavior* of the system.

Two $p \times q$ polynomial matrices $R(s)$ and $\tilde{R}(s)$ are called (row) equivalent if there is a unimodular matrix $U(s)$ with $\tilde{R}(s) = U(s)R(s)$. Clearly row equivalent matrices define the same behavior. On the other hand if the signal space is sufficiently rich, e.g. if $C^\infty(\mathbb{R}, \mathbb{R}^q) \subset \mathcal{H}$, then it is not hard to prove that polynomial matrices which are not row equivalent have also different behaviors. (Compare with [5, 14].) Motivated by those facts we define:

Definition 2.1 An equivalence class of full rank $p \times q$ polynomial matrices is called an autoregressive system.

The class of autoregressive systems generalizes the class of transfer functions in the following way: Consider a proper or improper $p \times m$ transfer function $G(s)$. Assume $G(s)$ has a left (polynomial) coprime factorization $D^{-1}(s)N(s) = G(s)$. If $\tilde{D}^{-1}(s)\tilde{N}(s) = G(s)$ is a second left coprime factorization then it is well known that the $p \times (m + p)$ polynomial matrices $(N(s)\ D(s))$ and $(\tilde{N}(s)\ \tilde{D}(s))$ are row equivalent. In other words $(N(s)\ D(s))$ defines an autoregressive system.

Finally we want to extend the notion of McMillan degree and the notion of observability indices to the class of autoregressive systems.

To do so recall the following well known Lemma, a proof of it can be found e.g. in [2, page 330].

Lemma 2.2 *Suppose given a $p \times q$ polynomial matrix $R(s)$. Then there exists a $p \times p$ unimodular matrix $U(s)$ and numbers $\nu_1 \geq \cdots \geq \nu_p$ only dependent on the equivalence class of $R(s)$ such that the i-th row degree of the matrix UP is given by $\partial_i(UP) = \nu_i$.*

Based on this Lemma we define the numbers $\nu = (\nu_1, \ldots, \nu_p)$ as the ordered observability indices (sometimes also called left Kronecker indices (compare with [10] or left Wiener Hopf indices (compare with [5])) of the autoregressive system $R(s)$. Finally we denote with the number $n := \sum_{i=1}^{p} \nu_i$ the McMillan degree of the system $R(s)$.

Above definitions naturally generalize the concept of McMillan degree and observability indices of a transfer function. For this consider a left coprime factorization $D^{-1}(s)N(s)$ of a proper or improper transfer function $G(s)$ with observability indices $\nu_1 \geq \cdots \geq \nu_p$ and McMillan degree n. Then it is well known that those indices correspond to the indices of $R(s) = (N(s)\,D(s))$ in the sense above. For more details concerning above relationships see e.g. [4, 10, 14] where also more references to the literature can be found.

3 Gap metric

In this section we first define the gap metric on the set of closed subspaces of a Hilbert space. This will then enable us to shortly review the notion of pointwise gap metric on the space of transfer functions. It will also serve as a preliminary for the next section, where we introduce a new metric on the space of autoregressive systems.

In the following let H be a real or complex Hilbert space with inner product $<,>$ and let $X, Y \subset H$ be two closed subspaces. Denote with P_X and P_Y the orthogonal projection on X respectively on Y. Then the gap between the subspaces X and Y is defined through

$$gap(X,Y) := \|P_X - P_Y\|. \tag{3.1}$$

Clearly $gap(X,Y) = gap(Y,X)$ and $gap(X,Y) = 0$ if and only if $P_X = P_Y$, i.e. $X = Y$. Finally the triangle inequality

$$gap(X,Z) \leq gap(X,Y) + gap(Y,Z) \tag{3.2}$$

follows directly from the corresponding triangle inequality of the corresponding operator norm. In short the function gap induces a metric on the set of closed subspaces of a Hilbert space H.

Define the angle between two nonzero vectors $x, y \in H$ through the requirement

$$\cos(x,y) := \frac{|<x,y>|}{\|x\| \cdot \|y\|}. \tag{3.3}$$

The following result shows that the gap between two lines X, Y in $\mathbb{R}^q$ corresponds to the sine of the angle between those lines. For similar results of this nature we refer to [1, 4, 11].

Lemma 3.1 *Assume* $x = (x_1,\ldots,x_q)^t$ *and* $y = (y_1,\ldots,y_q)^t$ *are two nonzero (column) vectors in* $\mathbb{R}^q$ *and* $X = \mathrm{span}(x)$ *and* $Y = \mathrm{span}(y)$. *Then one has*

$$gap(X,Y) = \sin(x,y) = \frac{\|x \wedge y\|}{\|x\| \cdot \|y\|}. \tag{3.4}$$

J. ROSENTHAL and X. WANG

Proof. The unique projection operator onto the line X is given through $P_X = ||x||^{-2}(x \cdot x^t)$. Similarly $P_Y = ||y||^{-2}(y \cdot y^t)$. In particular P_X is symmetric of rank one, $P_X^2 = P_X$ and the value of the trace is given by $\text{tr}(P_X) = 1$. Moreover the transformation $P_X - P_Y$ has rank 2 and zero trace. From this information follows that $P_X - P_Y$ has a zero eigenvalue of multiplicity $q - 2$, a positive eigenvalue λ_+ and a negative eigenvalue $\lambda_- = -\lambda_+$. Note that the norm of a finite operator is equal to the maximum eigenvalue, i.e. $||P_X - P_Y|| = \lambda_+$. Therefore one has:

$$\text{tr}((P_X - P_Y)^2) = 2\lambda_+^2 = 2||P_X - P_Y||^2. \tag{3.5}$$

The first equality now follows through

$$
\begin{aligned}
||P_X - P_Y||^2 &= \frac{1}{2}\text{tr}((P_X - P_Y)^2) & (3.6)\\
&= \frac{1}{2}(\text{tr}(P_X^2) + \text{tr}(P_Y^2) - 2\text{tr}(P_X P_Y)) & (3.7)\\
&= 1 - ||x||^{-2}||y||^{-2}\text{tr}(xx^t yy^t) & (3.8)\\
&= 1 - \frac{<x,y>^2}{||x||^2||y||^2} & (3.9)\\
&= 1 - \cos^2(x) = \sin^2(x). & (3.10)
\end{aligned}
$$

Finally the second equality follows from

$$\sin^2(x) = 1 - \cos^2(x) = \frac{||x||^2||y||^2 - <x,y>^2}{||x||^2||y||^2} \tag{3.11}$$

and the well known identity

$$||x \wedge y||^2 := \sum_{1 \leq i < j \leq q} (x_i y_j - x_j y_i)^2 = ||x||^2||y||^2 - <x,y>^2 \tag{3.12}$$

which completes the proof. $\qquad\square$

We are now in a position to define the pointwise gap metric as first introduced by Qiu and Davison in [7].

Consider two transfer functions $G(s)$ and $\tilde{G}(s)$ with left coprime factorizations

$$D^{-1}(s)N(s) = G(s) \quad \text{and} \quad \tilde{D}^{-1}(s)\tilde{N}(s) = \tilde{G}(s). \tag{3.13}$$

Then the pointwise gap between the transfer functions $G(s)$ and $\tilde{G}(s)$ is defined through (compare with [4, 7, 11])

$$d(G(s), \tilde{G}(s)) := \max_{s \in \mathbf{C}_+} gap(\text{rowsp}((N(s)\ D(s)), \text{rowsp}(\tilde{N}(s)\ \tilde{D}(s)). \tag{3.14}$$

DISTANCE BETWEEN AUTOREGRESSIVE SYSTEMS

Note that the gap function is continuous and the closed right half plane $\mathbb{C}_+$ is compact, i.e. the maximum is well defined. The following example taken from Qiu and Davison [7] illustrates some of the properties of the pointwise gap metric.

Example 3.2 Consider the family of transfer functions

$$G_\epsilon(s) := \begin{pmatrix} \frac{1+\epsilon}{s-1} & \frac{1}{s-1} \\ \frac{1}{s-1} & \frac{1}{s-1} \end{pmatrix}. \tag{3.15}$$

As shown in [7] one has for all $\epsilon > 0$

$$d(G_\epsilon(s), G_0(s)) = 1. \tag{3.16}$$

In particular $\lim_{\epsilon \to 0} G_\epsilon(s) \neq G_0(s)$ in the induced topology despite the fact that the convergence holds entrywise.

At this point it is not quite clear how to extend definition (3.14) or the definition of the gap metric [3, 15] or the graph metric [13] to the class of autoregressive systems. We would like to illustrate the anticipated problem on the following example.

Example 3.3 Consider two autoregressive systems

$$\Sigma_1: \quad \dot{u}_1(t) + u_1(t) = \dot{y}_1(t) + y_1(t) \tag{3.17}$$

and

$$\Sigma_2: \quad \dot{u}_2(t) - u_2(t) = \dot{y}_2(t) - y_2(t). \tag{3.18}$$

Taking the Laplace transform with zero initial condition one immediately sees that both transfer functions are the same, i.e.

$$\frac{\hat{y}_1(s)}{\hat{u}_1(s)} = \frac{s+1}{s+1} = \frac{s-1}{s-1} = \frac{\hat{y}_2(s)}{\hat{u}_2(s)} = 1. \tag{3.19}$$

This equality is mainly due to the fact that a pole zero cancellation occurred in both systems. Nonetheless the dynamics of both systems are as different as one can think. Indeed after adding an admissible static compensator

$$y(t) = \alpha u(t), \qquad \alpha \in \mathbb{R} \setminus \{1\} \tag{3.20}$$

the system Σ_1 is unstable where the second system is always stable! Due to this observation we propose in the next section a new metric which is capable of distinguishing between the systems Σ_1 and Σ_2.

4 A new metric on the set of autoregressive systems

In this section we will define a new metric on the space of autoregressive systems. For this consider the Hilbert space ℓ_2 consisting of all square integrable sequences. Denote with G the set of all finite dimensional subspaces of ℓ_2. Note that G is a metric space with respect to the gap metric.

Now we will describe an embedding of the space of all autoregressive systems of size $p \times q$ into G. Our embedding originates in our topological studies of the space of autoregressive systems as done in [8, 9, 12].

Consider an autoregressive system represented by a $p \times q$ polynomial matrix $R(s)$. Assume $R(s)$ is row reduced with row degree equal to $\nu_1 \geq \cdots \geq \nu_p$ and write

$$R(s) = R_0 + R_1 s + R_2 s^2 + \cdots. \qquad (4.1)$$

By assumption $R_k = 0$ if $k > \nu_1$. Consider now the following matrix with $(\nu_1 - \nu_p + 1)p$ rows and infinite many columns:

$$\tilde{R} := \left.\left[\begin{array}{cccccc} R_0 & R_1 & R_2 & \cdots & R_q & \cdots \\ 0 & R_0 & R_1 & \cdots & R_q & \cdots \\ \vdots & \vdots & \ddots & \cdots & \vdots & \cdots \\ 0 & \cdots & 0 & R_0 & R_1 & \cdots \end{array}\right]\right\} \nu_1 - \nu_p + 1 \text{ boxes} \qquad (4.2)$$

Note that elementary row operations on $R(s)$ correspond to elementary row operations on $\tilde{R}$. Identifying the system $R(s)$ with the rowspace of $\tilde{R}$ we get a well defined map from the set of autoregressive systems into the set G. Moreover one verifies that this map is even an embedding and induces therefore a metric on the set of autoregressive systems. We point out that this metric corresponds to the "coefficients" of the autoregressive system and that it fits into the topological results obtained by Nieuwenhuis and Willems [6] though this point has to be clarified in later work.

We conclude the paper with reviewing the two examples (3.2) and (3.3). In Example (3.2) one verifies that for all $\epsilon > 0$ $G_\epsilon(s)$ has a polynomial left coprime factorization

$$G_\epsilon(s) = D_\epsilon^{-1}(s)N_\epsilon(s) = \begin{pmatrix} s-1 & 0 \\ 0 & s-1 \end{pmatrix}^{-1} \begin{pmatrix} 1+\epsilon & 1 \\ 1 & 1 \end{pmatrix}. \qquad (4.3)$$

The corresponding autoregressive system is therefore given through

$$R_\epsilon(s) = \begin{pmatrix} 1+\epsilon & 1 & s-1 & 0 \\ 1 & 1 & 0 & s-1 \end{pmatrix} \qquad (4.4)$$

and associated to this matrix is the 2-plane in ℓ_2 defined by the row space of the matrix

$$\tilde{R}_\epsilon = \begin{pmatrix} 1+\epsilon & 1 & -1 & 0 & 0 & 0 & 1 & 0 & 0 & \cdots \\ 1 & 1 & 0 & -1 & 0 & 0 & 0 & 1 & 0 & \cdots \end{pmatrix}. \qquad (4.5)$$

On the other hand if $\epsilon = 0$, $G_\epsilon(s)$ has McMillan degree 1 and the associated autoregressive system is given by

$$R_0(s) = \begin{pmatrix} 1 & 1 & s-1 & 0 \\ 0 & 0 & 1 & -1 \end{pmatrix}. \tag{4.6}$$

The embedded plane in the set G is therefore of 4-dimensions, and therefore the process as ϵ approaches zero is not continuous either.

Finally in Example (3.3) the associated autoregressive system to Σ_1 is

$$r_1(s) = ((s+1) \ (-s-1)) \tag{4.7}$$

and associated to it is the vector

$$\tilde{r}_1 = (1 \ -1 \ 1 \ -1 \ 0 \ \ldots) \in \ell_2. \tag{4.8}$$

The corresponding situation for the system Σ_2 is:

$$r_2(s) = ((s-1) \ (-s+1)) \tag{4.9}$$

and associated to it is the vector

$$\tilde{r}_2 = (-1 \ 1 \ 1 \ -1 \ 0 \ \ldots) \in \ell_2. \tag{4.10}$$

Because the inner product $< \tilde{r}_1, \tilde{r}_2 >= 0$ we conclude using Lemma (3.1) that the distance between Σ_1 and Σ_2 is equal to 1.

References

[1] N.I.Achieser und I.M.Glasmann, *Theorie der Linearen Operatoren im Hilbert Raum*, Akademie-Verlag Berlin, 1954.

[2] H.Blomberg and R.Ylinen, *Algebraic Theory for Multivariable Linear Systems*, Academic Press, London, 1983.

[3] T.T.Georgiou and M.C.Smith, "Robust Stabilization in the Gap Metric: Controller Design for Distributed Plants," *IEEE Trans. Automat. Control*, vol.37, no.8, 1992, pp. 1133–1143.

[4] H.Glüsing-Lüerßen, *On Various Topologies for Finite-Dimensional Linear Systems*, report 273, University of Bremen, August 1992.

[5] M.Kuijper, *First-order Representation of Linear Systems*, Ph.D.–Thesis, Amsterdam, the Netherlands, 1992.

[6] J.W.Nieuwenhuis and J.C.Willems, "Continuity of Dynamical Systems: The Continuous Time Case," *Mathematics of Control, Signals and Systems,* vol.5, no.4, 1992, pp. 391–400.

[7] L.Qiu and E.J.Davison, "Pointwise Gap Metrics on Transfer Matrices," *IEEE Trans. Automat. Control,* vol.37, no.6, 1992, pp. 741–758.

[8] M.S.Ravi and J.Rosenthal, "A Smooth Compactification of the Space of Transfer Functions with Fixed McMillan Degree," preprint, 1992.

[9] M.S.Ravi, J.Rosenthal and X.Wang, "On Homogeneous Autoregressive Systems," *Proceedings of Symposium on Implicit and Nonlinear Systems,* Ft. Worth, Texas, 1992, pp. 229–235.

[10] J.Rosenthal, M.Sain and X.Wang, "Topological Considerations for Autoregressive Systems with Fixed Kronecker Indices," *Circuits Systems Signal Process.,* to appear.

[11] J.M.Schumacher, "A Pointwise Criterion for Controller Robustness," *Systems & Control Lett.,* 18, no. 1, 1992, pp. 1–8.

[12] X.Wang and J.Rosenthal, "A Cell Structure for the Set of Autoregressive Systems," preprint, 1992.

[13] M.Vidyasager, "The Graph Metric for Unstable Plants and Robustness Estimates for Feedback Stability," *IEEE Trans. Automat. Control,* vol. 29, 1984, pp. 403–418.

[14] J.C.Willems, "Paradigms and Puzzles in the Theory of Dynamical Systems," *IEEE Trans. Automat. Control,* vol.36, no.3, 1991, pp. 259–294.

[15] G.Zames and A.K.El-Sakkary, "Unstable Systems and Feedback: The Gap Metric." *Proc. of the 16-th Allerton Conference,* 1980, pp. 380–385.

SINC CONVOLUTION APPROXIMATE SOLUTION OF BURGERS' EQUATION

F. Stenger, * B. Barkey and R. Vakili

Department of Computer Science
University of Utah
Salt Lake City, Utah 84112

1   Introduction and Summary

Let $\mathcal{R}$ denote the real line, and let u_0 denote a given function defined on $\mathcal{R}$. We shall illustrate an integral equation procedure for solving the Burgers' equation problem

$$\frac{\partial}{\partial t} u(x,t) - \varepsilon \frac{\partial^2}{\partial x^2} u(x,t) \;=\; -\frac{1}{2}\frac{\partial}{\partial x} u^2(x,t), \quad x \in \mathcal{R}, \quad t > 0,$$

$$u(x,0) \;=\; u_0(x). \tag{1.1}$$

We accomplish this by first transforming the problem (1.1) into the equivalent integral equation problem

$$u(x,t) \;=\; \frac{1}{(4\pi\varepsilon t)^{1/2}} \int_{\mathcal{R}} \exp\left\{ -\frac{(x-\xi)^2}{4\varepsilon t} \right\} u_0(\xi)\, d\xi$$

$$+ \; \pi \int_0^t \int_{\mathcal{R}} \frac{x-\xi}{\{4\pi\varepsilon(t-\tau)\}^{3/2}} \exp\left\{ -\frac{(x-\xi)^2}{4\varepsilon(t-\tau)} \right\} u^2(\xi,\tau)\, d\xi\, d\tau, \tag{1.2}$$

which we then discretize via a novel Sinc collocation procedure, and the resulting discretized system is then solved via Neumann iteration. At the outset of the paper we briefly describe some notions of Sinc approximation, and we also briefly motivate the Sinc collocation procedure, without proofs, since a detailed proof of the error of Sinc approximation may be found in [1], [2], and a detailed proof of the error of Sinc collocation may be found in [2]. We then illustrate an explicit numerical procedure for solving (1.2).

We mention that excellent Sinc procedures have already been implemented in [1] for solving the heat equation, and these methods could also be used for solve the Burgers' problem (1.1). The procedure of the present paper has some advantages for purposes of parallel computation.

---

*Supported in part by IBM

2 Sinc Approximation

We begin with a summary of some properties *Sinc* methods for collocating differential and integral equations, the proofs of which can be found in [2]. *Sinc* methods excel for differential and integral equation problems whose solutions may have singularities, or infinite domains, or boundary layers.

Let (a, b) be an arbitrary interval, such as, e.g., the interval $(0, \infty)$, or the real line $\mathcal{R}$, and let ϕ denote a conformal map of a simply–connected domain $\mathcal{D}$ onto $\mathcal{D}_d$, with

$$\mathcal{D}_d = \{w \in C : |\Im w| < d\}, \tag{2.1}$$

where d is a positive number, such that ϕ is also a one - to - one map of (a, b) onto $\mathcal{R}$. We take $\alpha \in (0, 1]$, $\beta \in (0, 1]$, $d \in (0, \pi)$ and we define $M_{\alpha, \beta}(\mathcal{D})$ to be the family of all those functions $f \in \mathbf{Hol}(\mathcal{D})$, such that if g is defined by

$$g(z) = f(z) - \frac{f(a) + \rho(z)\, f(b)}{1 + \rho(z)}, \quad \rho = e^\phi, \tag{2.2}$$

then there exists a constant C_1, such that

$$|g(z)| \le C_1 \frac{|\rho(z)|^\alpha}{[1 + |\rho(z)|]^{\alpha + \beta}}, \quad z \in \mathcal{D}. \tag{2.3}$$

The set of all functions $g \in \mathbf{Hol}(\mathcal{D})$ which satisfy (2.3) is called $\mathbf{L}_{\alpha, \beta}(\mathcal{D})$. In this case there is no restriction on the size of the positive numbers α, β and d. If $\alpha \in (0, 1]$, $\beta \in (0, 1]$, and $d \in (0, \pi)$, then the class $\mathbf{L}_{\alpha, \beta}(\mathcal{D})$ is contained in the class $\mathbf{M}_{\alpha, \beta}(\mathcal{D})$.

For example, if (a, b) is a finite interval, we define ϕ by $w = \phi(z) = \log[(z - a)/(b - z)]$; this function ϕ provides a conformal transformation of the "eye – shaped" region $\mathcal{D} = \{z \in C : |\arg[(z - a)/(b - z)]| < d\}$ onto the strip $\mathcal{D}_d$ The same function ϕ also provided a one - to - one transformation of (a, b) onto the real line $\mathcal{R}$. A convenient space of functions for purposes of *Sinc* approximation is the space $\mathbf{M}_{\alpha, \beta}(\mathcal{D})$, which consists of all those functions f holomorphic in $\mathcal{D}$, i.e., $f \in \mathbf{Hol}(\mathcal{D})$, such that if g is defined by

$$g(z) = f(z) - \frac{(b - z)\, f(a) + (z - a)\, f(b)}{b - a} \tag{2.4}$$

then there exists a constant C_1 such that

$$|g(z)| \le C_1 |z - a|^\alpha |b - z|^\beta, \quad z \in \mathcal{D}. \tag{2.5}$$

It is easily seen that $\mathbf{M}_{\alpha, \beta}(\mathcal{D})$ includes all those functions $f \in \mathbf{Hol}(\mathcal{D})$ which are of class $\mathbf{Lip}_\alpha$ in that part of $\mathcal{D}$ within a distance $R \le (b - a)/2$ from a, and which are of class $\mathbf{Lip}_\beta$ in that part of $\mathcal{D}$ within a distance

R from b. The class $\mathbf{M}_{\alpha,\beta}(\mathcal{D})$ thus includes functions that are analytic in the interval (a,b), but which may have singularities at the end – points of (a,b).

For the case of $(a,b) = (0,\infty)$, the mapping $\phi(z) = \log(z)$ conformally maps the "sector" $\mathcal{D} = \{z \in \mathcal{C} : |\arg(z)| < d\}$ onto $\mathcal{D}_d$. This selection of ϕ enables us to consider problems whose solutions f are analytic in the "sector" $\mathcal{D}$, and which satisfy the relations $f(z) - f(0) = \mathcal{O}(|z|^\alpha)$ as $z \to 0$, and $f(z) - f(\infty) = \mathcal{O}(|z|^{-\beta})$ as $z \to \infty$. On the other hand, if we again have $(a,b) = (0,\infty)$, we may select $\phi(z) = \log\{\sinh(z)\}$, a transformation which conformally maps the "bullet – shaped" region $\{z \in \mathcal{C} : |\arg\{\sinh(z)\}| < d\}$ onto $\mathcal{D}_d$. This selection of ϕ enables us to consider problems whose solutions f are analytic and bounded in the "bullet - shaped" region $\mathcal{D}$, and which satisfy the asymptotic relations $f(z) - f(0) = \mathcal{O}(|z|^\alpha)$ as $z \to 0$, and $f(z) - f(\infty) = \mathcal{O}(e^{-\beta z})$ as $z \to \infty$. Similarly, several different mappings are possible for problems on $\mathcal{R}$.

The spaces $\mathbf{L}_{\alpha,\beta}(\mathcal{D})$ and $\mathbf{M}_{\alpha,\beta}(\mathcal{D})$ are motivated by the premise that most scientists and engineers use calculus to model differential and integral equation problems, and under this premise the solution to these problems are (at least piecewise) analytic. The spaces $\mathbf{L}_{\alpha,\beta}(\mathcal{D})$ and $\mathbf{M}_{\alpha,\beta}(\mathcal{D})$ house nearly all solutions to such problems, including solutions with singulalrities at end – points of (finite or infinite) intervals (or at boundaries of finite or infinite domains in more than one dimension). Although these spaces also house singularities, they are not as large as Sobolev spaces which assume the existence of only a finite number of derivatives in a solution, and consequently (see below) when *Sinc* methods are used to approximate solutions of differential or integral equations, they are usually more efficient than finite difference or finite element methods. In addition, *Sinc* methods are replete with interconnecting simple identities, including DFT (which is one of the *Sinc* methods, enabling the use of FFT), making it possible to use a *Sinc* approximation for nearly every type of operation arising in the solution of differential equations.

The spaces $\mathbf{L}_{\alpha,\beta}(\mathcal{D})$ and $\mathbf{M}_{\alpha,\beta}(\mathcal{D})$ are invariant, in the sense that if for $j = 1,2$ we have conformal mappings $\phi_j : \mathcal{D}_j \to \mathcal{D}_d$, and if $f \in \mathbf{L}_{\alpha,\beta}(\mathcal{D}_1)$ (resp., $f \in \mathbf{M}_{\alpha,\beta}(\mathcal{D}_1)$), then $f \circ \phi_1^{-1} \circ \phi_2 \in \mathbf{L}_{\alpha,\beta}(\mathcal{D}_2)$ (resp., $f \circ \phi_1^{-1} \circ \phi_2 \in \mathbf{M}_{\alpha,\beta}(\mathcal{D}_2)$).

Sinc approximation in $\mathbf{M}_{\alpha,\beta}(\mathcal{D})$ is defined as follows. Select a positive integer N, and define M by $M = [\beta N/\alpha]$, where $[\cdot]$ denotes the greatest integer function. Letting $\mathcal{Z}$ denote the set of all integers, set

$$
\begin{aligned}
\operatorname{sinc}(z) &= \frac{\sin(\pi z)}{\pi z}, \\
h &= \left(\frac{\pi d}{\beta N}\right)^{1/2}, \\
z_j &= \phi^{-1}(jh), \quad j \in \mathbb{Z} \\
\gamma_j &= \operatorname{sinc}\{[\phi - jh]/h\}, \quad j = -M, \cdots, N, \\
\omega_j &= \gamma_j, \quad j = -M+1, \cdots, N-1, \\
\omega_{-M} &= \frac{1}{1+\rho} - \sum_{j=-M+1}^{N} \frac{1}{1+e^{jh}}\gamma_j, \\
\omega_N &= \frac{\rho}{1+\rho} - \sum_{j=-M}^{N-1} \frac{e^{jh}}{1+e^{jh}}\gamma_j, \\
\varepsilon_N &= N^{1/2}e^{-(\pi d\beta N)^{1/2}}.
\end{aligned} \tag{2.6}
$$

We shall also define a norm by $\|f\| = \sup_{x \in (a,b)} |f(x)|$.

Theorem 2.1:[2] *If $f \in M_{\alpha,\beta}(\mathcal{D})$, then there exists a constant C_0 which is independent of N such that*

$$
\left\| f - \sum_{j=-M}^{N} f(z_j)\,\omega_j \right\| \leq C_0\varepsilon_N. \tag{2.7}
$$

The constants in the exponent in the definition of ε_N are the best constants for *Sinc* approximation. Hence best *Sinc* approximation of f is based on our being able to make good estimates on α, β, and d. If these constants cannot be accurately estimated, e.g., if instead of as in (2.6) above, we define h by $h = \gamma/N^{1/2}$, with γ a constant independent of N, then the right $-$ hand $-$ side of (2.7) is replaced by $C_1e^{-\delta N^{1/2}}$, where C_1 and δ are constants independent of N. Henceforth we shall take h as defined in (2.6).

Let us now remark that if the same function ϕ provides the conformal mappings $\phi : \mathcal{D}' \to \mathcal{D}_{d'}$, $\phi : \mathcal{D} \to \mathcal{D}_d$, with $0 < d < d'$, then $\mathcal{D} \subset \mathcal{D}'$. Moreover, it is readily shown in this notation, that if $f'/\phi' \in L_{\alpha,\beta}(\mathcal{D})$, then $f \in M_{\alpha,\beta}(\mathcal{D})$. Conversely, if $f \in M_{\alpha,\beta}(\mathcal{D}')$, then $f'/\phi' \in L_{\alpha,\beta}(\mathcal{D})$.

The following result, guarantees an accurate final *Sinc* approximation of f on (a, b), provided that we know a good approximation to f at the *Sinc* points.

Theorem 2.2:[2] *Let $f \in M_{\alpha,\beta}(\mathcal{D})$, and let the conditions of Theorem 1 be satisfied. Let $c_{-M}, \cdots, c_N$ be complex numbers such that*

$$
\left(\sum_{j=-M}^{N} |f(z_j) - c_j|^2\right)^{1/2} < \delta, \tag{2.8}
$$

where δ is a positive number. If C_0 and ϵ_N are defined as in (2.7), and if ω_j is defined as in (2.6), then

$$\left\| f - \sum_{j=-M}^{N} c_j \omega_j \right\| < C_0 \epsilon_N + \delta. \tag{2.9}$$

For purposes of describing *Sinc* indefinite integration and convolution, we use the notation

$$\begin{aligned} \sigma_k &= \int_0^k \operatorname{sinc}(x)\,dx, \quad k \in \mathcal{Z}, \\ e_k &= 1/2 + \sigma_k. \end{aligned} \tag{2.10}$$

We set $m = M + N + 1$, with M and N defined as in (2.6), and we define a Toeplitz matrix $I^{(-1)}$ of order m by $I^{(-1)} = [e_{i-j}]$, with e_{i-j} denoting the (i,j)th element of $I^{(-1)}$. If u is an arbitrary function defined on (a,b), we define a diagonal matrix $D(u)$ by $D(u) = \operatorname{diag}[u(z_{-M}), \cdots, u(z_N)]$. Again letting u be a function defined on (a,b), we define an operator V_m to convert u into a column vector, by the equation $V_m u = (u(z_{-M}), \cdots, u(z_N))^T$. Given the column vector $\mathbf{c} = (c_{-M}, \cdots, c_N)^T$, we define an operator Π_m by $\Pi_m \mathbf{c} = \sum_{j=-M}^{N} c_j \omega_j$, with ω_j defined as in (2.6). We also define matrices A_m and B_m, and operators $\mathcal{J}$, $\mathcal{J}'$, $\mathcal{J}_m$, and $\mathcal{J}_m'$ by

$$\begin{aligned} A_m &= h\,I^{(-1)}D(1/\phi'), & B_m &= h\,(I^{(-1)})^T D(1/\phi'), \\ (\mathcal{J}f)(x) &= \int_a^x f(t)\,dt, & (\mathcal{J}'f)(x) &= \int_x^b f(t)\,dt, \\ \mathcal{J}_m &= \Pi_m A_m V_m, & \mathcal{J}_m' &= \Pi_m B_m V_m, \end{aligned} \tag{2.11}$$

with $(I^{(-1)})^T$ denoting the transpose of $I^{(-1)}$. We can thus state the following theorem, the result of which enables us to collocate (linear or nonlinear, non $-$ stiff or stiff) initial value problems.

Theorem 2.3: *If $f/\phi' \in \mathbf{L}_{\alpha,\beta}(\mathcal{D})$, then there exists a constant C_1, which is independent of N, such that*

$$\begin{aligned} \|\mathcal{J}f - \mathcal{J}_m f\| &\le C_1 \epsilon_N, \\ \|\mathcal{J}'f - \mathcal{J}_m' f\| &\le C_1 \epsilon_N. \end{aligned} \tag{2.12}$$

Some partial differential equations can at times be conveniently transformed into convolution - type integral equations. Such integral equation representations can be collocated via *Sinc* formulas for collocating the indefinite convolution integrals

$$\begin{aligned} p(x) &= \int_a^x f(x-t)\,g(t)\,dt, \\ q(x) &= \int_x^b f(t-x)\,g(t)\,dt, \end{aligned} \tag{2.13}$$

where $x \in (a, b)$. Note, from (2.13), if f is the same in the definitions of p and q, then

$$p(x) + q(x) = \int_a^b f(|x - t|) \, g(t) \, dt, \qquad (2.14)$$

so that being able to collocate p and q also enables us to collocate $p+q$. The method which we shall describe below, of the *Sinc* collocation of p and q will thus provide a powerful technique for solving any initial or boundary value problem whose solution can be expressed in terms (definite or indefinite) convolution – type integral, or via a convolution–type integral equation, such as Abel's integral equation, or integral equations to which the Wiener – Hopf method is applicable.

Sinc collocation of p and q is possible under the following **Assumptions 2.4:** *We assume that the "Laplace transform",*

$$F(s) = \int_0^\infty f(t) \, e^{-t/s} \, dt \qquad (2.15)$$

exists for all $s \in \Omega^+ \equiv \{s \in C : \Re s > 0\}$, *and that* $F(s) = \mathcal{O}(s)$ *as* $s \to \infty$. *Let* $P(r, x)$ *be defined by*

$$P(r, x) = \int_a^x f(r + x - t) \, g(t) \, dt. \qquad (2.16)$$

We assume further, that

(i) $P(r, \cdot) \in M_{\alpha,\beta}(\mathcal{D}')$, *uniformly for* $r \in [0, b - a]$; *and that*

(ii) $P(\cdot, x)$ *is of bounded variation on* $(0, [b-a])$, *uniformly for* $x \in [a, b]$. Under these assumptions, we have

Theorem 2.5: *If the above assumptions are satisfied, then there exists a constant* C_1, *independent of* N, *such that*

$$\begin{aligned}
\|p - \Pi_m F(A_m) V_m g\| &\leq C_1 \varepsilon_N, \\
\|q - \Pi_m F(B_m) V_m g\| &\leq C_1 \varepsilon_N.
\end{aligned} \qquad (2.17)$$

Let us briefly motivate the proof of this theorem.

It suffices to consider only the case of $p(x)$ as defined in (2.13), since the case of $q(x)$ can be treated in exactly the same way. Letting $\mathcal{J}w$ be defined as above, with $w \in \mathbf{L}^1(a.b)$, it follows that for $n = 1, 2, \cdots$, we have

$$(\mathcal{J}^n w)(x) = \int_a^x \frac{(x - t)^{n-1}}{(n - 1)!} w(t) \, dt. \qquad (2.18)$$

We assume that the length of the interval (a, b) is finite at the outset. It is then convenient to take $\|w\| = (b - a)^{-1} \int_a^b |w(t)| \, dt$, since this choice of norm yields the simple inequality

$$\|\mathcal{J}w\| = \frac{1}{b-a} \int_a^b \left| \int_a^x w(t)\,dt \right| dx \le (b-a)\|w\|, \qquad (2.19)$$

which implies that

$$\begin{aligned}
\|\mathcal{J}^n\| &\le \frac{(b-a)^n}{n!} \\
\|\mathcal{J}^n\| &\le \|\mathcal{J}\|^n \le (b-a)^n.
\end{aligned} \qquad (2.20)$$

By using the Bromwich formula for the inversion of the Laplace transform, and then converting to the "Laplace transform" via replacement of s by $1/s$ (a transformation which transforms Ω^+ to itself) the expression (2.13) for $p(x)$ may be written in the alternate forms

$$\begin{aligned}
p &= -\int_a^x \frac{1}{2\pi i} \int_{-i\infty}^{i\infty} e^{(x-t)/s} F(s) s^{-2} ds\, g(t)\,dt \\
&= -\frac{1}{2\pi i} \int_{-i\infty}^{i\infty} w\,ds,
\end{aligned} \qquad (2.21)$$

where

$$w = \int_a^x s^{-2} e^{(x-t)/s} F(s)\, g(t)\,dt, \qquad (2.22)$$

and it thus follows from the definitions (2.11), that

$$\begin{aligned}
w &= \int_a^x \sum_{n=0}^{\infty} \frac{(x-t)^n}{n!\, s^{n+2}} g(t)\,dt \\
&= \left(\frac{\mathcal{J}}{s^2} \sum_{n=0}^{\infty} \frac{\mathcal{J}^n}{s^n} F(s) g \right)(x) \\
&= \left(\frac{\mathcal{J}}{s} (s - \mathcal{J})^{-1} F(s) g \right)(x).
\end{aligned} \qquad (2.23)$$

By analytic continuation as a function of s, it then follows that the identity

$$\int_a^x s^{-2} e^{(x-t)/s} F(s)\, g(t)\,dt = \left(\frac{\mathcal{J}}{s} (s - \mathcal{J})^{-1} F(s) g \right)(x) \qquad (2.24)$$

holds not only for all $s \in C$ such that $|s| > b-a$, but in the larger, *resolvent* set of $\mathcal{J}$, excluding the point $s = 0$. Here, the resolvent set of $\mathcal{J}$ is the set $\{s \in C : (s - \mathcal{J})^{-1} \text{ exists}\}$. The resolvent set of $\mathcal{J}$ can be more closely identified, upon setting

$$(u, v) \equiv \int_a^b u(x)\overline{v(x)}\,dx. \qquad (2.25)$$

It follows, in this notation, that

$$\Re(\mathcal{J}u, u) = \frac{1}{2}\left|\int_a^b u(x)\,dx\right|^2 \geq 0. \tag{2.26}$$

Hence, the resolvent set of $\mathcal{J}$ includes the set $\{s \in C : \Re s < 0\}$, as well as the set $\{\Re s \in C : |s| > b - a\}$.

Substitution of (2.24) into (2.21) yields the *Dunford* – type integral

$$p(x) = -\left(\frac{1}{2\pi i}\int_{-i\infty}^{i\infty} \frac{\mathcal{J}}{s}(s - \mathcal{J})^{-1} F(s)\,ds\,g\right)(x). \tag{2.27}$$

If (a, b) is a finite interval, then it follows that $F(s)$ is bounded for all sufficiently large $s \in C$. It is also well – known that the classical Laplace transform $\hat{f}(s) \equiv \int_0^c e^{-st} f(t)\,dt \to 0$ as $s \to \infty$, so that $F(0) = \hat{f}(\infty) = 0$. The expression (2.27) thus yields the formula

$$p(x) = (F(\mathcal{J})\,g)(x), \tag{2.28}$$

which is valid for bounded intervals (a, b).

Although we shall not do so here, it may be readily shown that the result (2.28) also holds for unbounded intervals (a,b).

Now, letting V_m, II_m and $\mathcal{J}_m$ be defined as in (2.11) above, it follows that

$$p(x) = (F(\mathcal{J})\,g)(x) \approx (F(\mathcal{J}_m)\,g)(x) = (\mathrm{II}_m F(A_m)V_m g)(x). \tag{2.29}$$

We remark here that it may be readily shown that every eigenvalue of the matrices A_m and B_m defined as in (2.11) above lies in the closure of the right half plane. Indeed, it has been shown by direct computation, that all eigenvalues of the matrices A_m and B_m lie in the open right half plane for $1 \leq m \leq 513$, and hence the matrices $F(A_m)$ and $F(B_m)$ are well defined for all such values of m, and may be evaluated in the usual way, via diagonalization of A_m and B_m.

Example 2.6 *Multidemensional convolution integrals* can also be effectively approximated by the above convolution procedure. Let us illustrate for the case of a two dimensional integral,

$$p(x_1, x_2) = \int_{x_2}^{b_2} \int_{a_1}^{x_1} f(x_1 - \xi_1, \xi_2 - x_2)\,g(\xi_1, \xi_2)\,d\xi_1 d\xi_2, \tag{2.30}$$

where an approximation is sought over the region $\mathbf{B} = \Pi_{i=1}^{2} \otimes (a_i, b_i)$, and with $(a_i, b_i) \subseteq \mathcal{R}$. In order to guarantee some accuracy in the final approximation, we shall simply assume, without going into detail, that the function p belongs to the class $\mathbf{M}_{\alpha,,\beta,}(\mathcal{D}'_j)$ with respect to each variable x_j, for all fixed values of the other variable in its interval of definition, $j = 1, 2$. We shall also assume that the mappings $\phi_j : \mathcal{D}'_j \to \mathcal{D}_{d'}$ have been determined. We furthermore assume that positive integers N_j and M_j as well as positive numbers h_j $(j = 1, 2)$ have been selected such that $[N_1\beta_1] = [N_2\beta_2]$, such that $M_j = [\beta_j N_j/\alpha_j]$, where $[\cdot]$ denotes the greatest integer function, and such that $h_j = \{\pi d/(\beta_j N_j)\}^{1/2}$. These definitions ensure that we get the same order of accuracy of approximation $\varepsilon = \mathcal{O}(N_1^{1/2} e^{-(\pi d\beta_1 N_1)^{1/2}})$ in each variable. We set $m_j = M_j + N_j + 1$, and we define the Sinc points by $z_\ell^{(j)} = \phi^{-1}(\ell h_j)$, for $\ell = -M_j, \cdots, N_j; j = 1, 2$. Next, we determine matrices A_j, X_j, S_j and X_j^{-1}, such that

$$
\begin{aligned}
A &= h_1 I_{m_1}^{(-1)} D(1/\phi_1') &= X_1 S_1 X_1^{-1}, \\
B &= h_2 \left(I_{m_2}^{(-1)}\right)^T D(1/\phi_2') &= X_2 S_2 X_2^{-1}.
\end{aligned}
\tag{2.31}
$$

In (2.31), $I_{m_j}^{(-1)}$ is defined as in (2.11), and the S_j are diagonal matrices,

$$
S_j = \text{diag}[s_{-M_j}^{(j)}, \cdots, s_{N_j}^{(j)}].
\tag{2.32}
$$

Arbitrarily taking $c_j \in [2(b_j - a_j), \infty]$, we set

$$
\begin{aligned}
F(s^{(1)}, y) &- \int_0^{c_1} f(x, y) e^{-x/s^{(1)}} dx, \\
G(s^{(1)}, s^{(2)}) &= \int_0^{c_2} F(s^{(1)}, y) e^{-y/s^{(2)}} dy.
\end{aligned}
\tag{2.33}
$$

We mention at this point, that the function F defined in (2.33) is introduced here solely for purposes of understanding the method of separation of variables described below. Only the function G defined in the last line of (2.33) is required in the final algorithm.

We now illustrate the method of separation of variables. To this end, we first rewrite (2.30) in the notationally more convenient form

$$
p(x, y) = \int_y^{b_2} \int_{a_1}^{x} f(x - \xi, \eta - y) g(\xi, \eta) \, d\xi \, d\eta.
\tag{2.34}
$$

Discretization with respect to x. We set

$$
\mathbf{g}(\eta) = \left(g(z_{-M_1}^{(1)}, \eta), \cdots, g(z_{N_1}^{(1)}, \eta)\right)^T,
\tag{2.35}
$$

and we then *define* a vector $\mathbf{p}(y)$ by

$$\mathbf{p}(y) = \int_y^{b_2} F(A, \eta - y)\mathbf{g}(\eta)\, d\eta, \qquad (2.36)$$

where A and F are defined in (2.31) and (2.33), respectively.
Using the diagonalization identity $A = X_1 S_1 X_1^{-1}$ given in (2.31), it now follows from (2.30) that

$$\mathbf{p}(y) = X_1 \int_y^{b_2} \begin{pmatrix} F(s_{-M_1}^{(1)}, \eta - y) & & \\ & \ddots & \\ & & F(s_{N_1}^{(1)}, \eta - y) \end{pmatrix} X_1^{-1} \mathbf{g}(\eta)\, d\eta. \qquad (2.37)$$

The expression (2.37) motivates the transformations

$$\mathbf{h}(\eta) = X_1^{-1}\mathbf{g}(\eta), \quad \mathbf{q}(y) = X_1^{-1}\mathbf{p}(y). \qquad (2.38)$$

Thus, denoting the components of $\mathbf{h}$ and $\mathbf{q}$ by h_i and q_i respectively, $i = -M_1, \cdots, N_1$, Equation (2.38) reduces to the *decoupled* set of scalar equations

$$q_i(y) = \int_y^{b_2} F(s_i^{(1)}, \eta - y)h_i(\eta)\, d\eta. \qquad (2.39)$$

Discretization with respect to y. We set

$$\mathbf{h}_i = \left(h_i(z_{-M_2}^{(2)}) \cdots, h_i(z_{N_2}^{(2)}) \right)^T, \qquad (2.40)$$

and we then *define* a vector $\mathbf{q}_i$ by

$$\mathbf{q}_i = G(s_i^{(1)}, B)\,\mathbf{h}_i, \qquad (2.41)$$

where B and G are defined in (2.31) and (2.33) respectively.
Using the diagonalization identity $B = X_2 S_2 X_2^{-1}$ given in (2.31), it now follows from (2.41) that

$$\mathbf{q}_i = X_2 \begin{pmatrix} G(s_i^{(1)}, s_{-M_2}^{(2)}) & & \\ & \ddots & \\ & & G(s_i^{(1)}, s_{N_2}^{(2)}) \end{pmatrix} X_2^{-1}\mathbf{h}_i. \qquad (2.42)$$

This last expression motivates the transformations

$$\mathbf{k}_i = X_2^{-1}\mathbf{h}_i, \quad \mathbf{r}_i = X_2^{-1}\mathbf{q}_i. \qquad (2.43)$$

Denoting the components of $\mathbf{r}_i$ and $\mathbf{k}_i$ by $r_{i,j}$ and $k_{i,j}$ respectively, $i = -M_1, \cdots, N_1$; $j = -M_2, \cdots, N_2$, Equation (2.43) reduces to the decoupled set of scalar equations

$$r_{i,j} = G(s_i^{(1)}, s_j^{(2)}) k_{i,j}. \tag{2.44}$$

By assumption, the $k_{i,j}$ are known at this point, and (2.44) then determines the $r_{i,j}$. The second equation in (2.43) next determines the vectors $\mathbf{q}_i$. The second equation in (2.38) is then used to determine the vectors $\mathbf{p}(y)$ at the Sinc points $y = z_j^{(2)}$. We can thus recover the complete array of values $p(x, y)$ at the set of Sinc points $(z_i^{(1)}, z_j^{(2)})$. The whole procedure is illustrated succinctly via the following algorithm. In this algorithm the we use the notation, e.g., $\mathbf{h}_{i,\cdot} = (h_{i,-M_2}, \cdots, h_{i,N_2})^T$. The computations are to be carried out in the order $g_{ij} \to h_{ij} \to k_{ij} \to r_{ij} \to q_{ij} \to p_{ij}$. We emphasize the obvious ease of adaptation of this algorithm to parallel computation.

Algorithm 2.7

1. Form the array $z_i^{(j)}$, and $\frac{d}{dx}\phi^{(j)}(x)$ at $x = z_i^{(j)}$ for $j = 1, 2$, and $i = -M_j, \cdots, N_j$, and then form the array $[g_{i,j}] = [g(z_i^{(1)}, z_j^{(2)})]$.
2. Determine A, B, S_j, X_j, and X_j^{-1} for $j = 1, 2$, as defined in (2.31).
3. Form   (See (2.38))   [DONE]

$$\mathbf{h}_{\cdot,j} = X_1^{-1}\mathbf{g}_{\cdot,j} \qquad\qquad \mathbf{p}_{\cdot,j} = X_1\mathbf{q}_{\cdot,j}$$

4. Form   (See (2.43))

$$\mathbf{k}_{i,\cdot} = X_2^{-1}\mathbf{h}_{i,\cdot}. \qquad\qquad \mathbf{q}_{i,\cdot} = X_2\mathbf{r}_{i,\cdot}.$$

5. Form   (See (2.44)

$$r_{i,j} = G(s_i^{(1)}, s_j^{(2)}) k_{i,j}.$$

Once the numbers $p_{i,j}$ have been computed, we can then use these numbers to approximate p on the region B via the use of a Sinc basis; upon setting $\rho^{(\ell)} = e^{\phi^{(\ell)}}$, we can define the functions

$$
\begin{aligned}
\gamma_i^{(\ell)} &= \mathrm{sinc}\{[\phi^{(\ell)} - ih]/h\}, \quad \ell = 1, 2; \quad i = -M_\ell, \cdots, N_\ell, \\
\omega_i^{(\ell)} &= \gamma_i^{(\ell)}, \quad \ell = 1, 2; \quad i = -M_\ell + 1, \cdots, N_\ell - 1, \\
\omega_{-M_\ell}^{(\ell)} &= \frac{1}{1 + \rho^{(\ell)}} - \sum_{j=-M_\ell+1}^{N_\ell} \frac{1}{1 + e^{jh_\ell}}\gamma_j^{(\ell)}, \\
\omega_{N_\ell}^{(\ell)} &= \frac{\rho^{(\ell)}}{1 + \rho^{(\ell)}} - \sum_{j=-M_\ell}^{N_\ell-1} \frac{e^{jh_\ell}}{1 + e^{jh_\ell}}\gamma_j^{(\ell)}.
\end{aligned}
\tag{2.45}
$$

We then get the approximation

$$p(x, y) \approx \sum_{i=-M_1}^{N_1} \sum_{j=-M_2}^{N_2} p_{i,j} \omega_i^{(1)}(x) \omega_j^{(2)}(y). \tag{2.46}$$

Assuming a one-dimensional error of the order of $\epsilon = N_1^{1/2} e^{-(\pi d \beta_1 N_1)^{1/2}}$, we may expect the approximation in (2.46) to have an error of the order of $\{\log(N_1)\}^2 \epsilon$.

It is also relatively simple to obtain an estimate of the complexity, i.e., the total amount of work required to achieve an error ϵ when carrying out the computations of the above algorithm on a sequential machine. By taking $\alpha_j = \beta_j = \alpha$, and then selecting $M_j = N_j = N$, it follows then, that the error in the final approximation is roughly of the order of $\epsilon = e^{-(\pi d \alpha N)^{1/2}}$. The amount of work required to factor the matrices A_j into the form $X_j S_j X_j^{-1}$ is of the order of $(2N+1)^3$, and the amount of work in the totality of the matrix – vector multiplications in steps 3 to 5 is of the order of $(2N + 1)^3$. Hence the complexity is $\mathcal{O}([2N + 1]^3) = \mathcal{O}([\log(\epsilon)]^6)$. The above algorithm extends readily to ν dimensions, where the complexity is $\mathcal{O}([\log(\epsilon)]^{2\nu+2})$.

Henceforth, we shall use the condensed notation for the Algorithm 2.8, namely,

$$[p_{ij}] = G(A, B, [g_{ij}]). \tag{2.47}$$

3 Application to Burgers' Problem

We now apply the above procedure to the solution of the Burgers' problem (1.2). To this end, we take

$$u_0(x) = a \exp\left\{-b(x - c)^2\right\} \tag{3.1}$$

This choice of u_0 enables an explicit expression for the first term on the right-hand side of (1.2), so that we can now rewrite (1.2) in the form

$$u(x,t) \;=\; v(x,t)$$

$$+ \; \pi \int_0^t \left[\int_{-\infty}^x \frac{x - \xi}{\{4\pi\epsilon(t - \tau)\}^{3/2}} \exp\left\{ -\frac{(x - \xi)^2}{4\epsilon(t - \tau)} \right\} u^2(\xi, \tau)\, d\xi \right.$$

$$\left. - \int_x^\infty \frac{\xi - x}{\{4\pi\epsilon(t - \tau)\}^{3/2}} \exp\left\{ -\frac{(x - \xi)^2}{4\epsilon(t - \tau)} \right\} u^2(\xi, \tau)\, d\xi \right] dt, \tag{3.2}$$

where

$$v(x,t) = \frac{1}{\{1 + 4b\varepsilon t\}^{1/2}} \exp\left\{ -\frac{b(x-c)^2}{1 + 4b\varepsilon t} \right\}. \tag{3.3}$$

Due to this explicit form of the function $v(x,t)$, the form (3.1) for u_0 makes it possible to approximate an arbitrary continuous function u_0 defined on $\mathcal{R}$ by use of the function $F_3(\beta, h)$ defined in [3].

We now proceed to discretize Equation (3.2) as outlined in Example 2.6. To this end we may note that it is possible to explicitly evaluate the "Laplace transform" of the convolution kernel in (3.2), i.e.,

$$\begin{aligned} G(s,\sigma) &= \int_0^\infty \int_0^\infty \exp\left\{ -\frac{x}{s} - \frac{t}{\sigma} \right\} \frac{x}{\{4\pi\varepsilon t\}^{3/2}} \exp\left\{ -\frac{x^2}{4\varepsilon t} \right\} dx\, dt \\ &= \frac{1}{4\varepsilon^{1/2}} \frac{s\, \sigma^{1/2}}{s + \varepsilon^{1/2}\sigma^{1/2}}. \end{aligned} \tag{3.4}$$

We now select $\varepsilon = 1/2$, $b = 1$, $c = 0$, $\phi_t(t) = \log\{\sinh(t)\}$, $\phi_x(x) = x$, $d_t = \pi/2$, $\alpha_t = \beta_t = 1/2$, $d_x = \pi/4$, $\alpha_x = \beta_x = 1$, and in this case it is convenient to take $M_t = N_t = M_x = N_x = x$. We thus form matrices

$$\begin{aligned} A_x &= h_x I^{(-1)} = X_t S_x X_t^{-1}, \quad A_x' = h_x (I^{(-1)})^T = X_t S_x X_t^{-1}, \\ B_t &= h_t I^{(-1)} D(1/\phi_t') = X_t S_t X_t^{-1}, \end{aligned} \tag{3.5}$$

where the superscpript "T" denotes the transpose, and where S_x and S_t are diagonal matrices, and then proceed as in Example 2.6 above, and the notation of Equation (2.47) to reduce the integral equation problem (3.2) to the nonlinear matrix problem

$$[u_{ij}] = G(A_x, B_t, [u_{ij}^2]) - G(A_x', B_t, [u_{ij}^2]) + [v_{ij}], \tag{3.6}$$

where the function v_{ij} may be evaluated a priori, via the formula $v_{ij} = v(ih_x, z_j)$, with $v(x,t)$ defined as in (3.3), and with

$$z_j = \log\left[e^{jh_t} + \left(1 + e^{2jh_t}\right)^{1/2} \right].$$

The system (3.6) may be solved by Neumann iteration, for a (defined as in (3.1)) sufficiently small. For example, with $a = 1/2$, and using the map $\phi_t(t) = \log[\sinh(t)]$ we achieved convergence in 4 iterations, for all values of N (between 10 and 30) that we attempted. We can also solve the above equation via Neumann iteration for larger values of a, if we restrict the time t to a finite interval, $(0, T)$, via the map $\phi_t(t) = \log\{Tt/(T-t)\}$.

The form of G given in (3.4) suggests that the solution to the problem (3.6) will become infinite as $\varepsilon \to 0$. However, due to our special form of A_x, it may be shown that this is, in fact, not the case.

References

[1] J. LUND and K.L. BOWERS, "Sinc Methods for Quadrature and Differential Equations," *SIAM*, Philadelphia, PA, 1992.

[2] F. STENGER, "Numerical Methods Based on Sinc and Analytic Functions," *Springer–Verlag*, New York, NY, 1993.

[3] F. STENGER, "Explicit Approximate Methods for Computational Control Theory," *Computation and Control*, edited by K. L. Bowers and J. Lund, *Birkhäuser, Basel* , 1989, pp. 299–316.

SINC-GALERKIN COLLOCATION METHOD FOR PARABOLIC EQUATIONS IN FINITE SPACE-TIME REGIONS

Marc Stromberg[*]
Department of Mathematics
Texas Tech University
Lubbock, Texas 79409

Xiaoning Li Gilliam[†]
Department of Mathematics
Texas Tech University
Lubbock, Texas 79409

1 Introduction

In this paper we develop the Sinc-Galerkin collocation method for approximating the solution of initial boundary value problems for the heat equation in a finite space-time cylinder. One important feature of the Sinc method is that a uniform exponential error bound of the form $O(e^{-c/h})$, where h is a step size, is typical for boundary data that is piecewise holomorphic with discontinuities at finitely many isolated values of t. This is allowed by calculating solutions in finite time intervals, which appears to be a new result and includes calculation of solutions that blow up at infinity or within finite time. The case of a finite time interval requires some care in evaluating the resulting inner products to obtain a sufficiently well-conditioned finite dimensional system of equations arising from discretization.

When the weights and nodes are taken to be the same in each space variable, the finite-dimensional system can be written in a particularly simple form involving tensor products. Using an eigenvalue decomposition method the system can be accurately and relatively quickly solved using parallel computation.

The ultimate goal of this paper is the numerical approximation of solutions of linear parabolic partial differential equations typified by the heat equation

$$u_t(x, y, t) - u_{xx}(x, y, t) - u_{yy}(x, y, t) = f(x, y, t); \quad (x, y, t) \in D$$

$$u(x, y, t) = g(x, y, t); \quad (x, y, t) \in \partial D, \ t \neq T \qquad (1.1)$$

$$D = \{(x, y, t) : 0 < x < 1; \ 0 < y < 1; \ 0 < t < T\}.$$

[*]Supported in part by the Texas ARP
[†]Supported in part by the first author

Our main result is the application of the Galerkin method in both space and time for the solution of the heat equation in a finite rectangular region in two spatial dimensions and on a finite time interval. To employ a Galerkin method, we use a Kronecker basis derived from sinc bases in x, y and t. These are derived from the sinc function

$$\text{sinc}(x) = \frac{\sin \pi x}{\pi x}$$

by translation and composition, by first defining

$$s(k, h)(z) = \text{sinc}(z/h - k).$$

In x the resulting basis is given by

$$b_i(x) = \begin{cases} (1 - x) - \displaystyle\sum_{q=-N_x}^{N_x} (1 - x_q)\, s(q, h) \circ \phi(x) & i = -N_x - 1 \\[2ex] s(i, h) \circ \phi(x) & i = -N_x, \cdots, N_x \\[2ex] x - \displaystyle\sum_{q=-N_x}^{N_x} x_q\, s(q, h) \circ \phi(x) & i = N_x + 1 \end{cases}$$

and similarly for y, where $\phi(x) = \log(x/(1 - x))$. In the t variable the basis elements are $\tilde{b}_k(t) = b_k(\tilde{t})$ where $\tilde{t} = t/T$ and T is fixed but otherwise arbitrary. We use a standard sinc quadrature to evaluate the inner products in x and y and a special half-step sinc quadrature to evaluate the inner products in t.

To apply a Galerkin scheme, we seek an approximate solution to this problem in the form

$$u(x, y, t) \approx \sum_{k=-N_t-1}^{N_t+1} \sum_{j=-N_y-1}^{N_y+1} \sum_{i=-N_x-1}^{N_x+1} u_{ijk} b_i(x) b_j(y) \tilde{b}_k(t) \tag{1.2}$$

requiring that

$$0 = \, <Lu - f, b_\ell(x) b_p(y) \tilde{b}_q(t)>$$

for $\ell = -N_x, \cdots, N_x$, $p = -N_y, \cdots, N_y$, and $q = -N_t, \cdots, N_t + 1$.

Introducing the notation

$$\sum = \sum_{k=-N_t-1}^{N_t+1} \sum_{j=-N_y-1}^{N_y+1} \sum_{i=-N_x-1}^{N_x+1}$$

we have

$$\sum \dot{u}_{ijk} < b_i(x), b_\ell(x) >< b_j(y), b_p(y) >< \frac{\partial}{\partial t}\tilde{b}_k(t), \tilde{b}_q(t) >$$

$$- \sum u_{ijk} < \frac{\partial^2}{\partial x^2}b_i, b_\ell >< b_j(y), b_p(y) >< \tilde{b}_k(t), \tilde{b}_q(t) > \qquad (1.3)$$

$$- \sum u_{ijk} < b_i(x), b_\ell(x) >< \frac{\partial^2}{\partial y^2}b_j, b_p >< \tilde{b}_k(t), \tilde{b}_q(t) >$$

$$= \; < f, b_\ell b_p \tilde{b}_q >,$$

which is the finite dimensional Galerkin system. However, numerical methods for solution of the Galerkin system (1.3) typically approximate the inner products by a quadrature scheme, and the way this is done can strongly affect the solvability of the resulting system, as we note below. To simplify notation, let

$$a_{i\ell}^0 =< b_i(x), b_\ell(x) >, \quad a_{i\ell}^2 =< \frac{\partial^2}{\partial x^2}b_i, b_\ell >,$$

$$b_{jp}^0 =< b_j(y), b_p(y) >, \quad b_{jp}^2 =< \frac{\partial^2}{\partial y^2}b_j, b_p >,$$

$$c_{kq}^0 =< \tilde{b}_k(t), \tilde{b}_q(t) >, \quad c_{kq}^1 =< \frac{\partial}{\partial t}\tilde{b}_k, \tilde{b}_q >, \quad \text{and} \quad f_{\ell pq} =< f, b_\ell b_p \tilde{b}_q > .$$

The Galerkin system is now expressed as

$$\sum u_{ijk}\{a_{i\ell}^0 b_{jp}^0 c_{kq}^1 - a_{i\ell}^2 b_{jp}^0 c_{kq}^0 - a_{i\ell}^0 b_{jp}^2 c_{kq}^0\} = f_{\ell pq}$$

in which the true unknowns are $\{u_{ijk}\}_{i=-N_x, j=-N_y, k=-N_t}^{N_x, N_y, k=N_t+1}$ and the remaining $\{u_{ijk}\}$ correspond to given initial and boundary values. Taking known quantities to one side of the equation, we have

$$\sum_{k=-N_t}^{N_t+1} \sum_{j=-N_y}^{N_y} \sum_{i=-N_x}^{N_x} u_{ijk}\{a_{i\ell}^0 b_{jp}^0 c_{kq}^1 - a_{i\ell}^2 b_{jp}^0 c_{kq}^0 - a_{i\ell}^0 b_{jp}^2 c_{kq}^0\} = RHS_{\ell pq}, \quad (1.4)$$

where $RHS_{\ell pq}$ is the data for the finite dimensional system, consisting of the coordinate-wise sum of the array f and the difference of the right side of (1.2) and the left side of (1.4).

For each fixed k, we introduce the notation $u_{ij}^k = u_{ijk}$ to represent a $(2N_x+1)\times(2N_y+1)$ matrix and similarly for fixed q set $RHS_{\ell p}^q = RHS_{\ell pq}$. Then for fixed q we can write (1.4) as

$$\sum_{k=-N}^{N+1} \left[a_{i\ell}^0\right]^T [u_{ij}^k] [b_{jp}^0] c_{kq}^1 \; - \; \left[a_{i\ell}^2\right]^T [u_{ij}^k] [b_{jp}^0] c_{kq}^0$$

$$- \left[a_{i\ell}^0\right]^T [u_{ij}^k] [b_{jp}^2] c_{kq}^0 = [RHS_{\ell p}^q],$$

where $[a_{i\ell}^0]$ is a $(2N_x+1)\times(2N_x+1)$ matrix and similarly for the remaining matrices.

Applying standard sinc quadrature in the x and y variables we have

$$< f,g >= \int_0^1 f(x)g(x)v(x)\,dx \approx h \sum_{\ell=-N_x}^{N_x} \frac{v(x_\ell)}{\phi'(x_\ell)} f(x_\ell)g(x_\ell) \qquad (1.5)$$

where $\{x_\ell\}$ are the sinc nodes

$$x_\ell = \frac{e^{\ell h}}{e^{\ell h}+1},\ \ell=-N_x,\ \cdots,\ N_x$$

and v is a weight function chosen primarily to allow inner products to be evaluated by sinc quadrature. If g is an arbitrary function, write $\mathcal{D}(g) = \text{diag}\,[g(x_\ell)]$. Then for the x and y inner products we obtain

$$[a_{i\ell}^0] = h\,\mathcal{D}\left(\frac{v}{\phi'}\right) = h\,\mathcal{D}_x,$$

$$- [a_{i\ell}^2]^T = h\,\mathcal{D}_x(A^{(2)})^T,$$

where

$$(A^{(2)})^T = -1/h^2\,\mathcal{D}\left((\phi')^2\right)\left[\delta_{ik}^{(2)}\right] + 1/h\,\mathcal{D}\,(\phi'')\left[\delta_{ik}^{(1)}\right]$$

and where

$$\delta_{jk}^{(1)} \equiv h\frac{d}{dx}[s(j,h)\circ\phi(x)]|_{x=x_k} = \left\{ \begin{array}{ll} 0, & j=k \\ (-1)^{k-j}/(k-j), & j\neq k \end{array} \right.$$

and

$$\delta_{jk}^{(2)} \equiv h^2\frac{d^2}{dx^2}[s(j,h)\circ\phi(x)]|_{x=x_k} = \left\{ \begin{array}{ll} -\pi^2/3, & j=k \\ -2(-1)^{k-j}/(k-j)^2, & j\neq k \end{array} \right.$$

with the same calculation giving the corresponding terms in y, namely $B^{(2)}$ and $\mathcal{D}_y$.

In the t variable we introduce a slightly more complicated quadrature formula to produce a uniquely solvable system. This amounts to approximately doubling the accuracy of quadratures in the t variable. The technique is based on the observation that ideally one would evalute the inner products in (1.1) exactly. Since this is not generally practicable, these must be evaluated by a quadrature method. However, if the quadrature formula (1.5) is used for inner products that involve $\tilde{b}_{N+1}(t)$ then the computed result is exactly zero, since these functions are zero at the quadrature points, and the system then becomes ill-conditioned. This reflects only the fact that the exact inner products are small (since they are zero to the error of sinc quadrature), not that they actually are zero. By evaluating the inner products at double accuracy in the t dimension we can more closely approximate their (nonzero) values; moreover, we obtain a system that is better conditioned (and has a unique solution). Thus we introduce the sinc quadrature based on a half-step,

$$< f, g >= \int_0^T f(t)g(t)w(t)\,dt \approx \frac{h}{2} \sum_{\ell=-4N_t}^{4N_t} \frac{w(t_\ell)}{\widetilde{\phi}'(t_\ell)} f(t_\ell)g(t_\ell)$$

where $\widetilde{\phi}(t) = \phi(\widetilde{t})$ and $\{t_\ell\}$ are the half-step sinc nodes defined by

$$t_\ell = \frac{Te^{\ell h/2}}{e^{\ell h/2} + 1} = \frac{Te^{\ell h/4}}{2\cosh(\ell h/4)}$$

for $\ell = -4N_t, \cdots, 4N_t$ and w is an appropriate weight function.

In order to describe the result of the half-step quadrature in the t inner products it is useful to define $i_k^m = m/2 - k$ and

$$
\begin{aligned}
[c_{k,q}^1] &= \left[< \frac{\partial}{\partial t}\widetilde{b}_k(t), \widetilde{b}_q(t) > \right] \\
&= \left[\frac{\text{sinc}'(i_k^m)}{\dfrac{h}{T\widetilde{\phi}'(t_m)}} - \sum_r \frac{t_r}{T}\text{sinc}'(i_r^m) \right] \mathcal{D}(w) \left[\text{sinc}(i_q^m)\ \widetilde{b}_{N_t+1} \right] \frac{1}{2} \\
&\equiv \frac{1}{2}\left[I_t^{(1)} \right],
\end{aligned}
$$

where $\mathcal{D}(w) = \text{diag}\,[w(t_m)]$ and $\widetilde{b}_{N_t+1} = t_m/T - \sum_{r=-N_t}^{N_t} \frac{t_r}{T}\text{sinc}(i_r^m)$. Next set

$$
\begin{aligned}
[c_{k,q}^0] &= \left[< \widetilde{b}_k(t), \widetilde{b}_q(t) > \right] \\
&= h/2 \left[\frac{\text{sinc}(i_k^m)}{\widetilde{b}_{N_t+1}} \right] \mathcal{D}\left(\frac{w}{\widetilde{\phi}'} \right) \left[\text{sinc}(i_q^m)\ \widetilde{b}_{N_t+1}(t_m) \right] \\
&= h/2 \left[I_t^{(0)} \right].
\end{aligned}
$$

Then for fixed q we have

$$\sum_{k=-N}^{N+1} h^3/2\,\mathcal{D}_x[u_{ij}^k]\,\mathcal{D}_y(1/h)c_{k,q}^1 + \sum_{k=-N}^{N+1} h^3/2\,\mathcal{D}_x\left(A^{(2)}\right)^T[u_{ij}^k]\,\mathcal{D}_y\,c_{k,q}^0$$

$$+ \sum_{k=-N}^{N+1} h^3/2 \, \mathcal{D}_x[u_{ij}^k] \, \mathcal{D}_y \, B^{(2)} c_{k,q}^0 = [RHS_{\ell p}^q].$$

Given a two-dimensional array (matrix of size $p \times q$) $M = [a_{i,j}]$, we represent M as a column vector $\mathrm{co}(M)$ by

$$\mathrm{co}(M) = [a_{1,1}, \cdots, a_{p,1}, \cdots, a_{1,q}, \cdots, a_{p,q}]^T.$$

Using this we define the "mat" notation for an array $\{u_{ijk}\}$ of size $p \times q \times r$ by

$$\mathrm{mat}(u_{ijk}) = [\mathrm{co}(u_{ij1}), \mathrm{co}(u_{ij2}), \cdots, \mathrm{co}(u_{ijr})]$$

which is a matrix of size $p \cdot q \times r$. In the present case the array of unknowns is three-dimensional of size $(2N+1) \times (2N+1) \times (2N+2)$. Expressing this in terms of the tensor product we have

$$h^3/2 \, \mathcal{D}_y \otimes \mathcal{D}_x \, \mathrm{mat}(u_{ijk}) 1/h \left[I_t^{(1)} \right]$$

$$+ \quad h^3/2 \, \mathcal{D}_y \otimes \mathcal{D}_x \left(A^{(2)} \right)^T \mathrm{mat}(u_{ijk}) \left[I_t^{(0)} \right]$$

$$+ \quad h^3/2 \, \mathcal{D}_y \left(B^{(2)} \right)^T \otimes \mathcal{D}_x \, \mathrm{mat}(u_{ijk}) \left[I_t^{(0)} \right] = \mathrm{mat}(RHS_{\ell p}^q).$$

Set $V \equiv \mathrm{mat}(u_{ijk}) = [\mathrm{co}(u_{ij}^{-N}), \cdots, \mathrm{co}(u_{ij}^{N+1})]$ then premultiply the preceding equation by $(2/h^3) \, \mathcal{D}_y^{-1} \otimes \mathcal{D}_x^{-1}$ and postmultiply by $\left[I_t^{(0)} \right]^{-1}$ to get

$$\frac{1}{h} V \left[I_t^{(1)} \right] \left[I_t^{(0)} \right]^{-1} + \left[I_y \otimes \left(A^{(2)} \right)^T + \left(B^{(2)} \right)^T \otimes I_x \right] V = H$$

where

$$H = 2/h^3 \, \mathcal{D}_y^{-1} \otimes \mathcal{D}_x^{-1} \, \mathrm{mat}(RHS_{\ell p}^q) \left[I_t^{(0)} \right]^{-1}.$$

Letting $F = (1/h) \left[I_t^{(1)} \right] \left[I_t^{(0)} \right]^{-1}$ and assuming $N_x = N_y$ and that the weight functions in x and y are the same, we have

$$E = (B^{(2)})^T = (A^{(2)})^T$$

which gives the more symmetric form

$$(I_y \otimes E + E \otimes I_x) V + V F = H.$$

Provided the matrix E is diagonalizable, so is the matrix $(I_y \otimes E + E \otimes I_x)$ and we have

$$(I_y \otimes E + E \otimes I_x) = [T \otimes T] \Lambda_E \left[T^{-1} \otimes T^{-1} \right]$$

where T diagonalizes E. If Q diagonalizes F, with $F = Q\Lambda_F Q^{-1}$, then

$$\left[(T \otimes T)\Lambda_E(T^{-1} \otimes T^{-1})\right] V + VQ\Lambda_F Q^{-1} = H.$$

Premultiplying $T^{-1} \otimes T^{-1}$ and postmultiplying by Q we arrive at

$$\Lambda_E(T^{-1} \otimes T^{-1})VQ + (T^{-1} \otimes T^{-1})VQ\Lambda_F = (T^{-1} \otimes T^{-1})HQ.$$

Letting

$$W = (T^{-1} \otimes T^{-1})VQ$$

we have

$$\Lambda_E W + W\Lambda_F = Z$$

where

$$Z = (T^{-1} \otimes T^{-1})HQ. \tag{1.6}$$

Since Λ_E and Λ_F are diagonal matrices we can easily solve for W_{ij} by

$$W_{ij} = \frac{Z_{ij}}{\Lambda_E(i,i) + \Lambda_F(j,j)}$$

and finally the solution is given by

$$V = (T \otimes T)WQ^{-1}.$$

The tensor product is very useful in representing the system of equations in the two-dimensional case but is not practical for computations because it is too large to store for realistic problems. Fortunately, its storage is not necessary since entries of the tensor product can be computed as required from far smaller matrices, and this is also true of the eigenvalue decompositions. For example, for $N = N_x = N_y = N_t = 4$, the matrix of the system (1.4) is of order 810, but if $N = 16$ it is of order 37026, so storage of this matrix is essentially impossible on most computers. For this reason we present a numerical procedure that runs slower but allows solution of systems of the large order required, and amounts to solving the tensor product system one y (or x) level at a time. Recall that the "mat" and "co" operations simply produced a stacked matrix. The first $(2N_x + 1) \times (2N_t + 2)$ block corresponding to the first y level (i.e., $j = -N_y$), the second block corresponding to the second y value, etc. Provided that we take $N_x = N_y$ this block structure can be exploited to reduce the computations considerably. The matrices in

$$\Lambda_E W + W\Lambda_F = Z \tag{1.7}$$

have a natural block structure. Let

$$W = \begin{bmatrix} W_1 \\ W_2 \\ \vdots \\ W_m \end{bmatrix}$$

where $m = (2N_x+1)$ and W_j is a $(2N_x+1) \times (2N_t+2)$ matrix corresponding to the jth block of $(2N_x + 1)$ rows of W. Similarly, the first matrix on the right is a block diagonal matrix

$$\Lambda_E = \operatorname{diag}[\Lambda_1, \cdots, \Lambda_m]$$

with

$$\Lambda_j = \Lambda_A + \Lambda_A(j,j)I_x$$

where $\Lambda_A = T(A^{(2)})^T T^{-1}$ consists of the eigenvalues of $(A^{(2)})^T$. The right-hand side of (1.7) is a $(2N_x + 1)^2 \times (2N_t + 2)$ matrix with a natural block structure of the form same as W, which we write as

$$Z = \begin{bmatrix} Z_1 \\ Z_2 \\ \vdots \\ Z_m \end{bmatrix}.$$

Our method is to exploit this block structure and solve for each jth block of unknowns one at a time without ever explicitly computing tensor products. In order for this to work we must consider the computation of the right-hand side. First note that the jth block of rows of size $(2N_x + 1)$ of the matrices $T \otimes T$ and $T^{-1} \otimes T^{-1}$ are easily computed from a knowledge of T and T^{-1}, respectively. Namely, the jth block of $(2N_x + 1)$ rows consists of m matrices of size $(2N_x + 1) \times (2N_x + 1)$. Then the jk block of $T \otimes T$ is given by

$$\mathbf{T}_{jk} = T_{jk}T$$

and the jkth block of $T^{-1} \otimes T^{-1}$ is

$$\mathbf{T}_{jk}^{-1} = T_{jk}^{-1}T^{-1}.$$

Therefore to compute the jth block of unknowns W_j, we compute the jth block of the right hand side from

$$Z_j = T^{-1}\left(\sum_{i=1}^{m} T_{ji}^{-1}\mathbf{H}_i\right)Q$$

where $\mathbf{H}_i$ is the ith block of $(2N_x + 1)$ rows of the known right-hand side H in (1.6). Thus we have to solve $m = 2N_x + 1$ equations of the form

$$\Lambda_j Z_j + Z_j \Lambda_f = W_j$$

which are easily computed as described above. Finally,

$$U = (T \otimes T)WQ^{-1}$$

can be computed using the same block matrix computations as used above. Namely, the jth block U_j of U is given by

$$U_j = T\left(\sum_{i=1}^{m} T_{ji} Z_i\right) Q^{-1}.$$

In order to apply these methods to the problem (1.1) in which boundary data is piecewise holomorphic, it is only necessary to calculate a solution in each interval of t for which the boundary data is holomorphic and to use the final values on that interval as the initial condition for the next interval of t in which the data is represented by a holomorphic function.

Finally, we note that the computations outlined above can be used in essentially the same way to calculate solutions of the more general parabolic problem

$$u_t - a(x)b(y)u_{xx} - c(x)d(y)u_{yy} = 0$$

on the same domain.

2 Sample Computations

The computations above were applied in the following examples, and the maximum error of the method computed on a uniform rectangular grid with step size .025, using the approximation (1.2). These errors were calculated at various time levels for display and are typical of those at all other time levels (the approximation (1.2) is uniform on the closure of D).

Example 1 In this example the true solution is given by

$$u(x, y, t) = \exp(t)\cos(2(x + y))$$

with

$$f(x, y, t) = 9(\exp(t)\cos(2(x + y)))$$

and boundary data given by

$$u(x, y, t) = \begin{cases} \exp(t)\cos(2y) & x = 0 \\ \exp(t)\cos(2(1 + y)) & x = 1 \\ \exp(t)\cos(2x) & y = 0 \\ \exp(t)\cos(2(x + 1)) & y = 1 \\ \cos(2(x + y)) & t = 0 \end{cases}.$$

The errors are of the following form.

M. STROMBERG AND X. GILLIAM

N	h_x	T	t level	mesh error
4	1.571	1	t=0.4	9.18E(-02)
8	1.111	1	t=0.4	4.18E(-03)
16	.785	1	t=0.4	2.11E(-04)

Example 2 The true solution is

$$u(x, y, t) = \exp(-t)\exp(x + y)$$

in which case the function

$$f(x, y, t) = -3(\exp(-t)\exp(x + y))$$

and the boundary data is given by

$$u(x, y, t) = \begin{cases} \exp(-t)\exp(y) & x = 0 \\ \exp(-t)\exp(1 + y) & x = 1 \\ \exp(-t)\exp(x) & y = 0 \\ \exp(-t)\exp(x + 1) & y = 1 \\ \exp(x + y) & t = 0 \end{cases}$$

The errors are:

N	h_x	T	t level	mesh error
4	1.571	1	t=0.4	1.31E(-02)
8	1.111	1	t=0.4	1.86E(-03)
16	.785	1	t=0.4	8.36E(-05)

Example 3 The true solution is

$$u(x, y, t) = \exp(-t)(x(1 - x))(y(1 - y)),$$

$$f(x, y, t) = -\exp(-t)(x(1 - x))(y(1 - y)) + 2\exp(-t)(x(1 - x) + y(1 - y))$$

and the boundary data is given by

$$u(x, y, t) = \begin{cases} 0 & x = 0,\ x = 1, \\ 0 & y = 0,\ y = 1 \\ x(1 - x)y(1 - y) & t = 0 \end{cases}$$

The errors are:

N	h_x	T	t level	mesh error
2	2.221	1	t=0.5	7.23E(-03)
4	1.571	1	t=0.5	9.82E(-04)
8	1.111	1	t=0.5	7.13E(-05)

Example 4   In our final example we consider the true solution given by

$$u(x, y, t) = 16 \exp(-t - K((x - 1/2)^2 + (y - 1/2)^2))x(1 - x)y(1 - y)$$

in which case the function

$$
\begin{aligned}
f(x, y, t) = \\
- \quad & 16 \exp(-t - K((x - 1/2)^2 + (y - 1/2)^2))x(1 - x)y(1 - y) \\
- \quad & 16 \exp(-t - K((x - 1/2)^2 + (y - 1/2)^2))\{-y(1 - y)(10Kx - 10Kx^2 \\
- \quad & 8K^2x^3 + 4K^2x^4 + 5K^2x^2 - K^2x - 2K + 2)) \\
+ \quad & (-x(1 - x)(10Ky - 10Ky^2 - 8K^2y^3 + 4K^2y^4 \\
+ \quad & 5K^2y^2 - K^2y - 2K + 2))\}
\end{aligned}
$$

where K is a constant and the boundary data is given by

$$
u(x, y, t) = \begin{cases} 0 & x(1 - x)y(1 - y) = 0 \\ 16 \exp(-K((x - 1/2)^2 + \\ (y - 1/2)^2))x(1 - x)y(1 - y) & t = 0 \end{cases}
$$

The errors for various values of K are:

N	h_x	K	t level	mesh error
2	2.221	0	t=0.2	1.19E(-01)
4	1.571	0	t=0.2	1.67E(-02)
8	1.111	0	t=0.2	1.39E(-03)

N	h_x	K	t level	mesh error
4	1.571	1	t=0.2	4.92E(-02)
8	1.111	1	t=0.2	4.75E(-03)
16	0.785	1	t=0.2	2.48E(-04)

N	h_x	K	t level	mesh error
4	1.571	5	t=0.2	2.66E(-01)
8	1.111	5	t=0.2	4.16E(-02)
16	0.785	5	t=0.2	2.80E(-03)

References

[1] D.L. LEWIS, J. LUND and K. L. BOWERS, "The Space-time Sinc-Galerkin Method for Parabolic Problems," *Internat. J. Numer. Methods Engrg.*, v. 24, 1987, pp 1629-1644.

[2] J. LUND, "Symmetrization of the Sinc-Galerkin Method for Boundary Value Problems," *Math. Comp.*, v. 47, 1986, pp. 571-588.

[3] J. LUND and K. L. BOWERS, *Sinc Methods for Quadrature and Differential Equations*, SIAM, Philadelphia, PA, 1992.

[4] F. STENGER, "Approximations via Whittakers Cardinal Function," *J. Approx. Theory*, v. 17, 1976, pp. 222-240.

[5] F. STENGER, "Numerical Methods Based on Whittaker Cardinal, or Sinc Functions," *SIAM Rev.*, v. 23, 1981, 165-224.

[6] M. STROMBERG,"Sinc Approximate Solution of Quasilinear Equations of Conservation Law Type," *Computation and Control*, Birkhäuser Boston, 1989, pp. 317-331.

A MODIFIED LEVENBERG-MARQUARDT ALGORITHM FOR LARGE-SCALE INVERSE PROBLEMS

C. R. Vogel *
Department of
Mathematical Sciences
Montana State University
Bozeman, MT 59717

J. G. Wade †
Institute for
Scientific Computation
Texas A&M University
College Station, TX 77843

1 Introduction

Distributed parameter estimation problems typically involve attempts to invert infinite dimensional nonlinear compact operators. In this case the derivative, or Jacobian, is a compact linear operator. Via the Hilbert-Schmidt Theorem one can construct, from a truncated spectral decomposition consisting of the largest eigenvalues and corresponding eigenfunctions, a uniformly convergent sequence of finite rank operator approximations to the Jacobian. This truncated spectral decomposition can be computed using a variety of iterative methods, including Subspace Iteration and the Lanczos method [5]. The approximate Jacobians can then be incorporated into a quasi-Newton scheme for solving the nonlinear problem. The purpose of this paper is to demonstrate that by combining Subspace Iteration with costate, or adjoint, ideas similar to those in [7], one can efficiently solve large-scale distributed parameter estimation problems.

To illustrate these ideas, consider the estimation of a spatially varying diffusion coefficient $\tilde{q}(x)$ in a one-dimensional linear diffusion/convection equation (with $x \in (0,1)$, $t \in (0, T_f)$)

$$u_t(x,t;\tilde{q}) = \Big(\tilde{q}(x)u_x(x,t;\tilde{q}) - \nu u(x,t;\tilde{q}) \Big)_x, \tag{1.1}$$

$$\tilde{q}(0)u_x(0,t;\tilde{q}) - \nu u(0,t;\tilde{q}) = b(t) \quad 0 < t < T_f,$$

$$\tilde{q}(1)u_x(1,t;\tilde{q}) + \kappa u(1,t;\tilde{q}) = 0 \quad 0 < t < T_f,$$

$$u(x,0;\tilde{q}) = 0 \quad 0 < x < 1,$$

given knowledge of the solution $u(x_i,t;\tilde{q})$ at "a few" locations $\{x_i\}_{i=1}^{n_{ob}}$. The inverse problem is cast in the standard output least squares formulation [1]. Given "data" $\bar{z} \in [L^2(0,T_f)]^{n_{ob}} \stackrel{\text{def}}{=} Z$ the goal is to minimize (a regularized

---

*Research was supported in part by NSF under Grant DMS-9106609.

†Research was supported in part by the Air Force Office of Scientific Research under grant AFOSR-90-0091 and by the Department of Energy under contract #SK966-19. Part of this work was carried out while the second author was a visitor at the University of Southern California, Los Angeles, CA.

version of) the objection functional

$$J(\tilde{q}) \stackrel{\text{def}}{=} \frac{1}{2}\|\mathbf{z}(\tilde{q}) - \bar{z}\|_Z^2 \tag{1.2}$$

over an admissible parameter space $\tilde{Q}_{AD} \subset L^\infty(0,1)$, the elements $\tilde{q}$ of which satisfy $\tilde{q}(x) \geq c$ for a positive constant c. The "output" variable $\mathbf{z}(\tilde{q})$ is defined by $z_i(t;\tilde{q}) = u(x_i, t; \tilde{q})$ with the $\tilde{q}$-dependent "state variable" $u(\cdot, \cdot; \tilde{q})$ defined as the solution of (1.1) where ν and κ are positive constants. Assuming that the boundary source term b above lies in $L^2(0, T_f)$ and that $\tilde{q} \in \tilde{Q}_{AD}$, this standard linear PDE has a unique solution $u(\cdot, \cdot; \tilde{q})$ in $H^1(0,1)$ which depends continuously but nonlinearly on $\tilde{q}$. Thus the map $\tilde{q} \mapsto z(\cdot; \tilde{q})$ is well-defined but nonlinear. Call this map $\mathcal{F}$, so that

$$\mathcal{F}(\tilde{q}) \stackrel{\text{def}}{=} \mathbf{z}(\tilde{q}) = \{u(x_i, \cdot; \tilde{q})\}_{i=1}^{n_{ob}}. \tag{1.3}$$

Thus the problem may be viewed as that of (approximately) inverting $\mathcal{F}$ via the least-squares method.

The numerical issues that arise in the minimization of J fall into three broad categories. First, since the problem is nonquadratic, attempts to solve it will usually be based on some quasi-Newton scheme. The Levenberg-Marquardt method [2] is one such scheme. It requires the derivative of $\mathcal{F}$, which will be denoted by $\mathcal{A}(\tilde{q})$. Second, since the spaces $\tilde{Q}_{AD}$ and Z are infinite dimensional, finite dimensional approximations of elements of $\tilde{Q}_{AD}$ and Z and a discretization of $\mathcal{F}$ and $\mathcal{A}(\tilde{q})$ are required. Thus a conceptual question arises: should one formulate the (infinite dimensional) optimization algorithm first, and then discretize, or should one discretize first, and then optimize? In this paper the former approach is taken. Third, the regularized optimization approach taken yields, after reduction to canonical form, a Hessian of the form $G + \gamma I$, where γ is small and positive, and the operator G is compact. With a high level of discretization, the exact evaluation and inversion of this Hessian is impractical. An iterative inversion method is developed in the next two sections based on a finite rank approximation to G constructed using Subspace Iteration. This approach requires only that for any $\tilde{q}, \delta\tilde{q} \in \tilde{Q}_{AD}$ and $z \in Z$ the quantities $\mathcal{F}(\tilde{q})$, $\mathcal{A}(\tilde{q})\delta\tilde{q}$ and $\mathcal{A}^*(\tilde{q})z$ can be evaluated. Then, in the context of this optimization scheme, numerical discretization issues (e.g., discretization of the PDE) are briefly considered. This is followed by a numerical example, after which concluding remarks are presented.

2 The Regularized Problem and α-Continuation

Let $\tilde{Q}_{AD}$ be a compact metric space contained in a Hilbert space $(Q, \langle \cdot, \cdot \rangle_Q)$ and let $\mathcal{R}$ be a self-adjoint, coercive linear operator with dense domain in

Q. Assume a factorization $\mathcal{R} = \mathcal{B}^* \mathcal{B}$ with $\mathcal{B}^{-1}$ compact and $\tilde{Q}_{AD} \subset \mathcal{B}^{-1}Q$. Let $(Z, \langle \cdot, \cdot \rangle_Z)$ be a Hilbert space and $\widetilde{\mathcal{F}} : \tilde{Q}_{AD} \mapsto Z$ a Fréchet-differentiable compact mapping. For a given $\bar{z} \in Z$ and $\tilde{q}_{RBF} \in \tilde{Q}_{AD}$ consider the problem of minimizing the regularized least squares functional

$$\widetilde{J_\alpha}(\tilde{q}) \overset{\text{def}}{=} \frac{1}{2}\|\widetilde{\mathcal{F}}(\tilde{q}) - \bar{z}\|_Z^2 + \frac{\alpha}{2}\langle(\tilde{q} - \tilde{q}_{RBF}), \mathcal{R}(\tilde{q} - \tilde{q}_{RBF})\rangle_Q \tag{2.1}$$

over $\tilde{Q}_{AD}$.

The positive scalar α is the regularization parameter, and the operator $\mathcal{R}$ is referred to as the regularization operator. For example, for the proto-type problem discussed in the introduction one might assume $\tilde{q} - \tilde{q}_{RBF}$ is smooth and vanishes at the endpoints of the interval $[0,1]$ and take $\mathcal{R} = -d^2/dx^2$ and $\mathcal{B} = \mathcal{R}^{1/2}$. The purpose of the regularization term in (2.1) is to "stabilize" the problem in the sense that for $\alpha > 0$, $\widetilde{J_\alpha}$ necessarily has a minimizer q^α which depends continuously on $\bar{z}$. This matter is discussed in detail in, for example, [1, 3].

It is convenient to recast (2.1) into canonical form with $\mathcal{R} = I$. This is achieved by the change of variable $q \overset{\text{def}}{=} \mathcal{B}\tilde{q}$ and the definition $\mathcal{F}(q) \overset{\text{def}}{=} \widetilde{\mathcal{F}}(\mathcal{B}^{-1}q)$. Then the problem of minimizing $\widetilde{J_\alpha}(\tilde{q})$ is equivalent to minimizing

$$J_\alpha(q) \overset{\text{def}}{=} \frac{1}{2}\|\mathcal{F}(q) - \bar{z}\|_Z^2 + \frac{\alpha}{2}\|q - q_{RBF}\|_Q^2, \tag{2.2}$$

over $Q_{AD} \overset{\text{def}}{=} \mathcal{B}^{-1}\tilde{Q}_{AD}$, where $q_{RBF} = \mathcal{B}\tilde{q}_{RBF}$. Henceforth it will be assumed that the problem is always in canonical form, as in (2.2).

Although it is not the main focus of this paper, the practical problem of choosing the value of α for (2.1) is important and must be addressed. Here the "L-curve" criterion [3] is used, in which one generates a sequence of α values and corresponding minimizers $\{q^\alpha\}$ of J_α, and then plots $\log(\|q^\alpha - q_{RBF}\|_Q^2)$ versus $\log(\|\mathcal{F}(q^\alpha) - \bar{z}\|_Z^2)$. This curve typically has an "L" shape, and the value of α corresponding to bend in the L tends to be optimal or nearly so. Assuming that some iterative optimization scheme (requiring an initial guess) is to be used for minimizing J_α for each α, an effective strategy is to use the following simple "α continuation". First pick $\alpha = \alpha_0$ for some α_0 which is large enough so that the (2.1) is relatively easy to solve (this step is discussed in detail below). Then reduce α by a constant factor and find a minimizer of J_α for the new α. Repeat this process successively until the slope of the L-curve exceeds some predefined threshold value. This yields a set of α values and corresponding q^α which can by used to obtain discrete points on the L-curve. The curve can then be approximated by a straightforward interpolation scheme and the "bend in the L" can be found by locating the point on the curve with maximum curvature.

3   The Optimization Scheme

The Levenberg-Marquardt Method. Among the most reliable methods for the solution of nonlinear least squares problems such as (2.2) is the Levenberg-Marquardt method, which is the Gauss-Newton method with an approximate trust region constraint (see [2]). These matters are now reviewed.

The basis of the Gauss-Newton method is successive linearization of the parameter to output map $\mathcal{F}(q)$ about the "current iterate" q_j^α, yielding $\mathcal{F}(q_j^\alpha + \delta q) \approx \mathcal{F}(q_j^\alpha) + \mathcal{A}(q_j^\alpha)\delta q$. The map $\mathcal{F}$ is replaced by this linearization in J_α and the next iterate q_{j+1}^α is determined by the solution of the resulting quadratic subproblem. With $\delta q_j^\alpha \overset{\text{def}}{=} q_{j+1}^\alpha - q_j^\alpha$, this leads to the following linear equation for δq_j^α:

$$\left(\mathcal{A}^*(q_j^\alpha)\mathcal{A}(q_j^\alpha) + \alpha I \right)\delta q_j^\alpha = -\mathcal{A}^*(q_j^\alpha)r(q_j^\alpha) - \alpha\,(q_j^\alpha - q_0). \qquad (3.1)$$

Note that the operator on the left side of this equation is positive definite for any $\alpha > 0$, so that these linear subproblems each have a stable solution.

The Gauss-Newton method has the advantage of quadratic convergence under certain conditions [2, p. 225]. However, it is not robust enough for general use because of the possibility of non-convergence (see [2]). A widely used approach to correcting this difficulty is the use of trust region techniques. These are based on the idea of constraining the linear subproblems so that $\|\delta q_j^\alpha\|_Q$ does not exceed a certain value r_j. This is equivalent (via the Kuhn-Tucker conditions) to adding a nonnegative μ_j to α on the left side of (3.1), yielding

$$\left(\mathcal{A}^*(q_j^\alpha)\mathcal{A}(q_j^\alpha) + \gamma I \right)\delta q_j^\alpha = -\mathcal{A}^*(q_j^\alpha)r(q_j^\alpha) - \alpha\,(q_j^\alpha - q_0), \qquad (3.2)$$

where $\gamma = \alpha + \mu_j$ is positive. This leads to the Levenberg-Marquardt (LM) class of methods. See [2] for strategies for choosing r_j and μ_j.

Solution of the Linear Subproblems. Equations (3.1) and (3.2) are of the form $(G + \gamma I)y = b$, where $G = \mathcal{A}(q_j^\alpha)^*\mathcal{A}(q_j^\alpha)$ is self-adjoint, positive definite, and compact. The Hilbert-Schmidt Theorem guarantees a spectral decomposition $G = \sum_{k=1}^\infty \lambda_k P_k$ where the λ_k's are positive eigenvalues, the P_k's are the orthogonal projections onto the spaces spanned by the corresponding eigenfunctions v_k, and $\lambda_k \to 0$ as $k \to \infty$. Assume that the eigenvalues are arranged in descending order and that $\|v_k\|_Q = 1$.

In severely ill-posed problems, the eigenvalues λ_k decay to zero very rapidly; for example,

$$\lambda_k \leq C_0 e^{-ck} \qquad (3.3)$$

for positive C_0 and c. Accordingly, G can be approximated reasonably well if all but the first "few", say, N_T, eigenvalues are replaced with zero. This leads to the "truncated spectral decomposition", $G_T \overset{\text{def}}{=} \sum_{k=1}^{N_T} \lambda_k P_k$, and the approximation

$$(G_T + \gamma I) y_T = b \tag{3.4}$$

to the linear system (3.2). Using the fact that $I = \sum_{k=1}^{\infty} P_k$, one can easily show that the solution to (3.4) is given by

$$y_T = \frac{1}{\gamma}\Big(I - \sum_{k=1}^{N_T} \frac{\lambda_k}{\lambda_k + \gamma} P_k \Big) b. \tag{3.5}$$

Suppose that for a given $N_T < \infty$, $\{\lambda_k, P_k\}_{k=1}^{N_T}$ is available. The goal is to choose N_T so as to ensure that y_T approximates y to within some error tolerance ϵ, so that $\|y - y_T\|_Q / \|y_T\|_Q < \epsilon$. From (3.5) one can show that this is guaranteed if

$$\lambda_{N_T} < \epsilon \alpha. \tag{3.6}$$

Computing the truncated spectral decomposition. It was assumed above that a truncated spectral decomposition was available for any given truncation index N_T. If the means by which to apply the operator G is available (i.e., if Gq can be computed given any $q \in Q$), then one may use Subspace Iteration [5, 6] to compute $\{\lambda_k, P_k\}_{k=1}^{N_T}$. But applying G requires application of the operators $\mathcal{A}(q_j^\alpha)$ and $\mathcal{A}^*(q_j^\alpha)$. How this can be done in the context of parameter estimation for PDE's is discussed in Section 4.

The conceptual basis of Subspace Iteration is as follows: Let $\{v_k^{(0)}\}_{k=1}^{N_T}$ be an arbitrary[1] orthonormal set in Q, and for $i \geq 0$ define the subspace $S_i \overset{\text{def}}{=} span\{G^i v_k^{(0)}\}_{k=1}^{N_T} \subset Q$. Let G_i denote the representation of G in S_i. Then, as $i \to \infty$ the spectrum of G_i converges to the first N_p components of the spectrum of G. The algorithm of [5, p. 293] is presented here for completeness.

Subspace Iteration algorithm

1. $z_k^{(i)} \leftarrow G v_k^{(i)}$ for $k = 1, 2, \ldots, N_T$.

2. Orthonormalize the $z_k^{(i)}$ to obtain $e_k^{(i)}$ for $k = 1, 2, \ldots, N_T$. Let R^T denote the $N_T \times N_T$ matrix representing this transformation. In the finite dimensional case this is a QR factorization.

---

[1] "Almost arbitrary". It must be true that none of the first N_T exact eigenvectors v_l is orthogonal to *all* of the N_T initial guesses $v_k^{(0)}$.

3. $U\Lambda W^T \leftarrow R^T$, compute the Singular Value Decomposition of R^T. The Ritz values (eigenvalue approximations) $\lambda_k^{(i+1)}$, $k = 1, \ldots, N_T$, are the diagonal entries of Λ.

4. $v_k^{(i+1)} \leftarrow \sum_{j=1}^{N_T} w_{jk} d_j$ for $k = 1, 2, \ldots, N_T$. Here the w_{jk} are the components of W in step 3.

Convergence rates may be found in [5] or, for the case in which singular values/vectors are being sought, [6]. Essentially the result is that $\lambda_k^{(i)}$ converges to λ_k at a rate which is asymptotically like $(\lambda_{N_T+1}/\lambda_k)^{2i}$. A similar results holds for the eigenvector approximations. This means that if there is good "relative separation" of the eigenvalues of G, then Subspace Iteration will converge very quickly. If an estimate of the form of (3.3) is available, then the rate of convergence is exponential:

$$|\lambda_k^{(i)} - \lambda_k| = \mathcal{O}(e^{-2ci(N_T+1-k)}) \qquad (3.7)$$

In practice, stopping criteria for these iterations must be considered. If (3.7) holds, then only a few iterations might be needed. Moreover, when Subspace Iteration takes place inside an "outer iteration" of an LM method, then acceptable performance can be obtained with even fewer of the inner (Subspace) iterations. In fact the authors have found that after the initialization stages of the algorithm, just one Subspace iteration per LM iteration is sufficient. There are two reasons for this. First, from one LM iteration to the next, the operator $G = \mathcal{A}^*(q_j^\alpha)A(q_j^\alpha)$ changes very little. Thus the trunctated eigensystem of $\mathcal{A}^*(q_{j-1}^\alpha)A(q_{j-1}^\alpha)$ provides a good initial guess for that of $\mathcal{A}^*(q_j^\alpha)A(q_j^\alpha)$. Second, eigencomponents corresponding to eigenvalues which are much larger than λ_{N_T} converge extremely rapidly (c.f., equation (3.7)), and hence, are accurately approximated after only one iteration. On the other hand, the eigencomponents corresponding to smaller eigenvalues are "filtered out" due to the term $\lambda_k/(\lambda_k+\gamma)$ in equation (3.5).

Finally, it is necessary to understand how to choose the truncation index N_T. Note that in light of (3.6), for a given error tolerance ϵ, the truncation index N_T should depend on α. Now recall from Section 2 that the plan is to use "α-continuation", starting with some α_0 which is yet to be specified. A simple strategy which the authors have found to be effective when an estimate such as (3.3) holds is as follows. Given an error tolerance ϵ and an initial guess $q_0 \in Q_{AD}$ (for example, $q_0 = q_{RBF}$), use Subspace Iteration to compute a truncated spectral decomposition for $\mathcal{A}^*(q_0)A(q_0)$ with, say, $N_T = 2$, and set $\alpha_0 = \lambda_2/\epsilon$. Then, as the iterations proceed, each time a linear system of the form (3.4) is to be solved, check whether (3.6) holds. If not, then perform a linear regression on $\{\log(\lambda_k)\}_{k=1}^{N_T}$ to estimate C_0 and c in (3.3). Extrapolate based on this to estimate a new N_T such that (3.6) holds.

4   Application to the model problem.

The weak form. For the parameter estimation problem discussed in the introduction, evaluation of $\mathcal{F}$, $\mathcal{A}$ and $\mathcal{A}^*$ involve solutions of (1.1). This PDE is discretized in x by the Galerkin method, which is most naturally discussed in terms of the weak, or variational, form of the PDE.

Denote by V and H the spaces $H^1(0,1)$ and $L^2(0,1)$, respectively, and by V^* the dual space of V with respect to the H inner product $\langle \cdot, \cdot \rangle$. Identify H with its dual in the usual way so that $V \hookrightarrow H = H^* \hookrightarrow V^*$. Also denote by $\langle \cdot, \cdot \rangle$ the usual extension of the H inner product to $V^* \times V$. See [8] for a more complete discussion of this standard "Gelfand triple" arrangement.

With $\mathbf{u}(t;q) \stackrel{\text{def}}{=} u(\cdot, t; q) \in H$, the weak form of (1.1) is written as

$$\langle \dot{\mathbf{u}}(t;q), \mathbf{v} \rangle + \sigma(q)\big(\mathbf{u}(t;q), \mathbf{v}\big) = \langle \mathbf{f}(t), \mathbf{v} \rangle, \tag{4.1}$$

Here, $\delta_{(x)}$ denotes the Dirac distribution with center at x and $\mathbf{f}(t) \stackrel{\text{def}}{=} -\delta_{(0)}b(t)$, and the sesquilinear form $\sigma(q)$ is defined by

$$\sigma(q)(\mathbf{u}, \mathbf{v}) \stackrel{\text{def}}{=} \langle \mathcal{B}^{-1}q\mathbf{u}_x - \nu\mathbf{u},\, \mathbf{v}_x \rangle + (\nu + \kappa)\langle \mathbf{uv}, \delta_{(1)} \rangle. \tag{4.2}$$

Via integration by parts and standard arguments one readily sees that (1.1) is equivalent to (4.1) under appropriate smoothness conditions. The sesquilinear form $\sigma(q)$ is easily seen to be bounded and coercive on V for each $q \in Q_{AD}$. Thus there exists a unique $L(q)$ such that

$$\sigma(q)(\mathbf{u}, \mathbf{v}) = \langle -L(q)\mathbf{u}, \mathbf{v} \rangle \tag{4.3}$$

for all $\mathbf{u}$ and $\mathbf{v}$ in V, and $L(q)$ is the infinitesimal generator of an analytic semigroup $S(t)$ on H.

The Jacobian and adjoint equations. The abstract weak form of (1.1) is given in (4.1), from which the action of $\mathcal{A}(q)$ and $\mathcal{A}^*(q)$ will be derived. However, (4.1) comprises a much wider class of PDE's than the one of current interest, and the following discussion is intended only to provide a framework for the application of the ideas of Section 3 to this wider class. Thus the discussion proceeds formally; the technicalities underlying this formalism will be considered in future work.

The solution of (4.1) can be written in terms of a "solution operator" $\mathcal{S}(q)$:

$$\mathbf{u}(t;q) = [\mathcal{S}(q)\mathbf{f}](t) \stackrel{\text{def}}{=} \int_{\tau=0}^{t} S(\tau)\mathbf{f}(t-\tau)\,d\tau \tag{4.4}$$

As a generalization of the pointwise evaluation in (1.3), let $\mathcal{C}$ be a linear operator from $\mathcal{H} \stackrel{\text{def}}{=} L^2(0, T_f; H)$ to a Hilbert space Z such that $\|\mathcal{C}\mathcal{S}(q)\mathbf{f}\|_Z \le$

$K < \infty$ for all $q \in Q_{AD}$. This C is referred to as the "observation operator". The map $\mathcal{F}$ is now of the form $\mathcal{F}(q) = CS(q)\mathbf{f}$.

Assume that the Fréchet derivatives of $\sigma(q)$ and $S(q)$ exist, and for $p \in Q_{AD}$ denote by δ_p the Gateaux derivative in the direction p. Assume that $d(\delta_p \mathbf{u})/dt = \delta_p(\dot{\mathbf{u}}(t; q))$. Formally applying δ_p to both sides of (4.1) then yields

$$\langle \frac{d(\delta_p \mathbf{u})}{dt}(t; q), \mathbf{v} \rangle + \sigma(q)(\delta_p \mathbf{u}(t; q), \mathbf{v}) = (\delta_p \sigma(q))(\mathbf{u}(t; q), \mathbf{v}). \qquad (4.5)$$

This will be referred to as the "Jacobian equation", because to evaluate $\mathcal{A}(q)p$, one must solve (4.5). In terms of $S(q)$, this means that

$$\mathcal{A}(q)p = \delta_p\Big(C\big(S(q)\mathbf{f}\big)\Big) = CS(q)\big(\delta_p L(q)\big)\mathbf{u}(\cdot; q). \qquad (4.6)$$

For evaluation of the adjoint, it is enough to know how to evaluate $\langle \mathcal{A}^*(q)z, p \rangle_Q$ for any $z \in Z$ and $p \in Q_{AD}$. So consider the inner product of $\mathcal{A}(q)p$ against any $z \in Z$. From (4.6),

$$\begin{aligned}
\langle z, \mathcal{A}(q)p \rangle_Q &= \Big\langle z, CS(q)\big(\delta_p L(q)\big)\mathbf{u} \Big\rangle_Z \\
&= \langle\!\langle S^*(q)C^*z, \big(\delta_p L(q)\big)\mathbf{u}(\cdot; q) \rangle\!\rangle.
\end{aligned}$$

Here, $\langle\!\langle \cdot, \cdot \rangle\!\rangle$ denotes the $\mathcal{H} \stackrel{\text{def}}{=} L^2(0, T_f; H)$ inner product. For a given $z \in Z$ define $\mathbf{y}^z \stackrel{\text{def}}{=} S^*(q)C^*z$, so that $\langle \mathcal{A}^*(q)z, p \rangle_Q = \langle\!\langle \mathbf{y}^z, \big(\delta_p L(q)\big)\mathbf{u}(\cdot; q) \rangle\!\rangle$. The operator $S^*(q)$ is given, for $\mathbf{g} \in \mathcal{H}$, by

$$[S^*(q)\mathbf{g}](t) = \int_t^{T_f} S^*(s)\mathbf{g}(t - s)\, ds.$$

The differential equation which $\mathbf{y}^z$ satisfies is

$$\begin{aligned}
\langle -\dot{\mathbf{y}}^z(t), \mathbf{v} \rangle &= \sigma(q)(\mathbf{v}, \mathbf{y}^z(t)) + \langle (C^*z)(t), \mathbf{v} \rangle, \qquad (4.7) \\
\mathbf{y}^z(T_f) &= 0.
\end{aligned}$$

Equation (4.7) is referred to as the adjoint equation.

Thus, given the solution $\mathbf{u}(\cdot; q)$ of the (4.1), in order to evaluate $\mathcal{A}^*(q)\mathcal{A}(q)p$ for a $p \in Q_{AD}$, one must:

1. Solve the Jacobian equation (4.5) for $\delta_p \mathbf{u}$.
2. Form $z = C\delta_p \mathbf{u}$.
3. Form $C^*z \in \mathcal{H}$.
4. Solve the adjoint equation (4.7) for $\mathbf{y}^z$.
5. Compute the inner product

$$\langle \mathcal{A}^*(q)z, p \rangle = \langle\!\langle \mathbf{y}^z, \big(\delta_p L(q)\big)(q)\mathbf{u} \rangle\!\rangle. \qquad (4.8)$$

Note that this process requires the solution of two linear PDE's, the Jacobian equation and the adjoint equation. Thus each Subspace Iteration of Section 3 requires the solution of $2 \times N_T$ PDE's.

One other note is in order concerning (4.8) for the specific case at hand, in which $\sigma(q)$ is defined in (4.2). From (4.2) and (4.3) it is clear that $\left(\delta_p L(q)\right) \mathbf{u} = (B^{-1} p \, \mathbf{u}_x)_x$. Therefore, integrating by parts and interchanging the order of dx and dt in the integrals in (4.8) leads to $\langle \mathcal{A}^*(q)z, p \rangle_Q = \langle w, p \rangle_Q$, with

$$w \overset{\text{def}}{=} \int_{t=0}^{T_f} (B^{-1} y^z)_x(x, t) u_x(x, t; q) \, dt \tag{4.9}$$

This has proven much more efficient in practice than (4.8).

A Numerical Example.

As a test of the method of Section 3, equations (4.1), (4.5) and (4.7) were discretized with a standard linear spline/Galerkin method based on n subintervals in x and the trapazoidal time marching scheme based on m time steps between 0 and $T_f \overset{\text{def}}{=} 2$. Elements of Q_{AD} were approximated by piecewise constant functions on the same mesh as the linear spline basis elements. The values of ν and κ were 1 and 1/3, respectively. The Subspace Iteration/Levenberg-Marquardt (SILM) method outlined above was compared to the standard LM method on identical runs except that the full LM Jacobian $\mathcal{A}(q)$ was computed by the solution of (4.5) for each element of the basis for the (now finite dimensional) parameter space $\tilde{Q}_{AD}$. (This is standard practice. See for example [1, 4]).

The boundary forcing term $b(t)$ in (1.1) was the following smooth approximation of a step function with step at $T_f/2$:

$$b(t) = 1/2 - \frac{1}{\pi} \tan^{-1}\left(\frac{t - T_f/2}{0.02}\right).$$

The observation points x_i in (1.3) were taken as $x_i = i/4$, $i = 0, \ldots, 4$, and "synthetic" data $\bar{z}$ were generated by defining the "exact parameter" to be $\bar{q}(x) = 1 - 0.8 \exp(-100(x - 1/3)^2)$ and then solving the forward problem (1.1), with n and m set to twice their respective values which were used in the inverse problem. The "reference parameter" was set as $q_{REF} \equiv 1$.

The sequence of α values $\{\alpha_l\}$ was generated as discussed at the end of Section 2, and the threshold L-curve slope discussed there was defined to be 1000 times the slope between α_0 and α_1. For each α, the LM stopping criterion $\|q_{j-1}^{\alpha} - q_j^{\alpha}\| / \|q_j^{\alpha}\| < 10^{-2}$ was employed. The value of ϵ in (3.6) was 10^{-1}.

To compare the efficiency of SILM with that of LM, two separate runs were made with each method, one with $n = m = 40$ and one with $n = m = 80$. For both methods, most of the computational effort was spent in the solution of the PDE's — evaluating $\mathcal{F}, \mathcal{A}$ and $\mathcal{A}^*$. Thus a reasonable index for comparison is a count of the total number of PDE solutions required. PDE solution counts are tabulated below.

Method	$n = m = 40$	$n = m = 80$
SILM	660	566
LM	1121	1954

Figure 1 below gives a graphical representation of the results. Denote by q^α the parameter obtained from the SILM run discussed above with $n = m = 80$. (The result of the corresponding LM run was essentially identical. They had a relative difference of $< 5 \times 10^{-3}$). Figure 2 shows a semilog plot of the singular values σ_k of the Jacobian at the exact parameter $\bar{q}$. These are related to the eigenvalues of $\mathcal{A}^*(\bar{q})\mathcal{A}(\bar{q})$ by $\sigma_k = \sqrt{\lambda_k}$. Note that (3.3) clearly appears to hold.

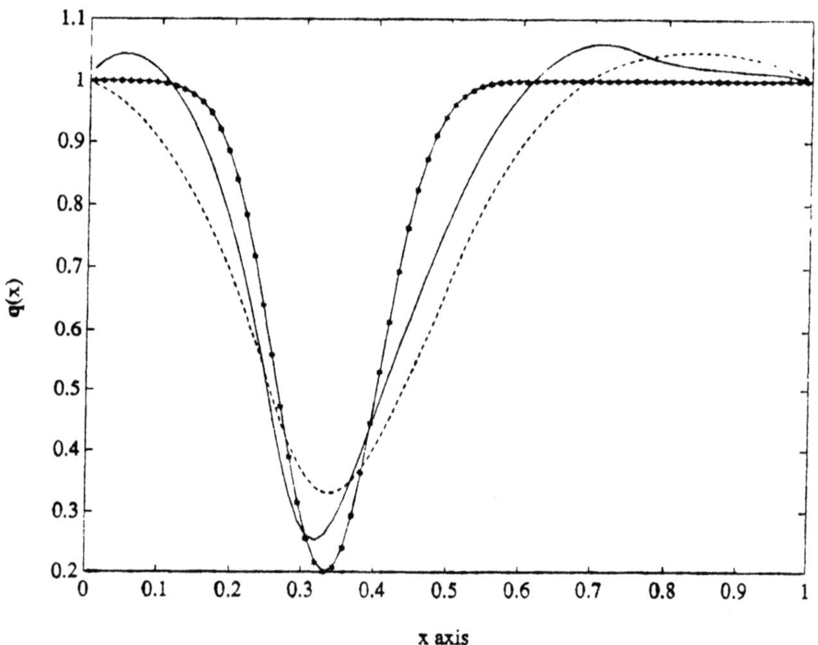

Figure 1. Plot of exact parameter $\bar{q}(x)$ (Solid line with stars) and approximate parameters $q^\alpha(x)$ for $\alpha = 3.3622 \times 10^{-8}$ (dashed line) and $\alpha = 4.9812 \times 10^{-9}$ (solid line).

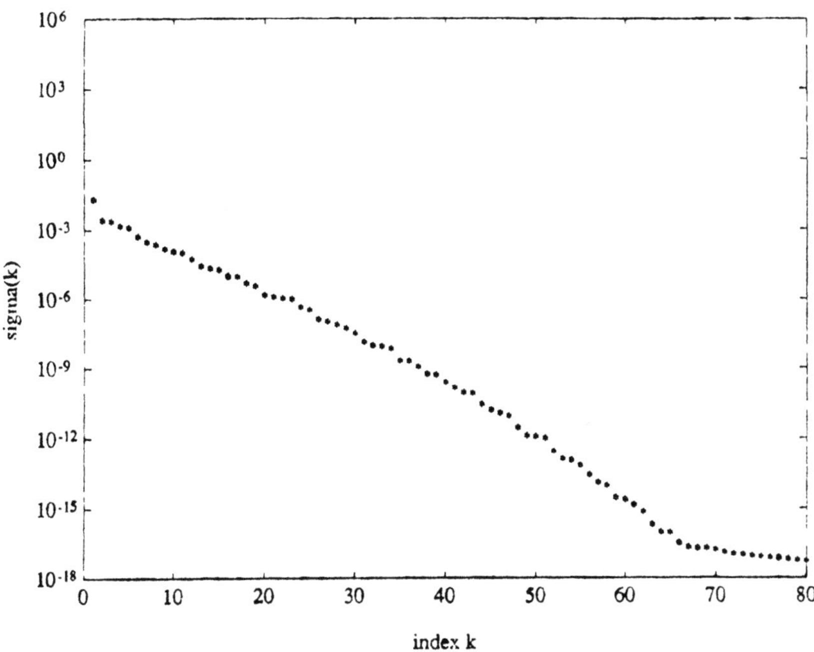

Figure 2. Semilog plot of the singular values of the Jacobian.

Conclusions and future directions. The numerical example above suggests that Subspace Iteration holds great promise for large scale inverse problems. Allowing the number of degrees of freedom in q to be as large as possible and restricting updates to lie in the dominant eigenspaces can be viewed roughly as "allowing the problem to choose its own discretization". This is a distinct advantage over imposing a fixed (usually low dimensional) discretization on q and allowing the state space discretization to be arbitrarily large, as is usually the practice. Note that in the SILM method, even though the dimension of Q is in principle unbounded, the number of PDE solutions appears to be mesh-independent, whereas with the standard LM method the number of PDE solutions grows like $dim(Q)$.

Finally, it should be noted that the Subspace Iteration technique lends itself readily to parallel implementation. In Step 1 of the algorithm outlined in Section 3, each of the evaluations $Gv_k^{(i)}$ is completely independent of the others, so that these can done in parallel. This facilitates a simple master/slave arrangement yielding a very effective coarse-grain parallelism.

References

[1] H. T. Banks and K. Kunisch, *Estimation Techniques for Distributed Parameter Systems*, Birkhäuser, 1989.

[2] J. E. Dennis and R. B. Schnabel, *Numerical Methods for Unconstrained Optimization and Nonlinear Equations*, Prentice Hall, Englewood Cliffs, 1983.

[3] P.C. Hansen, "Analysis of discrete ill-posed problems by means of the L-curve", *SIAM Review*, **34** (1992), pp. 561-580.

[4] J. Lund and C.R. Vogel, "A fully-Galerkin method for the numerical solution of an inverse problem in a parabolic partial differential equation", *Inverse Problems*, **6** (1990), pp. 205–217.

[5] B. Parlett, *The Symmetric Eigenvalue Problem*, Prentice Hall, 1980.

[6] C.R. Vogel and J.G. Wade, "Iterative SVD-based method for ill-posed problems", submitted to *SIAM J. Scientific Computing* (1992).

[7] C.R. Vogel and J.G. Wade, "Analysis of Costate Discretizations in Parameter Estimation for Linear Evolution Equations", submitted to *SIAM J. Control and Optimization* (1992).

[8] J. Wloka, *Partial Differential Equations*, Cambridge University Press, 1987.

A LOCAL SAMPLING SCHEME FOR INVARIANT EVOLUTION EQUATIONS ON A COMPACT SYMMETRIC SPACE, ESPECIALLY THE SPHERE

Dorothy I. Wallace

Department of Mathematics and Computer Science
Dartmouth College
Hanover, New Hampshire 03755

1 Introduction

The theorems in this paper deal with the problem of constructing a sampling scheme for an invariant evolution equation on a compact symmetric space. In [5] we showed that there exist many ways to take samples of such a space so that the system is observable as the number of observations increases. That is, given a system

$$u_t(x,t) = D_x u(x,t) \qquad *$$

where $x \in G/K$, G a compact Lie group and D_x an invariant differential operator on $L^2(G/K)$, we wish to use a series of measurements, $u(x_i,t_j) = d_{ij}$ to construct an ever improving approximation to the exact solution having these measurements, with initial condition

$$\lim_{t \to 0} u(x,t) = b(x).$$

If this can be done we say that the triple $(X, *, S)$ is *observable*, where

$$\begin{aligned}
X &= G/K \\
* : u_t &= D_x u \\
S &= \{(x_i,t_j)\} \text{ where } u(x_i,t_j)
\end{aligned}$$

In a subsequent paper, [6] we investigated the acuity of certain observation schemes for equations of this type. This means that for certain sampling schemes we are able to say at what rate the approximate solution, based upon n measurements, tends to the correct solution as n increases. The answer depends on the spectrum of D, the Sobolev class of the initial condition $b(x)$, and also on the operator norms of the inverses of certain matrices of the form

$$M = \begin{bmatrix} ^T C \, \pi_j(g_i) \end{bmatrix}$$

where π_j is some irreducible representation of G occurring in $L^2(G/K)$, and $^T C$ is a row vector of all ones. One of the questions this paper attempts to address is how one chooses a basis for π_j and appropriate x_i so that the matrix M is easy to invert.

Now, the problem described above is just a special case of the problem of observing a representation of the group G. That is, given a representation $\pi : G \to GL(v)$ and a set of measurements of the form

$$
\begin{aligned}
d_i &= {}^T\!C\,\pi(g_i)v_0 \\
S &= \{g_i, 1 \le i \le \dim V\}
\end{aligned}
$$

we would like to know if it is possible to reconstruct v_0 from the data. That is, we would like to know when the triple $(\pi, {}^T\!C, S)$ is observable. In Wolf, [7] , it is shown that if π is the sum of finite dimensional irreducibles π_i, each occurring with multiplicity m_i and if for all i, $m_i \le \deg \pi_i$, then there exist ${}^T\!C$ and S such that $(\pi, {}^T\!C, S)$ is observable. The necessary condition for ${}^T\!C$ is that it be a *cocyclic* vector for π. This result also holds for a representation of a Lie algebra, $\mathfrak{g}$. Solving the above problem explicitly amounts to inverting

$$
M = [{}^T\!C\,\pi(g_i)]_{i=1}^{\dim v}
$$

where each row of M is ${}^T\!C\,\pi(g_i)$ for some i. Again, choosing an appropriate basis of V_π and suitable g_i will make M invertible and, in fact, well conditioned and easy to invert numerically.

Thus this paper also addresses the point placement problem for representations of a Lie algebra. In fact, it is necessary to do the algebra case first in order to do the problem for a representation of the group and ultimately the representation in question, namely the left regular representation of G on $L^2(G/K)$, necessary for sampling the equation $*$.

The ideas of [7], on which this paper is based, are inspired by a result due to Gilliam, Li and Martin, [2]. The acuity problem for their example is worked out in DeStefano, Kaliszewski and Wallace, [4] and provides a model for the acuity theorems in [6]. This paper and some of the techniques in it are motivated by a more recent paper by Gilliam, Lund and Martin, [3]. The reader is urged to survey these references for a better insight as to why the proofs in this paper work the way they do.

In asking where to place sensors on a compact manifold in order to observe solutions to, say, the heat equation, we can ask at least three types of questions. First, we might not have any control over where the sensors are placed. In this case we can only say a few things about the recovery process, the most useful being the acuity estimates in [6]

Second, we might be able to place sensors anywhere on the manifold. The techniques used in this paper do not lend themselves particularly well to optimizing this sort of scheme. In fact we will see that the methods used herein might restrict points of observation to a horosphere within the manifold.

The kind of sampling problem our machinery is built to answer is a local one. We have some x_0, a basepoint in M. We wish to take samples in a neighborhood of x_0, the size of which we specify in advance. We sample at points we get to by following the paths of a family of horocycles in M. How far we go along a path is restricted by the size of the neighborhood in which the points must stay.

The last section of the paper deals with local point placement on the sphere. Using the theorems we prove as well as special properties of the 2-sphere, we are able to construct (at least abstractly) a sampling scheme which is observable and, even better than that, numerically easy to compute.

2 Some General Theorems

For the purposes of this section G will be a compact Lie group, $\mathfrak{g}$ its Lie algebra, $\mathfrak{g}_\mathbb{C}$ the complexification of $\mathfrak{g}$ and ψ will be the extension of $d\pi$ to $\mathfrak{g}_\mathbb{C}$ where π is the left regular representation of G on $L^2(G/K)$. Then ψ lifts to $U(\mathfrak{g}_\mathbb{C})$, the universal enveloping algebra of $\mathfrak{g}_\mathbb{C}$, in the obvious way by sending a monomial

$$x_1 x_2 x_3 \in U(\mathfrak{g}_\mathbb{C})$$

to $\psi(x_1)\psi(x_2)\psi(x_3) \in gl(V)$. The invariant subspaces for π are also invariant under ψ and so the representations decompose in the same way into irreducibles of multiplicity one.

Fix forever a Cartan subalgebra $\mathfrak{a}$ of $\mathfrak{g}_\mathbb{C}$. Write $\mathfrak{g}_\mathbb{C}$ as

$$\oplus_j \mathcal{X}_j \oplus \mathfrak{a} \oplus_j \mathcal{Y}_j,$$

where each $\mathcal{X}_k = \text{span } X_k$ and $\mathcal{Y}_k = \text{span } Y_k$, are positive and negative root spaces for $\mathfrak{a}$. We can choose a basis for V_i of weight vectors for $\mathfrak{a}$

$$V_i = \text{span } \{W_{i,1} W_{i,2} \ldots, W_{i,deg\ \psi_i}\}$$

in such a way that

1. $W_{i,1}$ is the highest weight for ψ_i, i.e. if $M(X_1, \ldots, X_k)$ is any nonconstant monomial in $X_1, \ldots, X_k$, in $U(\mathfrak{g}_\mathbb{C})$, then

$$\psi\big(M(X_1, \ldots, X_k)\big) W_{i,1} = 0$$

2. $W_{i,deg\ \psi_i}$ is a lowest weight vector, i.e., if $(Y_1, \ldots, Y_k)$ is a nonconstant monomial in Y_1 to Y_k in $U(\mathfrak{g}_\mathbb{C})$ then

$$\psi\big(N(Y_1, \ldots, Y_k)\big) W_{i,deg\ \psi_i} = 0$$

3. Each $W_{i,j} = \psi(M_j(Y_1, \ldots, Y_k))W_{i,1}$ for some monomial

$$M_j(Y_1, \ldots, Y_k) \in U(\mathfrak{g}_{\mathbb{C}}).$$

It is well known (see Wolf, [7], for example) that $W = W_{1,1} \otimes W_{2,1} \otimes \ldots \otimes W_{k,1} \otimes \ldots$, the product of the highest weight vectors in each V_k, is both cyclic and cocyclic for π. In fact, any vector with nontrivial projection onto each V_k is cyclic. The point of this section is to give an algorithm for selecting algebra elements which will actually give a basis when applied to W and do so in an orderly fashion. By *orderly*, what we mean is that the projection of the first n basis vectors onto $V_1 \oplus \ldots \oplus V_r$ (where n is the dimension of this space) gives a basis for $V_1 \oplus \ldots \oplus V_r$. This will allow us to actually use the algorithm described in Wallace & Wolf, [6], to reconstruct coefficients. We will now identify ψ_k with π_k.

3 A partial ordering on highest weights

Let λ_k be the highest weight of π_k and denote λ_k by the s-tuple

$$(b_{1,k}, \ldots, b_{s,k})$$

where

$$b_{i,k} = \frac{< \lambda_k, \beta_i >}{< \beta_i, \beta_i >} = (\lambda_k, \beta_i)$$

and β_i is the ith simple root of $\mathfrak{g}_{\mathbb{C}}$. The following facts will prove useful:

A λ_k is determined uniquely by $(b_{1,k} \ldots, b_{s,k})$,

B $b_{j,k}$ is the length of the β_j-string through $\lambda_{k,1}$

C Any of the monomials $\pi_k(Y_j^{b_{j,k}+1})$ kills $W_{k,1}$ (the highest weight vector)

D Any monomial $X_j^a Y_j^a$ where $a \le b_{j,k}$ satisfies

$$W_{k,1} = \pi_k(X_j^a Y_j^a)W_{k,1}.$$

We now put the standard partial ordering on π_k by setting $j < k$ if for all $1 \le i \le s$

$$b_{i,j} \le b_{i,k}.$$

Notice that any ordering satisfying this partial order has the property that

E $j < k \Rightarrow b_{i,j} < b_{i,k}$ for some i.

Theorem 3.1 *Let π be a representation each of whose irreducibles is multiplicity free. Then, $W_{1,1} \oplus \cdots \oplus W_{i,1} + \cdots = W$ is a cyclic (cocyclic) vector for π and a basis B_i for V satisfying*

$$\left(\bigcup_{i=1}^{\dim \tilde{V}_r} \mathrm{Proj}_{W_r} B_i \right) \text{ spans } V_1 \oplus \cdots \oplus V_r = \tilde{V}_r \text{ and} \qquad *$$

can be constructed explicitly as $B_i = \pi(M_i)W$ where the M_i are monomials in $X_1, \ldots, X_s, Y_1, \ldots, Y_s$ (given below).

Proof: (By induction) First we construct a basis for V_1 of weight vectors. It is known this can be done and

$$W_{1,j} = \pi_1\big(M_{1,j}(Y_1, \ldots, Y_s)\big)W_{1,1}.$$

It follows that if
$$B_i = \pi\big(M_{1,j}(Y_1, \ldots, Y_s)\big)W$$

then
$$\mathrm{Proj}_{\tilde{V}_1}(B_i) = W_{1,j}$$

and hence satisfies *.

Now, suppose we have $\tilde{V}_r = V_1 \oplus \cdots \oplus V_r$, dim $\tilde{V}_r = n$, and vectors $B_1, \ldots, B_n$ such that

$$\{\mathrm{Proj}_{\tilde{V}_r}(B_i)\} \text{ is a basis for} \tilde{V}_r.$$

We will construct vectors $B_{n+1}, \ldots, B_{n+m}$ where $m = \dim V_{r+1}$ such that

1. $\mathrm{Proj}_{\tilde{V}_r} B_k = 0 \qquad n+1 \leq k \leq n+m$

2. $\{\mathrm{Proj}_{V_{r+1}}(B_k)\}_{k=n+1}^{n+m} \qquad$ spans V_r.

In fact the projections in 2) will be weight vectors for π_{r+1}. Conditions 1) and 2) will provide the inductive step

Suppose $\lambda_{r+1} = (b_1, \cdots, b_s)$ and consider the monomial

$$M = X_s^{b_s} Y_s^{b_s} \quad . Y_2^{b_2} X_1^{b_1} Y_1^{b_1}.$$

By fact **D**, $\pi_{r+1}(M)W_{1,r+1} = W_{1,r+1}$. But by fact **E**, $\pi_k(M)W_{1,k} = 0$ for all $k < r+1$. (This is because the first b_j that exceeds (λ_k, β_j) has the property that

$$\pi_k(Y^{b_j})W_{1,k} = 0. \,)$$

Therefore $\mathrm{Proj}_{\tilde{V}_r}\big(\pi(M)W\big) = 0$. Now we need only apply monomials in $Y_1, \ldots, Y_s$ (as for V_1) to produce an entire basis for V_r. That is, there exist monomials $M_{r+1,i}$ in the Y's such that

$$\{\pi_{r+1}(M_{r+1,i})W_{1,r+1}\}_{i=1}^{deg\ \Pi_{r+1}}$$

form a basis for V_{r+1}. Then we have

$$V_1 = \pi(M_{r+1,i})\pi(M)W$$

satisfying $\mathrm{Proj}_{\tilde{V}_r} V_i = 0$ and

$$\{\mathrm{Proj}_{V_{r+1}} V_i\}_{i=1}^{deg\ V_{r+1}} \text{ spans } V_{r+1}.$$

The proof follows by induction.

The construction of the $M_{r+1,i}$ is known and can be found in the appendix to Dynkin, [1].

It is easy to generalize this theorem to the following case.

Theorem 3.2 *Let $U_1 \ldots U_k + \cdots = U$ be a vector such that*

$$Proj_{W_{1,1}} U_i \neq 0.$$

Then U is cyclic for π. (Again, summands of π have multiplicity one.)

Proof: The proof proceeds the same way except that at the inductive step, replace U_k with $\pi_k(R_k L_k)U_k = cW_{1,k}$ where L_k is a monomial in $Y_1, \ldots, Y_s$ which lowers the highest weight for π_k to the lowest. This kills off all lower weight components of $U_k \cdot R_k$ then raises the lowest weight to the highest. Annihilating $V_1 + \cdots + V_{k-1}$ and then constructing a basis of V_k proceeds as before.

In fact we can do better than this.

Theorem 3.3 *If π is multiplicity free then any vector V whose projection onto V_i is nonzero for all i is cyclic.*

The proof is essentially the same as before, choosing some weight vector for each π_i whose component from V is nonzero, raising it to the highest weight vector for π_i and proceeding as before. Since this is not very constructive unless one knows V more specifically, we leave the proof to the reader.

Lastly we note that one can prove a theorem analogous to Theorem 3.1 for representations with multiplicity providing

$$\mathrm{mult}(\pi_i) \leq deg\ (\pi_i)$$

although there are no analogous results for Theorems 3.2 and 3.3 is completely false for this situation. We give the theorem and construction of basis vectors but leave the reader to fill in the proof.

Theorem 3.4 *Let π be a representation of $\mathfrak{g}$ decomposing as $\oplus m_i \pi_i$ such that $m_i \leq \deg \pi_i$. Let $W_{1,i}, \ldots, W_{m_i,i}$ be a collection of m_i distinct weight vectors for π_i with $W_{1,i}$ the highest, and where*

$$W_{k,i} = \pi(m_{k,i}(Y_1, \ldots, y_n))W_{i,i}.$$

Order the $W_{k,i}$ by the length of the shortest string of roots between it and $W_{1,i}$. (This is a partial ordering.) Then one can construct a basis $\{B_i\}$ of V_π satisfying:

1. $B_i = \otimes_i \left[\overset{m_i}{\underset{k=1}{\otimes}} W_{k,i} \right]$ *is cyclic for π*

2. $\mathrm{Proj}_{\widetilde{V}_r} B_i$ *is a basis for $\widetilde{V}_r$ for $i = 1$ to $\deg \widetilde{V}_r$.*

Construction is again done so that

1. $\mathrm{Proj}_{\widetilde{V}_r} B_i = 0 \quad n+1 \leq i \leq n+m, \quad n = \deg \widetilde{V}_r, \quad m = \deg V_{r+1}.$

2. $\{\mathrm{Proj}_{V_{r+1}} B_i\}_{i=n+1}^{n+m}$ is a basis for V_{r+1}.

Here V_{r+1} is the representation space for π_{r+1}, counted with multiplicity. Again we proceed inductively. Suppose we have $B_1, \ldots, B_n$ so that

$$\{\mathrm{Proj}_{\widetilde{V}_r} B_i\}_{i=1}^{n} \text{ is a basis for } \widetilde{V}_r.$$

Suppose $\mathrm{Proj}_{V_{r+1}} W = W_{k,i}$, that is the kth occurrence of π_i. There is a monomial $R_{k,i}$ in the X_j's so that $\pi_i(R_{k,i})W_{k,i} = W_{1,i}$. Note that $\pi_i(R_{k,i})W_{j,i} = 0$ if $j < k$. Thus we have killed of the previous $W_{1,1}$ to $W_{1,j}$ in the representation spaces of $\underbrace{\pi_i \oplus \cdots \oplus \pi_i}_{k-1}$. We then proceed as in Theorem 3.1, using the monomials

$$\pi(X_s^{b_s} \ldots X_1^{b_1} Y_1^{b_1})$$

to kill off $W_{j,h}$ for all $h < j$. To handle the case of equal weights, just note that distinct weight vectors correspond to different monomials in the Y_i. Details are left to the reader.

In summary, what we have done in this section is to characterize cyclic vectors for π and to give a construction of monomials (via Theorem 3.2) in $U(\mathfrak{g}_{\mathbb{C}})$ which, for most choices of cyclic vector, generate under multiplicity-free π a basis for V_π. Next we need to adjust the argument so as to pass from $U(\mathfrak{g}_{\mathbb{C}})$ to $U(\mathfrak{g})$.

4 Point Placement for Local Sampling on the Sphere

The problem we are addressing in this section is, given a function on S (the 2-sphere), and a basepoint x_0, find points in a small region U around x_0 so that reconstructing the function is "easy". We will use a basis ϕ_{ij} for the space of functions on S which is tailored to make use of the theorems in the preceeding section. We will allow ourselves to oversample in the interest of numerical ease. A scheme similar to the one we propose was studied by Gilliam, Lund and Martin [3], in a very different situation, namely sampling on the interval.

Recall that, for some point $x_k \in U$,

$$
\begin{aligned}
f(x_k) \;&=\; f(g_k x_0) \qquad\qquad g_k \in SO(3) \\[4pt]
&=\; \sum_{i,j} c_{ij}\phi_{ij}(g_k x_0) \\[4pt]
&=\; (1,1,\ldots\ldots,1,\ldots) \\
&\qquad\times
\begin{pmatrix}
\pi_1(g_k^{-1}) & & \\
& \pi_2(g_k^{-1}) & \\
& & \ddots
\end{pmatrix} \\
&\qquad\times
\begin{pmatrix}
c_{i,1}\phi_{i,1}(x_0) \\
\hline
c_{i,2}\phi_{i,2}(x_0) \\
\hline
\vdots
\end{pmatrix}
\end{aligned}
$$

where $\{\phi_{ij}\}$ are arranged in blocks according to the irreducible representations of $SO(3)$ from smallest up. The j subscript denotes the jth representation, i denotes the ith coordinate in the jth representation, so $1 \le i \le \deg \pi_j$. So finding f amounts to inverting

$$
M =
\begin{pmatrix}
(1,\ldots,1,\ldots)\ \pi(g_1^{-1}) \\
\hline
(1,\ldots,1,\ldots)\ \pi(g_2^{-1}) \\
\hline
\vdots
\end{pmatrix}
$$

or, at least, a truncated version of it. Inspired by Gilliam, Lund and Martin [3] we wish to make the rows of M as close to orthogonal as possible.

Drawing on the last section, we pass immediately to the complexified lie algebra of $SO(3)$, and its enveloping algebra $\mathfrak{U}(\mathfrak{g})$.

To do this, break π into irreducibles π_j and for each j choose a basis for the invariant subspace V_j so that $\mathrm{Proj}_{V_j}{}^T C$ is a left highest weight vector, (call it ${}^T C_j$). That is, in the notation of the last section,

$$\begin{aligned}
{}^T C_j \tilde{\pi}_j(x) &= 0 \\
{}^T C_j \tilde{\pi}_j(a) &= \lambda_j {}^T C
\end{aligned}$$

Then ${}^T C_j \tilde{\pi}_j(Y^n)$ are all left weight vectors and because of the special structure of $SO(x)$, they are an orthonormal basis of V_j.

Having ordered the π_j by degree, (this is a multiplicity-free representation), and since we know ${}^T C_j \tilde{\pi}_j(Y^{n_j}) = 0$ for some n_j, to get the next algebra element, use ${}^T C \tilde{\pi}(Y^{n_j+1} X^{n_j+1})$. This will kill all components of ${}^T C$ in $V_1, \ldots, V_j$ and leave a highest weight vector in V_{j+1}.

As a result of this algorithm, the matrix

$$\begin{pmatrix} {}^T C \ \tilde{\pi}(z_1) \\ {}^T C \ \tilde{\pi}(z_2) \\ \vdots \end{pmatrix}$$

where the z_i are the chosen algebra elements, now has the form

$$\begin{pmatrix} \widetilde{M_1} & * & * & * \\ 0 & \widetilde{M_2} & * & * \\ & & \ddots & \\ 0 & 0 & \widetilde{M_3} & \\ \vdots & \vdots & & \ddots \end{pmatrix}$$

That is, the $\widetilde{M}$ in question is block upper triangular and the blocks M_j, corresponding to $\tilde{\pi}_j$, have orthogonal rows. Thus, any truncated version of $\widetilde{M}$, such as

$$\widetilde{M}_T = \begin{pmatrix} M_1 & * \\ 0 & M_2 \end{pmatrix}$$

is easy to invert.

Fortunately, $\mathfrak{U}(\mathfrak{g})$ also acts on the same space as G, that is the space of functions on S. It does this via directional derivatives. Let $z = X^t Y^s$, for example.

$$\tilde{\pi}(z)f(x_0)\underbrace{\tilde{\pi}(x)\ldots\tilde{\pi}(x)}_{t}\underbrace{\tilde{\pi}(y)\ldots\tilde{\pi}(y)}_{s}f(x_0)$$

where

$$\tilde{\pi}(z)f(x_0) = \frac{d}{du}f(exp\ (uY)x_0)|_{u=0}.$$

But each directional derivative is approximated in U by some finite linear combination of values $f(x_l)$. So,

$$\tilde{\pi}(z_l) \cong \sum_k a_{l,k}\pi(g_{l,k})$$

so our measurements are of the form

$$^TC(\sum_k a_{l,k}\pi(g_{l,k})) \begin{pmatrix} c_{11}\phi_{11}(x_0) \\ \vdots \\ c_{ij}\phi_{ij}(x_0) \end{pmatrix}$$

thus

$$M = \begin{pmatrix} ^TC\ \sum_k a_{1,k}\ \pi(g_{1,k}) \\ \hline ^TC\ \sum_k a_{2,k}\ \pi(g_{2,k}) \\ \hline \vdots \end{pmatrix}$$

is "close to" $\widetilde{M}$.

Just for sanity's sake, we'll state our results in the form of a theorem.

Theorem 4.1 *Using the notation above, let $\epsilon > 0$ be given and let M_T be some truncated version of $\widetilde{M}$. Then there exist points $x_{lk} \in U$ such that, if we are given measurements*

$$\sum_k a_{l,k}f(x_{l,k}) = d_l$$

we can reconstruct the approximate Fourier coefficients C_{ij} of f by inverting M_T, the upper left hand truncation of M corresponding to $\widetilde{M}_T$. Further, the $x_{l,k}$ can be chosen so that

$$\sum_{ij}[M_T - \widetilde{M}_T]^2_{ij} < \epsilon.$$

In other words, with only a finite number of directional derivatives to approximate, we can do so within any tolerance.

5 Conclusions and Questions

Any scheme like the one we just described raises certain questions. For example, to get a good approximation to a high order derivative, one is tempted to place sample points close together. This will affect the conditioning of M_T because the rows may approach orthogonal but their norms may become small. Since U is a small neighborhood one will run into this problem eventually, especially between irreducible representations where the order of the derivative jumps dramatically.

Two methods of dealing with this problem leap to mind. One is to develop some sort of quadrature technique for derivatives of functions in some finite subspace. Another is to use symbolically computed derivatives of the appropriate ϕ_{ij} to construct M_T, rather than the corresponding approximation. Actually, the results in this paper await some computational work to see how far they go toward fixing the conditioning of M when applied in a naive fashion.

Finally, we want to thank John Lund and Ken Bowers for organizing a wonderful conference and for welcoming fairly theoretical results such as these along with their more immediately useful brethren.

References

[1] DYNKIN. Maximal subgroups of the classical groups. *AMS Translations*, **Series 2, Vol. 6.**

[2] GILLIAM, LI and MARTIN. "Discrete observability of the heat equation on bounded domains," *Int. J. Control*, **48, 2,** (1988), pp. 755-780.

[3] GILLIAM, LUND and MARTIN. "Inverse parabolic problems and discrete orthogonality," *to appear.*

[4] DESTEFANO, KALISZEWSKI and WALLACE. "Acuity of observation for the heat equation on a bounded domain," *to appear.*

[5] WALLACE and WOLF. "Observability evolution equations for invariant differential operators," *J. Math. Systems, Estimation, and Control,"* **Vol. 1,** (1990).

[6] WALLACE and WOLF. "Acuity of observation for invariant evaluation equations," *Computation and Control II*, (Proceedings, Bozeman, (1990). Birkhäuser, "Progress in Systems and Control Theory 11."

[7] WOLF. "Observability and group representation theory," *Computation and Control I*, (Proc. Bozeman, 1988), Birkhäuser, "Progress in Systems and Control Theory 1."

HASSE DIAGRAM AND DYNAMIC FEEDBACK OF LINEAR SYSTEMS

Xiaochang Wang[*]
Department of Mathematics
Texas Tech University
Lubbock, TX 79409

Joachim Rosenthal[†]
Department of Mathematics
University of Notre Dame
Notre Dame, IN 46556

1 Introduction

The output feedback pole placement problem of linear systems with static or dynamic compensators belongs to the classical problems in control theory and many theoretical and numerical research papers were already devoted to this problem. Although the considered systems are linear, the problem requires the solution of an over determined system of quadratic polynomial equations.

A lot of progress has been made on the static feedback problem. It was Brockett and Byrnes [1] who explained first the pole placement problem with static compensators as an intersection problem in a compactified set of static compensators, the Grassmannian $\mathrm{Grass}(p, m + p)$. In making the connection to the classical Schubert calculus they were able to show that there are

$$d(p, m) = \deg \mathrm{Grass}(p, m + p) = \frac{1!\,2! \cdots (p - 1)!\,(mp)!}{m!\,(m + 1)! \cdots (m + p - 1)!} \qquad (1.1)$$

complex static output feedback laws which assign each set of poles for a nondegenerate m-input, p-output linear system of McMillan degree $n = mp$. In particular if the number $d(p, m)$ is odd, pole assignment by real static feedback is always possible. It follows that the generic system has the arbitrary pole assignability by static output feedback if $d(p, m)$ is odd and $mp \geq n$. Although this is not true for even $d(p, m)$ as showed by Willems and Hesselink [13], the first author proved in [11] using geometric techniques that a real solution exists for the generic system as soon as $mp > n$.

People have been looking for similar results for the dynamic pole placement problem for a long time. Willems and Hesselink [13] proved in 1978 that

$$q(m + p - 1) + mp \geq n \qquad (1.2)$$

is a necessary condition for generic pole placement by output feedback with dynamic compensator of order q. It had been a conjecture for a long

---

[*]Supported in part by the Research Enhancement Fund from College of Art and Sciences, Texas Tech University.

[†]Supported in part by NSF grant DMS-9201263.

time that (1.2) is also sufficient over $\mathbf{C}$. This conjecture was proved by the second author recently in [8]. In [7,8] the second author explained the pole placement problem with dynamic compensators again as an intersection problem in a compactified space of dynamic compensators which we like to denote with $K^q_{p,m}$. It was also proved in [8] that if a plant has McMillan degree $n = q(m + p - 1) + mp$ and is q-nondegenerate then there exist

$$d(p, m, q) = \deg K^q_{p,m} \tag{1.3}$$

complex dynamic feedback compensators of order q which places a set of $n + q$ closed loop poles at a desired location. It follows that if $d(p, m, q)$ is odd and $q(m + p - 1) + mp \geq n$ the generic system has the arbitrary pole assignability by dynamic feedback of order q.

Unlike the Grassmannian which has been studied over a century, $K^q_{p,m}$ is little known. The formula for $\deg K^q_{p,m}$ and therefore the number $d(p, m, q)$ has not been found in general.

The main result of this paper is that the variety $K^q_{p,m}$ has associated a Hasse diagram, a certain partially ordered set determined by m, p and q. In this diagram the number $d(p, m, q)$ can be identified with the number of totally ordered subsets.

2   Pole Placement Map and the Compactification $K^q_{p,m}$

Consider a linear system

$$\dot{x} = Ax + Bu, \quad y = Cx \tag{2.1}$$

where u, x and y are m, n and p-vectors. Denote with z a q-vector. Then

$$\dot{z} = Fz - Gy, \quad u = Hz - Ky \tag{2.2}$$

describes a dynamic compensator of order q. The closed loop system then becomes

$$\begin{bmatrix} \dot{x} \\ \dot{z} \end{bmatrix} = \begin{bmatrix} A - BKC & BH \\ -GC & F \end{bmatrix} \begin{bmatrix} x \\ z \end{bmatrix} \quad y = Cx. \tag{2.3}$$

If the realizations of system and compensator are irreducible, the closed loop characteristic polynomial can be written as [6,7,8]

$$\begin{aligned}
p(s) &= \det \begin{bmatrix} sI - A + BKC & -BH \\ GC & sI - F \end{bmatrix} \\
&= \det(D(s)F_2(s) + N(s)F_1(s)) \\
&= \det[F_1(s) \ F_2(s)] \begin{bmatrix} N(s) \\ D(s) \end{bmatrix}
\end{aligned}$$

where $D^{-1}(s)N(s) = C(sI-A)^{-1}B$ and $F_1(s)F_2^{-1}(s) = K + H(sI-F)^{-1}G$ are left and right coprime fractions of the transfer functions of the system and compensator, respectively.

Notice that $[F_1(s)\ F_2(s)]$ is a curve in $\mathrm{Grass}(p, m+p)$. Let

$$i = (i_1, \ldots, i_p), \quad 1 \le i_1 < \cdots < i_p \le m+p$$

and denote with

$$f_i(s) = z_i^q s^q + z_i^{q-1} s^{q-1} + \cdots + z_i^0$$

the full size minor of $[F_1(s)\ F_2(s)]$ consisting of the i_1th through i_pth columns. Note that for each number s, $f_i(s)$ is exactly the i-th Plücker coordinate of the plane $\mathrm{rowsp}[F_1(s)\ F_2(s)]$. In addition the set of Plücker coordinates $\{f_i(s)\}$ satisfies a set of quadratic equations defining $\mathrm{Grass}(p, m+p)$ in $\mathbf{P}^{\binom{m+p}{p}-1}$ [3] for all s. Equating polynomial coefficients we obtain a set of quadratic equations satisfied by the coordinates $\{z_i^d\}$. As shown in [8] this set of equations defines a projective variety in $\mathbf{P}^N$,

$$N = (q+1)\binom{m+p}{p} - 1, \tag{2.4}$$

which we will denote with $K_{p,m}^q$.

Example 2.1 [7] $\mathrm{Grass}(2,4)$ is defined by

$$x_{12}x_{34} - x_{13}x_{24} + x_{14}x_{23} = 0 \tag{2.5}$$

in $\mathbf{P}^5$. Let $f_{ij}(s) = z_{ij}^1 s + z_{ij}^0$ and

$$f_{12}(s)f_{34}(s) - f_{13}(s)f_{24}(s) + f_{14}(s)f_{23}(s) = 0. \tag{2.6}$$

The defining equations of $K_{2,2}^1$ in $\mathbf{P}^{11}$ are then given by three quadratic equations

$$z_{12}^1 z_{34}^1 - z_{13}^1 z_{24}^1 + z_{14}^1 z_{23}^1 = 0 \tag{2.7}$$

$$z_{12}^1 z_{34}^0 - z_{13}^1 z_{24}^0 + z_{14}^1 z_{23}^0 + z_{12}^0 z_{34}^1 - z_{13}^0 z_{24}^1 + z_{14}^0 z_{23}^1 = 0 \tag{2.8}$$

$$z_{12}^0 z_{34}^0 - z_{13}^0 z_{24}^0 + z_{14}^0 z_{23}^0 = 0. \tag{2.9}$$

Because $\dim K_{2,2}^1 = 8$ one sees in this case that $K_{2,2}^1$ is a complete intersection. From Bézout's theorem then follows that $\deg K_{2,2}^1 = 8$.

The closed loop characteristic polynomial $p(s)$ can be written as

$$p(s) = \sum_i f_i(s) g_i(s) = \sum_i \sum_d z_i^d g_i(s) s^d \tag{2.10}$$

where $g_i(s)$ is the full size minor of $\begin{bmatrix} N(s) \\ D(s) \end{bmatrix}$ consisting of the i_1th through i_pth rows. Let $z = (z_i^d) \in \mathbf{P}^N$ and define

$$\chi(z) = \sum_i \sum_d z_i^d g_i(s) s^d. \tag{2.11}$$

Then $\chi : K_{p,m}^q - E \to \mathbf{P}^{n+q}$ is a central projection [9] where

$$E = \{z \in \mathbf{P}^N | \chi(z) = 0\} \tag{2.12}$$

is the center of χ and a polynomial $b_0 + b_1 s + \cdots + b_{n+q} s^{n+q}$ is identified with the point $(b_0, b_1, \ldots, b_{n+q}) \in \mathbf{P}^{n+q}$.

Definition 2.1 [8] A system is called q-nondegenerate if

$$K_{p,m}^q \cap E = \emptyset. \tag{2.13}$$

It is proved in [8] that the generic system of degree n is q-nondegenerate if

$$q(m + p) + mp = n + q. \tag{2.14}$$

For a q-nondegenerate system of McMillan degree $n = q(m+p-1)+mp$, the central projection $\chi : K_{p,m}^q \to \mathbf{P}^{n+q}$ is a finite morphism [9], and therefore is onto and there are $\deg K_{p,m}^q$ points (counted with multiplicity) in $\chi^{-1}(y)$ for each $y \in \mathbf{P}^{n+q}$. Furthermore, since $g_{m+1,m+2, \ldots ,m+p}(s)$ is the only minor of $\begin{bmatrix} N(s) \\ D(s) \end{bmatrix}$ having degree n, all the points in $\chi^{-1}(y)$ must have the form $[F_1(s), F_2(s)]$ with $\det F_2(s)$ a polynomial of degree q if y is a polynomial of degree $n + q$. Therefore there are $\deg K_{p,m}^q$ dynamic compensators assigning each characteristic polynomial of degree $n + q$. If the degree is an odd number, then we can always find a real compensator.

3   Main Result

We re-index the coordinates of $\mathbf{P}^N$ first.

Definition 3.1 For any $i = (i_1, \ldots, i_p)$, $1 \leq i_1 < \cdots < i_p \leq m + p$ and any positive integer d, $d = kp + r$, $0 \leq r < p$, define

$$z_i^d = z_\alpha$$

where $\alpha = (\alpha_1, \ldots, \alpha_p)$ with

$$\alpha_l = \begin{cases} k(m + p) + i_{l+r}, & \text{for } l = 1, 2, \ldots, p - r \\ (k + 1)(m + p) + i_{l-p+r}, & \text{for } l = p - r + 1, \ldots, p. \end{cases} \tag{3.1}$$

Example 3.1

$$\begin{aligned}
z_i^0 &= z_i. \\
z_i^1 &= z_{(i_2,\ ,i_p,i_1+m+p)}. \\
z_i^2 &= z_{(i_3,\ ,i_p,i_1+m+p,i_2+m+p)}. \\
z_i^p &= z_{(i_1+m+p,\ ,i_p+m+p)}. \\
z_i^{p+1} &= z_{(i_2+m+p,\ ,i_p+m+p,i_1+2(m+p))}.
\end{aligned}$$

Definition 3.2

$$I(p,m) = \{\alpha = (\alpha_1,\ldots,\alpha_p) \mid 1 \le \alpha_1 < \cdots < \alpha_p,\ \alpha_p - \alpha_1 < m + p\}. \tag{3.2}$$

On the set $I(p,m)$ we define a partial order:

$$(\alpha_1,\ldots,\alpha_p) \le (\beta_1,\ldots,\beta_p)$$

if and only if

$$\alpha_l \le \beta_l \text{ for all } l.$$

Denote in the following with $Z(\alpha)$ the algebraic set defined by

$$Z(\alpha) = \{z \in K_{p,m}^q \mid z_\beta = 0 \text{ for all } \beta \not\le \alpha\} \tag{3.3}$$

and let

$$|\alpha| = \sum_l (\alpha_l - l). \tag{3.4}$$

Using the cell decomposition of the set of autoregressive systems introduced in [12], one can prove:

Proposition 3.1 *$Z(\alpha)$ is a subvariety of dimension $|\alpha|$.*

Let

$$L_\alpha = \{z \in \mathbf{P}^N \mid z_\alpha = 0\} \tag{3.5}$$

be a linear subspace of $\mathbf{P}^N$. Then one has the following Proposition.

Proposition 3.2

$$Z(\alpha) \cap L_\alpha = \bigcup_{\beta \le \alpha,\ |\beta|=|\alpha|-1} Z(\beta) \tag{3.6}$$

and the intersection multiplicity

$$\iota(Z(\alpha) \cap L_\alpha; Z(\beta)) = 1. \tag{3.7}$$

From Bézout's theorem [2,10] then follows

Proposition 3.3

$$\deg Z(\alpha) = \sum_{\beta \leq \alpha, \ |\beta| = |\alpha| - 1} \deg Z(\beta) \qquad (3.8)$$

Let $\alpha(p, m, q) = (\alpha_1, \ldots, \alpha_p)$ be defined by

$$\alpha_l = \begin{cases} k(m+p) + m + l + r, & \text{for } l = 1, 2, \ldots, p - r \\ (k+1)(m+p) + m + l - p + r, & \text{for } l = p - r + 1, \ldots, p \end{cases} \qquad (3.9)$$

where k and r are the two integers such that

$$q = kp + r, \quad 0 \leq r < p. \qquad (3.10)$$

Then

$$K_{p,m}^q = Z(\alpha(p, m, q)). \qquad (3.11)$$

Let $I(p, m, q)$ be the subset of $I(p, m)$ defined by

$$I(p, m, q) = \{\alpha \in I(p, m) | \alpha \leq \alpha(p, m, q)\}. \qquad (3.12)$$

Then by Proposition 3.3 it follows:

Theorem 3.4 *The degree of $K_{p,m}^q$ is equal to the number of maximal totally ordered subsets of $I(p, m, q)$.*

4 The partially ordered set $I(p, m, q)$

In this section we illustrate the counting procedure for the number of totally ordered subsets on two examples. Recall first some standard definitions in combinatorics:

Definition 4.1 [5] α is said to cover β if $\alpha > \beta$ and there exists no γ such that $\alpha > \gamma > \beta$.

Definition 4.2 [5] The Hasse diagram of a partially ordered set is a directed graph whose vertices are elements of the set and whose directed edges $\alpha \to \beta$ are precisely those ordered pairs such that α covers β.

Consider the Hasse diagram of $I(p, m, q)$. It has a unique maximum element $\alpha(p, m, q)$ and a unique minimum element $(1, \ldots, p)$. If we label the vertices in such a way that the number on $\alpha(p, m, q)$ is 1 and the number on the vertex α is the sum of the numbers on the vertices which cover α. Then the number on $(1, 2, \ldots, p)$ is $d(p, m, q)$. The following examples show, how it is possible to calculate the degree of the varieties $K_{2,2}^1$ and $K_{2,3}^1$ using elementary counting methods. We want to mention at this point that those numbers agree with the results obtained by the second author using software programs in commutative algebra. (Compare with [7]).

Example 4.1 The Hasse diagram of $K_{2,2}^1$ is given by

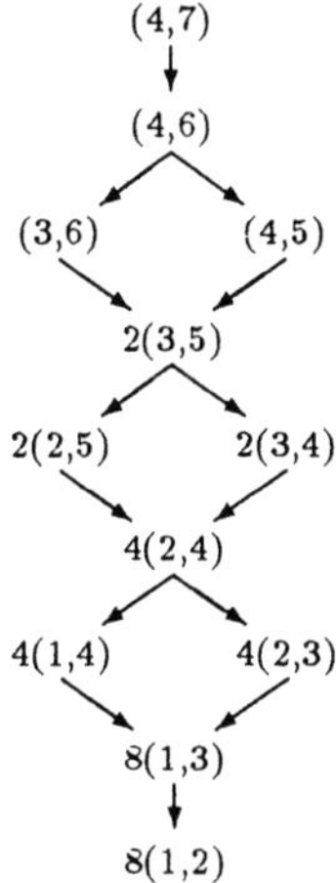

i.e. $\deg K_{2,2}^1 = 8$.

Example 4.2 The Hasse diagram of $K_{2,3}^1$ is given by

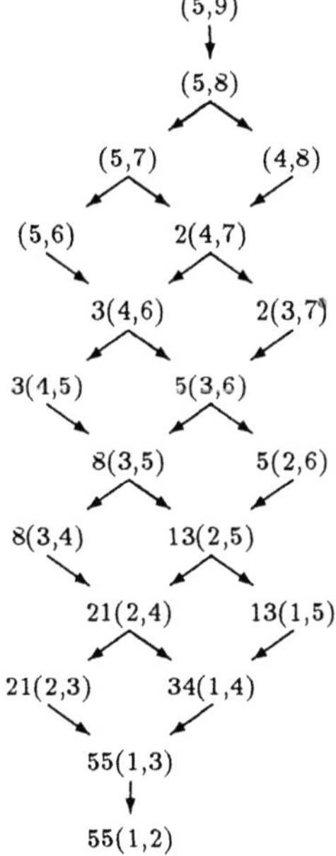

i.e. $\deg K_{2,3}^1 = 55$.

References

[1] R.W.BROCKETT and C.I.BYRNES, "Multivariable Nyquist Criteria, Root Loci and Pole Placement: A geometric viewpoint," *IEEE Trans. Aut. Control* AC-26, 1981, pp. 271–284.

[2] R.HARTSHORNE, *Algebraic Geometry*, Springer-Verlag, Berlin, 1977.

[3] W.V.D.HODGE and D.PEGOE, *Methods of Algebraic Geometry, Vol. II*, University Press, Cambridge, 1952.

[4] S.L.KLEIMAN, "Problem 15. Rigorous Foundation of Schubert's Enumerative Calculus," *Proceedings of Symposia in Pure Mathematics*, vol. 28, 1976, pp. 445–482.

[5] V.KRISHNAMURTHY, *Combinatorics: Theory and Applications*, Ellis Horwood Limited, England, 1986.

[6] J.ROSENTHAL, *Geometric Methods for Feedback Stabilization of Multivariable Linear Systems*, Ph.D.–Thesis, Arizona State University, 1990.

[7] J.ROSENTHAL, "On Minimal Order Dynamical Compensators of Low Order Systems," *Proc. of European Control Conference*, Grenoble, France, 1991, pp. 374 – 378.

[8] J.ROSENTHAL, "On Dynamic Feedback Compensation and Compactification of Systems," *SIAM J. Control Optim.*, to appear.

[9] I.R.SHAFAREVICH, *Basic Algebraic Geometry*, Springer-Verlag, Berlin, New York, 1974.

[10] W.VOGEL, *Results on Bézout's Theorem*, Tata Institute of Fundamental Research, Springer-Verlag, Berlin, New York, 1981.

[11] X.WANG, "Pole Placement by Static Output Feedback," *Journal of Math. Systems, Estimation, and Control*, vol.2 no.2, 1992, pp. 205–218.

[12] X.WANG and J.ROSENTHAL, "A Cell Structure for the Set of Autoregressive Systems", *preprint*.

[13] J.C.WILLEMS and W.H.HESSELINK, "Generic Properties of the Pole Placement Problem", *Proc. of the IFAC*, Helsinki, Finland, 1978, pp. 1725–1729.

POINT PLACEMENT FOR OBSERVATION OF
THE HEAT EQUATION ON THE SPHERE

Joseph A. Wolf

Department of Mathematics, University of California, Berkeley CA 94720

0 Abstract

In [14] Wallace and I studied discrete observability for invariant evolution equations on compact homogeneous spaces, e.g. for the heat equation on the sphere. The observations there were given by simultaneous measurements, corresponding to function evaluations. The initial data was observed as a limit of truncations deduced from a finite number of measurements. That procedure naturally involves two types of errors. Observations in the form of evaluation functionals are restricted to a finite part of the Fourier Peter Weyl expansion; that restriction implicitly involves a convolution. This discretization error occurs in the head of the approximation. The other error is the truncation error; that is the error in the tail of the approximation. In a later paper [15], we showed that the head (discretization) error depends linearly on the tail (truncation) error, and we investigated the consequences of various spectral growth conditions on the rate of vanishing of the tail error. Here I return to the rate of dependence of the head error on the tail error. This is a matter of placing the observation points, and the results to date are just the outcome of numerical experiments.

1 Background

Consider a riemannian homogeneous space $X = G/K$ where G is a compact Lie group. For example, if X is the n–dimensional sphere S^n, we can take G to be the rotation group $SO(n + 1)$ on the ambient euclidean

Research partially supported by N.S.F. Grant DMS 91 00578.

space $\mathbb{R}^{n+1}$, and then K will be the group $SO(n)$ of rotations about a base point $x_0 \in X$.

Function theory on X corresponds to representation theory for G by means of the Fourier–Peter–Weyl decomposition

$$L^2(X) = \sum_{\pi \in \widehat{G}} V_\pi \otimes \mathbb{C}^{m(\pi)} \tag{1.1}$$

as follows ([1], [2], [5]). $\widehat{G}$ is the set of (equivalence classes of) irreducible unitary representations of G. Given $\pi \in \widehat{G}$ we write V_π for the representation space; it is finite dimensional because G is compact. Each $\pi \in \widehat{G}$ defines the subspace E_π of $L^2(G)$ spanned by the matrix coefficients $f_{u,v}(g) = \langle u, \pi(g)v \rangle_{V_\pi}$. Here G acts from the left, $\ell(g)f(g') = f(g^{-1}g')$, by $\ell(g)f_{u,v} = f_{\pi(g)u,v}$. Similarly G acts from the right, $r(g)f(g') = f(g'g)$, by $r(g)f_{u,v} = f_{u,\pi(g)v}$. So $E_\pi \cong V_\pi \otimes V_\pi^*$ under the combined left and right action of G. Now let $m(\pi) = m(\pi, G, K)$ denote the dimension of the space of $\pi(K)$–fixed vectors in V_π, which is the multiplicity of the trivial representation 1_K in the restriction $\pi|_K$. Then the space of functions defined on $X = G/K$ that come from E_π, in other words the subspace of E_π consisting of functions invariant under the right action of K, as a left G–module, is just $V_\pi \otimes \mathbb{C}^{m(\pi)}$. Since $L^2(G)$ is the Hilbert space completion of the (orthogonal) direct sum of the E_π (that's the Peter–Weyl Theorem), we have the decomposition (1.1) of $L^2(X)$ as a unitary G–module.

Look at the case where $X = S^n$. For each integer $m \geq 0$ let V_m denote the span of the harmonic polynomials of degree m on $\mathbb{R}^{n+1}$. The $G = SO(n+1)$ acts irreducibly on V_m by the representation π_m whose highest weight is m times the highest weight of the usual (vector) representation on $\mathbb{C}^{n+1}$. One can check that $\dim V_m = \deg \pi_m = \frac{n-1+2m}{n-1} \prod_{k=1}^{n-2} \left(\frac{m+k}{k} \right)$. Also, if $\pi \in \widehat{G}$ then either π is (unitarily equivalent to) one of the π_m and $m(\pi) = 1$, or $\pi \not\cong \pi_m$ for any m and $m(\pi) = 0$. Thus, for the sphere, (1.1) specializes to

$$L^2(S^n) = \sum_{m=0}^{\infty} V_m . \tag{1.2}$$

This corresponds to the expansion of a function $f \in L^2(S^n)$ in terms of hyperspherical polynomials. See [3, Vol. II, §11.2], where those hyperspherical polynomials are described in terms of Geggenbauer polynomials. If $n = 2$ that is the classical expansion in terms of the bases of the V_m given by the spherical functions $Y_{m,k}(-m \leq k \leq m)$, as in [13, p. 129].

Let Δ denote the intrinsic Laplacian on X for the riemannian metric given by the negative of the Killing form of G. If $\pi \in \widehat{G}$ has highest weight

ν we write π_ν for π. Write ρ for half the sum of the positive roots. Then $\Delta f = (||\nu + \rho||^2 - ||\rho||^2)f$ for all $f \in V_{\pi_\nu} \otimes \mathbb{C}^{m(\pi_\nu)} \subset L^2(X)$. That gives a complete, precise description of the action of Δ. For example, if $X = S^n$ then Δ has eigenvalue $\frac{(n-1)m+m^2}{2n-2}$ on V_m. Note that we are using the convention in which Δ is a positive operator.

Now consider the heat equation on X with initial data $b \in L^2(X)$,

$$\Delta_x f(x : t) + \frac{\partial}{\partial t} f(x : t) = 0 \quad (x \in X, t \geq 0) \text{ with } f(x : 0) = b(x). \quad (1.3)$$

Let $d(\pi) = \dim V_\pi$ and $n(\pi) = \dim V_\pi \otimes \mathbb{C}^{m(\pi)} = d(\pi)m(\pi)$. Fix an orthonormal basis $\{\phi_{\pi,\imath}\}_{1 \leq \imath \leq n(\pi)}$ for each $V_\pi \otimes \mathbb{C}^{m(\pi)}$. Then we have the orthonormal basis $\{\phi_{\pi,\imath}\}_{\pi \in \widehat{G}, 1 \leq \imath \leq n(\pi)}$ of $L^2(X)$. If the initial data of (1.3) has expansion

$$b(x) = \sum_{\pi \in \widehat{G}} \sum_{\imath=1}^{n(\pi)} a_{\pi,\imath} \phi_{\pi,\imath}(x) \quad (1.4a)$$

then the corresponding solution to the heat equation has expansion

$$f(x : t) = \sum_{\pi \in \widehat{G}} \sum_{\imath=1}^{n(\pi)} a_{\pi,\imath} e^{-t(||\nu + \rho||^2 - ||\rho||^2)} \phi_{\pi,\imath}(x). \quad (1.4b)$$

The problem: effectively compute the initial data function b from a sequence of simultaneous observations (point evaluations) of f. In other words, given numbers $f(x_k : t) = \sum_{\pi \in \widehat{G}} \sum_{\imath=1}^{n(\pi)} a_{\pi,\imath} e^{-t(||\nu + \rho||^2 - ||\rho||^2)} \phi_{\pi,\imath}(x_k)$, solve for the $a_{\pi,\imath}$ in an accurate and efficient manner.

The solution is given in principle as follows ([14], [15]). Truncate $\widehat{G}$ and $L^2(X)$ by defining

$$\widehat{G}_r = \{\pi_\nu \in \widehat{G} \mid ||\nu|| < r\} \text{ and } L_r^2(X) = \sum_{\pi \in \widehat{G}_r} V_\pi \otimes \mathbb{C}^{m(\pi)}. \quad (1.5)$$

Denote $n_r = \dim L_r^2(X)$. Enumerate the orthonormal basis of $L^2(X)$ used in (1.4) as a complete orthonormal sequence $\{\phi_1, \phi_2, \ldots\}$ that is the increasing union of sets

$$\Phi_r = \{\phi_1, \ldots, \phi_{n_r}\} . \text{ basis of } L_r^2(X). \quad (1.6a)$$

Then $\Phi_r \subset \Phi_{r'}$ whenever $r \leq r'$. Dually, we can choose

$$S_r = \{s_1, \ldots, s_{n_r}\} \subset G \text{ with } S_r \subset S_{r'} \text{ whenever } r \leq r' \quad (1.6b)$$

so that corresponding evaluation functionals

$$\psi_j : C(X) \to \mathbb{C} \text{ given by } \psi_j(\phi) = \phi(s_j^{-1}x_0) \text{ for } 1 \leq j \leq n_r \qquad (1.7a)$$

map by restriction to form a basis $\{\psi_{r,1}, \psi_{r,2}, \ldots, \psi_{r,n_r}\}$ of the linear dual space $L_r^2(X)^*$. The corresponding matrices

$$M_r = (m_{i,j}) \text{ given by } m_{i,j} = \psi_i(\phi_j) \qquad (1.7b)$$

plays a key role in solving our initial value problem. Specifically, if $\phi = \sum a_j \phi_j \in L_r^2(X)$ and $b_i = \psi_i(\phi)$, then

$$\vec{b} = M_r \vec{a} \text{ so } \vec{a} = M_r^{-1} \vec{b}. \qquad (1.8)$$

Thus we want to choose the s_j so that the matrices M_r are reasonably well conditioned, and so that their conditioning does not degenerate too badly as r increases. This is the point placement problem.

2 Geometric Approach for the Sphere S^n, $n > 2$

Here I'll only describe some work on the sphere S^n, $n > 2$. In the next Section we'll look at S^2. Properly interpreted, all this works for riemannian symmetric spaces G/K of rank 1.

I want to use the nested sequence $S^1 \subset S^2 \subset S^3 \ldots$, so I'll use the spherical coordinates $\{r, \theta_1, \theta_2, \ldots \theta_n\}$ on $\mathbb{R}^{n+1}$ related to the linear coordinates $\{x_1, \ldots, x_{n+1}\}$ by

$$
\begin{aligned}
x_{n+1} &= r \sin \theta_n \\
x_n &= r \cos \theta_n \sin \theta_{n-1} \\
x_{n-1} &= r \cos \theta_n \cos \theta_{n-1} \sin \theta_{n-2} \\
&\cdots \\
x_2 &= r \cos \theta_n \cos \theta_{n-1} \cos \theta_{n-2} \cdots \cos \theta_2 \sin \theta_1 \\
x_1 &= r \cos \theta_n \cos \theta_{n-1} \cos \theta_{n-2} \cdots \cos \theta_2 \cos \theta_1
\end{aligned}
\qquad (2.1)
$$

where $-\frac{\pi}{2} \leq \theta_k \leq \frac{\pi}{2}$ for $k \geq 2$ and $-\pi \leq \theta_1 \leq \pi$. We'll view S^n as the sphere $r = 1$ and use the θ_k as coordinates on that sphere.

Since we are going to deal with nested spheres, we should be more specific about the sphere under consideration in (1.2), as follows.

$V_m = V_m(S^n) : SO(n+1)$–submodule of $L^2(S^n)$ generated by x_{n+1}^m,

$\pi_m = \pi_{SO(n+1),m} :$ action of $SO(n+1)$ on $V_m(S^n)$.

$$(2.2)$$

Then $\pi_{SO(n+1),m}$ is an irreducible unitary representation of $SO(n+1)$ and its restriction to $SO(n)$ is a direct sum of corresponding representations:

$$V_m(S^n) = \sum_{j=0}^{m} V_j(S^{n-1}) \cdot x_{n+1}^{m-j} \quad \text{and} \quad \pi_{SO(n+1),m}|SO(n) = \sum_{j=0}^{m} \pi_{SO(n),j}. \quad (2.3)$$

Choose $m+1$ points along the x_{n+1}-axis, for example the Chebyshev points

$$t_{m,k} = \cos(\alpha_{m,k}), \quad \alpha_{m,k} = \tfrac{2k+1}{2m+2}\pi, \quad 0 \leq k \leq m. \quad (2.4)$$

The hyperplanes $x_{n+1} = t_{m,k}$ intersect S^n in $m+1$ horizontal sub–spheres

$$S_k^{n-1} = \{x \in S^n \mid \theta_n = \tfrac{\pi}{2} - \alpha_{m,k}\} \quad (2.5)$$

equally spaced at $\tfrac{\pi}{2} - \theta_n = \tfrac{\pi}{2m+2}, \tfrac{3\pi}{2m+2}, \cdots, \tfrac{(2m+1)\pi}{2m+2}$ along any great circle arc $\theta_j = \text{constant}$, $1 \leq j \leq n-1$.

Following the decomposition of $V_m(S^n)$ in (2.3), we "allocate" one of the $V_j(S^{n-1}) \cdot x_{n+1}^{m-j}$ to each of the S_k^{n-1}. Here we will take $\dim V_j(S^{n-1})$ observation points on the associated sphere S_k^{n-1}, so of course we associate the largest of the $V_j(S^{n-1}) \cdot x_{n+1}^{m-j}$ (given by $j = m$) to the largest horizontal sphere S_k^{n-1} (given by minimizing $|\tfrac{\pi}{2} - \alpha_{m,k}|$), the next largest to the next largest, etc. For example, if $m = 3$ the α are $\alpha_{3,0} = \tfrac{\pi}{8}, \alpha_{3,1} = \tfrac{3\pi}{8}, \alpha_{3,2} = \tfrac{5\pi}{8}$ and $\alpha_{3,3} = \tfrac{7\pi}{8}$, corresponding to $\theta_n = \tfrac{3\pi}{8}, \tfrac{\pi}{8}, \tfrac{-\pi}{8}$ and $\tfrac{-3\pi}{8}$, so we associate $V_3(S^{n-1}) \cdot x_{n+1}^0$ to S_1^{n-1}, $V_2(S^{n-1}) \cdot x_{n+1}^1$ to S_2^{n-1}, $V_1(S^{n-1}) \cdot x_{n+1}^2$ to S_0^{n-1}, and $V_0(S^{n-1}) \cdot x_{n+1}^3$ to S_3^{n-1}. More generally,

2.6. Lemma. *If $m = 2\ell$ then the $\alpha_{m,k}$ are $\tfrac{\pi}{4\ell+2}, \tfrac{3\pi}{4\ell+2}, \cdots, \tfrac{(4\ell+1)\pi}{4\ell+2}$, corresponding to the values $\tfrac{\ell}{2\ell+1}\pi, \tfrac{\ell-1}{2\ell+1}\pi, \cdots, \tfrac{1}{2\ell+1}\pi, 0, -\tfrac{1}{2\ell+1}\pi, \cdots, -\tfrac{\ell-1}{2\ell+1}\pi, -\tfrac{\ell}{2\ell+1}\pi$ for θ_n. Then we associate $V_{2j}(S^{n-1}) \cdot x_{n+1}^{2\ell-2j}$ to $S_{2\ell-j}^{n-1}$ for $0 \leq j \leq \ell$, $V_{2j+1}(S^{n-1}) \cdot x_{n+1}^{2\ell-2j-1}$ to S_j^{n-1} for $0 \leq j < \ell$.*

2.7. Lemma. *If $m = 2\ell - 1$ then the $\alpha_{m,k}$ are $\tfrac{\pi}{4\ell}, \tfrac{3\pi}{4\ell}, \cdots, \tfrac{(4\ell-1)\pi}{4\ell}$, corresponding to the values $\tfrac{2\ell-1}{4\ell}\pi, \tfrac{2\ell-3}{4\ell}\pi, \cdots, \tfrac{1}{4\ell}\pi, -\tfrac{1}{4\ell}\pi, \cdots, -\tfrac{2\ell-3}{4\ell}\pi, -\tfrac{2\ell-1}{4\ell}\pi$ for θ_n. Then we associate $V_{2j}(S^{n-1}) \cdot x_{n+1}^{2\ell-2j}$ to $S_{2\ell-1-j}^{n-1}$ for $0 \leq j < \ell$, $V_{2j+1}(S^{n-1}) \cdot x_{n+1}^{2\ell-2j-1}$ to S_j^{n-1} for $0 \leq j < \ell$.*

We modify the notation (1.5) to fit our current situation. Denote the part of $L^2(S^n)$ given by the restrictions of polynomials of degree $\leq m$ on $\mathbb{R}^{n+1}$ by

$$L_m^2(S^n) = \sum_{k=0}^{m} V_k(S^n). \quad (2.8)$$

As an $SO(n)$–module it is given by

$$L^2_m(S^n) = \sum_{k=0}^{m}\sum_{j=0}^{k} V_j(S^{n-1}) \cdot x_{n+1}^{k-j} \cong \sum_{k=0}^{m} L^2_k(S^{n-1}). \qquad (2.9)$$

Suppose that we can work with the induction hypothesis that we have a good choice of $\dim L^2_k(S^{n-1})$ observation points on the S^{n-1} allocated to $V_k(S^{n-1}) \cdot x_{n+1}^{m-k}$. Then, putting these together as k runs from 0 to m, we can hope to have a good choice of $\dim L^2_m(S^n)$ points on S^n to observe functions in $L^2_m(S^n)$. In principle this would reduce the point placement problem from S^n to S^{n-1}, then from S^{n-1} to S^{n-2}, etc., and finally from S^3 to S^2.

There are, however, two points that must be addressed in the scheme just described. First, the good choices of observation points on the horizontal S^{n-1} must be coordinated. Second, we still have to look at S^2. We will address those two points together in the next Section.

3   Geometric Approach and Numerical Experiments for S^2

Compute $\dim V_k(S^2) = 2k+1$ so $\dim L^2_m(S^2) = \sum_{k=0}^{m}(2k+1) = (m+1)^2$. So we distribute $m+1$ observation points along each of the horizontal circles $\theta_2 = \frac{m-2s}{2m+2}\pi$, $0 \leq s \leq m$.

The easiest way to distribute these points would be to use a product Chebyshev point array relative to the coordinate (θ_1, θ_2). That would correspond to the product array of Shepp [10] for quadrature. Thus we would use the observation points

$$x_{r,s} = (\cos\theta_2 \cos\theta_1, \cos\theta_2 \sin\theta_1, \sin\theta_2) \qquad (3.1a)$$

for

$$\theta_1 = \frac{2r+1}{m+1}\pi \text{ and } \theta_2 = \frac{m-2s}{2m+2}\pi \text{ with } 0 \leq r, s \leq m. \qquad (3.1b)$$

But for $m = 1$ this gives

$$M_1 \text{ is a multiple of } \begin{pmatrix} 1 & 1 & 1 & 1 \\ 0 & 0 & 0 & 0 \\ 1/\sqrt{2} & 1/\sqrt{2} & -1/\sqrt{2} & -1/\sqrt{2} \\ 1 & -1 & 1 & -1 \end{pmatrix}, \qquad (3.2)$$

which is singular. Thus the product Chebyshev point array scheme fails.

Next, we introduce a twist in the product Chebyshev point array, rotating successive horizontal circles by $\frac{1}{m+1}$ of the θ_1–difference between successive points. Thus we use observation points

$$x_{r,s} = (\cos\theta_2 \cos\theta_1, \cos\theta_2 \sin\theta_1, \sin\theta_2) \qquad (3.3a)$$

for

$$\theta_1 = \left(\frac{2r+1}{m+1} + \frac{2s}{(m+1)^2}\right)\pi \text{ and } \theta_2 = \frac{m-2s}{2m+2}\pi \qquad (3.3b)$$

with $0 \leq r, s \leq m$. Then the matrices M_m , m small, have condition $\left|\frac{\text{max eigenvalue}}{\text{min eigenvalue}}\right|$ given as follows.

m	|max eigenvalue|	|min eigenvalue|	condition of M_m	
1	0.62516399438	0.34549414947	1.80947780254	
2	1.00693716448	0.41451582176	2.42918873447	
3	1.34513407701	0.14001836426	9.60684038917	
4	1.45657947728	0.08972517943	16.23378728748	
5	2.03402663182	0.01941004919	104.79245115255	(3.4)
6	2.20069969803	0.01354934360	162.42112996351	
7	2.46770320771	0.00031113486	7931.29773271503	
8	2.67137747418	0.00018666035	14311.43525272416	
9	3.03503554382	0.00009838771	30847.71134154818	
10	3.24067602389	0.00006048654	53576.81091075991	

Thus the conditioning for the M_m degenerates quite rapidly for the twisted Chebyshev point array (3.3).

Now consider a variation on the twisted Chebyshev point array (3.3), in which we introduce a parameter t as follows. We use observation points

$$x_{r,s} = (\cos\theta_2 \cos\theta_1, \cos\theta_2 \sin\theta_1, \sin\theta_2) \qquad (3.5a)$$

for

$$\theta_1 = \left(\frac{2r+1}{m+1} + \frac{2st}{(m+1)^2}\right)\pi \text{ and } \theta_2 = \frac{m-2s}{2m+2}\pi \qquad (3.5b)$$

with $0 \leq r, s \leq m$. Then the product Chebyshev point array (3.1) is given by $t = 0$ and the twisted Chebyshev point array (3.3) is given by $t = 1$. Let's examine the case that is most spread out, the case $t = \frac{m+1}{2}$ where (3.5b) becomes

$$\theta_1 = \frac{2r+s+1}{m+1}\pi \text{ and } \theta_2 = \frac{m-2s}{2m+2}\pi \qquad (3.6a)$$

so the observation points are spread out relative to the coordinate (θ_1, θ_2) in a diamond pattern

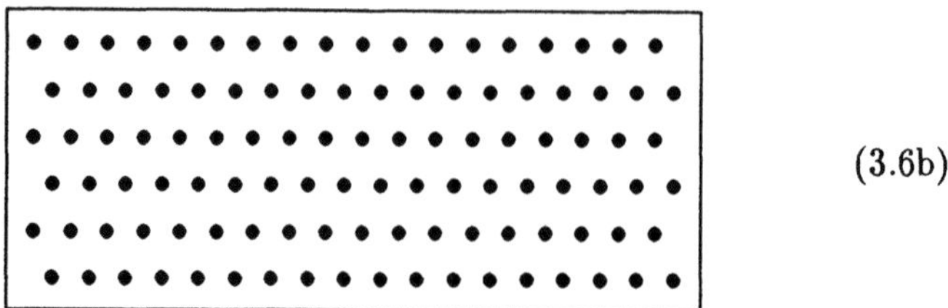

$$(3.6\text{b})$$

This diamond Chebyshev point array satisfies Grünbaum's criterion [4] of minimizing electrostatic potential. It correspond to the twisted product array of Louis and Shepp [8] for quadrature on S^2. The results are a lot better than for the twisted Chebyshev point array (3.3):

m	|max eigenvalue|	|min eigenvalue|	condition of M_m
1	0.62516399438	0.48860251190	1.27949402458
2	0.97577539125	0.64003105603	1.52457506875
3	1.20133247489	0.57853020692	2.07652506389
4	1.60723665467	0.60986420125	2.63540088332
5	2.21953966920	0.32288458362	6.87409613770
6	2.17402518607	0.34221284331	6.35284510386
7	2.36290298753	0.09803985534	24.10145322419
8	2.75156743888	0.03763120749	73.11929704824
9	3.14406231216	0.02254098010	139.48205884540
10	3.26083727533	0.00600301339	543.20006707837
11	4.11806808579	0.00162975400	2526.80348934114
12	4.07180887546	0.00021370678	19053.25101923603
13	4.51278799019	0.00003580171	126049.4988601687*
14	4.79115431289	0.00003283345	145922.9519870381*

$$(3.7)$$

The defect in even this improved scheme is that the observation points are close at the poles and far at the equator. There are a few obvious ways of evening this out, based on methods of numerical integration. In principle, one should average over a finite group of isometries of S^2 that mixes things up thoroughly (i.e. acts irreducibly), as in the group equilibrium quadrature schemes of Konayev [6], Lebedev [7], McLaren [9], Sobolev [11] and Tam [12]. I am now in the process of implementing that bit of group theory.

References

1. R. S. Cahn and J. A. Wolf, *Zeta functions and their asymptotic expansions for compact symmetric spaces of rank one*, Comm. Math. Helv. **51** (1976), 1–21.

2. É. Cartan, *Sur la détermination d'un système orthogonal complet dans un espace de Riemann symétrique clos*, Rendiconti Palermo **53** (1929), 217–252.

3. A. Erdélyi, Editor, *Higher Transcendental Functions, Volume II (Bateman Manuscript Project)*, Robert E. Krieger Publ. Co., Malabar, Fl., 1981.

4. F. A. Grünbaum, *Mathematical Aspects of Computerized Tomography (Proceedings, Oberwolfach, 1980)*, Springer–Verlag, 1981.

5. S. Helgason, *Differential Geometry and Symmetric Spaces*, Academic Press, 1962.

6. S. I. Konayev, *Quadratures of Gaussian type for a sphere invariant under the icosahedral group with inversion*, Math. Note **25** (1979), 629–634.

7. V. I. Lebedev, *Quadratures on a sphere*, Zh. Vychisl. Mat. Fiz. **16** (1976), 293–306.

8. A. K. Louis, *Optimal sampling in nuclear magnetic resonance zeugmatography*, preprint, Fachb. Angew. Math. u Informatik, Saarbrücken (1981).

9 A D McLaren, *Optimal numerical integration on a sphere*, Math. Comput. **17** (1963), 361–383.

10. L. A. Shepp, *Computerized tomography and nuclear magnetic resonance*, J Computerized Tomography **4** (1980), 94–107.

11. S. L. Sobolev, *On mechanical quadrature formulae on the surface of a sphere*, Siberskii Mat. Zh. **3** (1962), 769–796.

12. A. M. Tam, *Optimal Choice of Directions for the Reconstruction of an Object from a Finite Number of its Plane Integrals*, thesis, University of California, Berkeley, 1982.

13. N. Ya. Vilenkin, *Special Functions and the Theory of Group Representations (Transl. Math. Monographs, Vol. 22)*, American Math. Society, 1968

14. D. I. Wallace and J. A. Wolf, *Observability of evolution equations for invariant differential operators*, J. Math. Systems, Estimation, and Control **1** (1991), 29–44.

15. D. I. Wallace and J. A. Wolf, *Acuity of observation for invariant evolution equations*, in "Computation and Control II," (Proceedings, Bozeman, 1990), Birkhäuser, Progress in Systems and Control Theory **11** (1991), 325–350.

16. J. A. Wolf, *Observability and group representation theory*, in "Computation and Control," (Proceedings, Bozeman, 1988), Birkhäuser, Progress in Systems and Control Theory **1** (1989), 385–391.

Progress in Systems and Control Theory

Series Editor

Christopher I. Byrnes
Department of Systems Science and Mathematics
Washington University
Campus P.O. 1040
One Brookings Drive
St. Louis, MO 63130-4899

Progress in Systems and Control Theory is designed for the publication of workshops and conference proceedings, sponsored by various research centers in all areas of systems and control theory, and lecture notes arising from ongoing research in theory and applications control.

We encourage preparation of manuscripts in such forms as LATEX or AMS TEX for delivery in camera-ready copy which leads to rapid publication, or in electronic form for interfacing with laser printers.

Proposals should be sent directly to the editor or to: Birkhäuser Boston, 675 Massachusetts Avenue, Cambridge, MA 02139, U.S.A.

PSCT1   Computation and Control: Proceedings of the Bozeman Conference, Bozeman, Montana, 1988
K. Bowers and J. Lund

PSCT2   Perspectives in Control Theory: Proceedings of the Sielpia Conference, Sielpia, Poland, 1988
B. Jakubczyk, K. Malanowski, and W. Respondek

PSCT3   Realization and Modelling in System Theory: Proceedings of the International Symposium MTNS-89, Volume 1
M. A. Kaashoek, J. H. van Schuppen, and A. C. M. Ran

PSCT4   Robust Control of Linear Systems and Nonlinear Control: Proceedings of the International Symposium MTNS, Volume II
M. A. Kaashoek, J. H. van Schuppen, and A. C. M. Ran

PSCT5   Signal Processing, Scattering and Operator Theory, and Numerical Methods: Proceedings of the International Symposium MTNS-89, Volume III
M. A. Kaashoek, J. H. van Schuppen, and A. C. M. Ran

PSCT6   Control of Uncertain Systems: Proceedings of an International Workshop, Bremen, West Germany, June 1989
D. Hinrichsen and B. Mårtensson

PSCT7   New Trends in Systems Theory: Proceedings of the Università di Genova - The Ohio State University Joint Conference, July 9-11, 1990
G. Conte, A. M. Perdon, and B. Wyman

PSCT8   Analysis of Controlled Dynamical Systems: Proceedings of a Conference
          in Lyon, France, July 1990
          B. Bonnard, B. Bride, J. P. Gauthier and I. Kupka

PSCT9   Nonlinear Synthesis: Proceedings of a IIASA Workshop held in Sopron,
          Hungary, June 1989
          Christopher I. Byrnes and Alexander Kurzhansky

PSCT10   Modeling, Estimation and Control of Systems with Uncertainty: Proceedings
          of a Conference held in Sopron, Hungary, September 1990
          Giovanni B. Di Masi, Andrea Gombani and Alexander B. Kurzhansky

PSCT 11   Computation and Control II: Proceedings of the Second Conference at
          Bozeman, Montana, August 1990
          Kenneth Bowers and John Lund

PSCT 12   Systems, Models, and Feedback: Theory and Applications. Proceedings
          of a U.S.-Italy Workshop in honor of Professor Antonio Ruberti, held in
          Capri, Italy, 15-17, June 1992
          A. Isidori and T.J. Tarn

PSCT 13   Discrete Event Systems: Modelling and Control. Proceedings of a Joint
          Workshop held in Prague, Poland, August 1992
          S. Balemi, P. Kozak, and R. Smedinga

PSCT 14   Perspectives in Control.
          H. L. Trentelman and J. C. Willems

PSCT 15   Computation and Control III: Proceedings of the Third
          Bozeman Conference, Bozeman, Montana, August 1992
          K. L. Bowers and J. Lund